WITHDRAWN-UNL

DIGIT-SERIAL COMPUTATION

THE KLUWER INTERNATIONAL SERIES IN ENGINEERING AND COMPUTER SCIENCE

VLSI, COMPUTER ARCHITECTURE AND DIGITAL SIGNAL PROCESSING
Consulting Editor
Jonathan Allen

Other books in the series:

FORMAL SEMANTICS FOR VHDL, Carlos Delgado Kloos
    ISBN: 0-7923-9552-2
ON OPTIMAL INTERCONNECTIONS FOR VLSI, Andrew B. Kahng, Gabriel Robins
    ISBN: 0-7923-9483-6
SIMULATION TECHNIQUES AND SOLUTIONS FOR MIXED-SIGNAL COUPLING IN INTEGRATED CIRCUITS, Nishath K. Verghese, Timothy J. Schmerbeck, David J. Allstot
    ISBN: 0-7923-9544-1
MIXED-MODE SIMULATION AND ANALOG MULTILEVEL SIMULATION, Resve Saleh, Shyh-Jye, A. Richard Newton
    ISBN: 0-7923-9473-9
CAD FRAMEWORKS: Principles and Architecutres, Pieter van der Wolf
    ISBN: 0-7923-9501-8
PIPELINED ADAPTIVE DIGITAL FILTERS, Naresh R. Shanbhag, Keshab K. Parhi
    ISBN: 0-7923-9463-1
TIMED BOOLEAN FUNCTIONS: A UNIFIED FORMALISM FOR EXACT TIMING ANALYSIS, William K. C. Lam, Robert K. Brayton
    ISBN: 0-7923-9454-2
AN ANALOG VLSI SYSTEM FOR STEREOSCIPIC VISION, Misha Mahowald
    ISBN: 0-7923-9444-5
ANALOG DEVICE-LEVEL LAYOUT AUTOMATION, John M. Cohn, David J. Garrod, Rob A. Rutenbar, L. Richard Carley
    ISBN: 0-7923-9431-3
VLSI DESIGN METHODOLOGIES FOR DIGITAL SIGNAL PROCESSING ARCHITECTURES, Magdy A. Bayoumi
    ISBN: 0-7923-9428-3
CIRCUIT SYNTHESIS WITH VHDL, Roland Airiau, Jean-Michel Berge, Vincent Olive
    ISBN: 0-7923-9429-1
ASYMPOTIC WAVEFORM EVALUATION, Eli Chiprout, Michel s. Nakhla
    ISBN: 0-7923-9413-5
WAVE PIPELINING: THEORY AND CMOS IMPLEMENTATION, C. Thomas Gray, Wentai Liu, Ralph K. Cavin, III
    ISBN: 0-7923-9398-8
CONNECTIONIST SPEECH RECOGNITION: A Hybrid Appoach, H. Bourlard, N. Morgan
    ISBN: 0-7923-9396-1
BiCMOS TECHNOLOGY AND APPLICATIONS, SECOND EDITION, A.R. Alvarez
    ISBN: 0-7923-9384-8
TECHNOLOGY CAD-COMPUTER SIMULATION OF IC PROCESSES AND DEVICES, R. Dutton, Z. Yu
    ISBN: 0-7923-9379
VHDL '92, THE NEW FEATURES OF THE VHDL HARDWARE DESCRIPTION LANGUAGE, J. Bergé, A. Fonkoua, S. Maginot, J. Rouillard
    ISBN: 0-7923-9356-2
APPLICATION DRIVEN SYNTHESIS, F. Catthoor, L. Svenson
    ISBN:0-7923-9355-4

DIGIT-SERIAL COMPUTATION

by

Richard Hartley
General Electrical CRD

Keshab K. Parhi
University of Minnesota

KLUWER ACADEMIC PUBLISHERS
Boston / Dordrecht / London

Distributors for North America:
Kluwer Academic Publishers
101 Philip Drive
Assinippi Park
Norwell, Massachusetts 02061 USA

Distributors for all other countries:
Kluwer Academic Publishers Group
Distribution Centre
Post Office Box 322
3300 AH Dordrecht, THE NETHERLANDS

Library of Congress Cataloging-in-Publication Data

A C.I.P. Catalogue record for this book is available
from the Library of Congress.

Copyright © 1995 by Kluwer Academic Publishers

All rights reserved. No part of this publication may be reproduced, stored in a retrieval system or transmitted in any form or by any means, mechanical, photo-copying, recording, or otherwise, without the prior written permission of the publisher, Kluwer Academic Publishers, 101 Philip Drive, Assinippi Park, Norwell, Massachusetts 02061

Printed on acid-free paper.

Printed in the United States of America

Contents

1 **Digit-Serial Architecture**   1
   1.1 Data-Flow Architectures 1
   1.2 Synchronous Circuits 2
   1.3 Bit-Serial Architecture 4
       1.3.1 Comparison of Bit-Serial and Bit-Parallel Computation . 7
       1.3.2 Bit-Serial Circuit Architecture 8
       1.3.3 Scheduling 9
       1.3.4 Timing 10
       1.3.5 Previous Values 11
   1.4 Digit-Serial Architecture 14
       1.4.1 Example Digit-Serial Circuit 16
   1.5 Digit-Serial Operators 18
       1.5.1 Arithmetic Operators 18
       1.5.2 Comparison Operators 19
       1.5.3 Logic Operators 20
       1.5.4 Constants 20
       1.5.5 Delay Operators 21
       1.5.6 Shift Operators 22
       1.5.7 Converters 22
       1.5.8 Timing Signal Generation 22
       1.5.9 Special Cells 23
   1.6 Layout Overview 23
       1.6.1 Bit-Slicing the Standard Cells 24
       1.6.2 Complete Chip Layout 24

2 **Digit-Serial Cell Design**   27
   2.1 Digit-Serial Layout 27
       2.1.1 The Standard Template 27
       2.1.2 Delay Cells 28
       2.1.3 Deviation From The Standard Template 30
   2.2 The Add/Subtract/Compare Cell 32
       2.2.1 Division 34

	2.3		Digit-Serial Shifting	35
	2.4		Cordic Operators	37
3	**Multipliers**		**43**
	3.1		Multipliers	43
	3.2		Parallel Array Multiplier	43
	3.3		Bit-Serial multiplication	44
		3.3.1	Signed Multiplication	47
		3.3.2	Serial Multiplication with Constant Word Length	51
		3.3.3	Size of the Multiplier	52
		3.3.4	Constant Multiplication	53
		3.3.5	Serial-Serial Multiplication	53
		3.3.6	Software Support	54
	3.4		Digit-Serial Multiplication	54
	3.5		Low Latency Multiplication	56
4	**Digit-Serial Input Language**		**63**
	4.1		Cells	64
		4.1.1	Leaf Cells	65
		4.1.2	Stack Cells	67
		4.1.3	Composite Cells	68
		4.1.4	Symbolic Cells	68
	4.2		Function Calls	75
		4.2.1	Adding New Functions	78
	4.3		Control Structures	80
	4.4		Standard Libraries	82
		4.4.1	Data Conversion	82
		4.4.2	Parallel/Serial Conversion	83
		4.4.3	Serial/Parallel Conversion	84
		4.4.4	Latches	84
		4.4.5	ROMs	85
		4.4.6	Static and Periodic Signals	86
	4.5		Examples	88
		4.5.1	Square Root	88
		4.5.2	Complex Arithmetic	90
5	**Layout of Digit-Serial circuits**		**95**
	5.1		Digit-Serial Cell Layout Conventions	95
		5.1.1	Layout of the Complete Chip	97
		5.1.2	Linear Layout	97
		5.1.3	Two-Dimensional Placement	100
	5.2		Chip Examples	102

CONTENTS

6 Scheduling — 107
- 6.1 Scheduling — 107
- 6.2 Solving The Programming Problem — 113
- 6.3 Previous Values and Feedback Loops — 117
  - 6.3.1 Alternative Scheduling Method For z^{-1} Operators — 118
- 6.4 Earliest-Possible Scheduling — 119
- 6.5 Swapping Multiplier Inputs — 123
- 6.6 Trees of Associative and Commutative Operators — 124
- 6.7 Algorithm for Optimizing Single Adder Trees — 126
- 6.8 Rearranging General Data-Flow Graphs — 129
- 6.9 Handling Subtractors — 131
- 6.10 Examples — 131
- 6.11 Why not other optimizations ? — 134

7 Digit-Serial Performance — 137
- 7.1 Ripple-Carry Operators — 138
- 7.2 Look-Ahead Operators — 140
- 7.3 Improved Performance Through Unfolding — 141
- 7.4 An Example — 144

8 Bit-Level Unfolding — 147
- 8.1 Introduction — 147
- 8.2 Description of the Technique — 147
  - 8.2.1 Operators With Latency — 154
- 8.3 Collapsing Outputs — 157
- 8.4 Automatic Unfolding — 158
- 8.5 Unfolding to Arbitrary Digit Sizes — 159
- 8.6 Parallel to Digit-Serial Converter — 160

9 The Folding Transformation — 165
- 9.1 Introduction — 165
- 9.2 Digit-Serial Design Using Folding — 165
  - 9.2.1 Pipelining and Retiming For Folding — 167
  - 9.2.2 Folding of Bit-Parallel to Bit-Serial and Digit-Serial Architectures — 169
- 9.3 Folding of Regular Data-Flow Graphs — 171
- 9.4 Digit-Serial Architectures by Unfolding — 175

10 Wavelet Transform Architectures — 183
- 10.1 Introduction — 183
- 10.2 The Wavelet Computation — 184
- 10.3 The Analysis Wavelet Architecture — 185
  - 10.3.1 Life-Time Analysis — 186
  - 10.3.2 Register Allocation — 191

10.3.3 Digit-Serial Building Block	192
10.4 The Synthesis Wavelet Architecture	193

11 Digit-Serial Systolic Arrays — 195
- 11.1 Introduction 195
- 11.2 Digit Serial Systolic Arrays 197
- 11.3 Design Examples 198
  - 11.3.1 Convolution 198
  - 11.3.2 Band Matrix Multiplication 201
- 11.4 The General Case - Theory 203
  - 11.4.1 Cell Specification 203
  - 11.4.2 Timing of the Whole Array 205
  - 11.4.3 Generating Digit-Serial Arrays 207
  - 11.4.4 Expanding the Period 208
- 11.5 Finding the Minimum Digit-Size 211

12 Canonic Signed Digit Arithmetic — 217
- 12.1 Canonic Signed Digit Format 217
- 12.2 CSD Multiplication 220
  - 12.2.1 Pushing Subtractions Towards the Root 220
  - 12.2.2 Size of Adders 223
  - 12.2.3 Error Computation 224
  - 12.2.4 Bypassing Adders 225
- 12.3 Sub-Expression Sharing 226
  - 12.3.1 Finding Common Sub-expressions 227
  - 12.3.2 Choice of Sub-expression 230
  - 12.3.3 An Example 236
  - 12.3.4 Routability 237
  - 12.3.5 Using the Two Most Common Sub-expressions 238
- 12.4 Carry-Save Arithmetic 240
  - 12.4.1 Carry-Save Subtraction 242
  - 12.4.2 Speed of a Carry-Save Multiplier Circuit 243
  - 12.4.3 Filters using Carry-Save Arithmetic 244
- 12.5 Asymptotic Occurrence of Pairs. 246
  - 12.5.1 Number of n-bit CSD integers. 246
  - 12.5.2 Number of non-zero bits in CSD integers. 247
  - 12.5.3 Restricting the range of CSD integers 248
  - 12.5.4 Asymptotic frequency of non-zero bits. 249
  - 12.5.5 Frequency of pairs. 249
  - 12.5.6 Asymptotic frequency of pairs. 251

13 Online Arithmetic   **253**
  13.1 Redundant Data Formats 253
  13.2 Reduction of the Range of Digits 255
  13.3 Addition 261
  13.4 An Alternative Implementation 262
  13.5 Data Range and Overflow 264
        13.5.1 Reduction without Overflow Inhibition 266
        13.5.2 Reduction with Overflow Inhibition 267
        13.5.3 Overflow with the Alternative Reduction Scheme 267
  13.6 Radix-2 Carry-Free Addition 268
  13.7 Data Format Conversion 271
        13.7.1 Radix-2 Format Conversion 275
  13.8 Radix-2 Multiplication Architectures 275
        13.8.1 Doubly-Redundant Multiplier Architectures 276
        13.8.2 Singly-Redundant Multiplier Architectures 278
  13.9 Radix-4 and Higher Radix Multiplication 280
        13.9.1 Constant Multiplication 280
        13.9.2 Generalization to Higher Radices 283
        13.9.3 Variable-by-Variable Multiplication 288

Preface

Digital signal processing (DSP) is used in a wide range of applications such as speech, telephone, mobile radio, video, radar and sonar. The sample rate requirements of these applications range from 10 KHz to 100 MHz. Real time implementation of these systems requires design of hardware which can process signal samples as these are received from the source, as opposed to storing them in buffers and processing them in batch mode. Efficient implementation of real-time hardware for DSP applications requires study of families of architectures and implementation styles out of which an appropriate architecture can be selected for a specified application. To this end, the digit-serial implementation style is proposed as an appropriate design methodology for cases where bit-serial systems cannot meet the sample rate requirements, and bit-parallel systems require excessive hardware. The number of bits processed in a clock cycle is referred to as the digit-size. The hardware complexity and the achievable sample rate increase with increase in the digit-size. As special cases, a digit-serial system is reduced to bit-serial or bit-parallel when the digit-size is selected to equal one or the word-length, respectively. A family of implementations can be obtained by changing the digit-size parameter, thus permitting an optimal trade-off between throughput and size.

Because of their structured architecture, digit-serial designs lend themselves to automatic compilation from algorithmic descriptions. An implementation of this design methodology, the Parsifal silicon compiler was developed at the General Electric Corporate Research and Development laboratory. Parsifal constitutes a complete design environment, including capabilities for simulation, fault-simulation, chip layout, layout verification and timing verification. However, this book limits itself to a description of the digit-serial architecture and design and layout synthesis methods used in Parsifal.

The book goes beyond a description general digit-serial architectures (as embodied for instance in Parsifal) to discuss some wider-ranging issues in digit-serial design. We discuss systematic methods for the design of digit-serial architectures and design elements in chapters on "folding" and "unfolding". In addition we include chapters on systolic arrays, canonic-signed-digit number representation and carry-save arithmetic from the viewpoint of digit-serial arithmetic.

The emphasis of the book is on least-significant-digit (LSD) first digit-serial computation. However, much of the material applies as well to most-significant-digit (MSD) first arithmetic. MSD-first digit-serial arithmetic is often called on-line arithmetic, and has been a popular area of research. In on-line arithmetic, a redundant number representation is necessary to allow computation to proceed in a most-significant-bit first manner. The carry-free property of redundant number based architectures makes them ideal for high-speed implementation of DSP algorithms involving feedback loops, where minimization of loop latency is important. Without attempting a comprehensive survey we include a chapter on on-line arithmetic as a way of introducing the reader to this interesting area.

The book is organized as follows. The first five chapters relate most closely to the Parsifal design system. Chapter 1 introduces the digit-serial architecture and implementation methodology. Issues related to the design of digit-serial computational units, or cells are discussed in Chapter 2, leaving to Chapter 3 the discussion of design of digit-serial multipliers. The bit-serial language "BSL" used for describing DSP circuits in Parsifal is described in Chapter 4, and Chapter 5 describes the method of layout generation. Chapter 6 describes the scheduling algorithms used in Parsifal. These algorithms are also generally applicable to a wider class of synchronous designs.

The remaining chapters are concerned with more general aspects of digit-serial and related architectures. Chapter 7 analyzes the performance of digit-serial circuits. Chapters 8 and 9 explore formal approaches to the design of digit-serial circuits using architecture transformations. They present methods for systematic design of digit-serial circuits from bit-serial and bit-parallel designs using unfolding and folding techniques, respectively. Chapter 10 presents a case study where the digit-serial methodology is shown to be ideal for implementation of discrete wavelet transforms. This chapter also considers approaches to register minimization in architectures using life-time analysis. Systolic arrays are considered in Chapter 11 where it is shown that they offer special advantages over bit-parallel implementations. In particular, complete hardware utilization may be achieved by replacing bit-parallel processing elements with digit-serial ones. This also reduces the area by the hardware under-utilization factor. Chapter 12 presents canonic signed digit representation and carry-save arithmetic as techniques for design of DSP circuits which consume less area and power. Finally, Chapter 13 addresses on-line arithmetic.

Acknowledgment The authors gratefully acknowledge their colleagues who have contributed to the development of the Parsifal compiler or helped with the preparation of this book.

The GE bit-serial compiler (BSSC, or Parsifal I) was developed at the GE Corporate Research and Development center by Sharbel Noujaim, Richard Hartley and Jeffrey Jasica with layout help particularly from Dan Locker. The Parsifal compiler is an extension of BSSC implementing the digit-serial design methodology. It was developed by Richard Hartley and Peter Corbett with design assistance from Steve Karr, Phil Jacobs and Kathy Panton. Richard Hartley thanks all these people for their help, but wishes to single out Sharbel Noujaim, Peter Corbett and Jeff Jasica for particular thanks, and acknowledgment of their contribution to the development of the ideas presented in this book. Indeed much of the material of the book is derived from our joint publications. In addition, the section on tree-height reduction is derived from joint work with Albert Casavant. Thanks are also gratefully given to managers Ron

Jerdonek and Manuel d'Abreu for their support of this project, and to the many users of BSSC and Parsifal, particularly Fathy Yassa, for their suggestions and feedback.

Keshab Parhi acknowledges Ching-Yi Wang's contributions to the unfolding algorithm and his help in preparation of the index. He is also grateful to Tracy Denk who generated many figures used in this book. Finally, we would like to thank Bob Holland and Kluwer Academic Publishers for their encouragement and help in producing this book.

Chapter 1

Digit-Serial Architecture

1.1 Data-Flow Architectures

Digital signal processing (DSP) is used in a wide range of applications such as speech, telephone, mobile radio, satellite communications, biomedical, image, video, radar and sonar. Different applications of signal processing impose different constraints on hardware architectures. This book is concerned with design of *real-time* architectures where signal samples are processed as soon as these are received from the signal source. This is in contrast with batch-mode processing where signal samples are first stored in buffers and then processed in batch. Real-time DSP architectures are designed to match the hardware speed to the application sample speed. Since different real-time applications require different sample rates, a DSP algorithm needs to be implemented in different styles for various applications. Thus, design of real-time architectures for these applications requires study of families of architectures.

This book will be concerned with the compilation of computational circuits suitable for Digital Signal Processing (DSP) applications, and most specifically the compilation of data-flow architectures. Data-flow architectures can be implemented using different implementation styles such as bit-parallel [1][2], bit-serial [3][4][5][6] and digit-serial [7][8][9][10]. The bit-parallel architectures process all bits of a sample in one clock cycle and bit-serial architectures process one bit of the sample every clock cycle. On the other hand, digit-serial architectures process more-than-one but not-all bits of the sample every clock cycle. The number of bits processed in a clock cycle is referred to as the *digit-size*. For example, a 12-bit sample would be processed in 1 cycle in bit-parallel, 12 cycles in bit-serial, and 4 cycles in digit-serial with digit-size 3. By changing the digit-size, it is possible to design a family of architectures using the digit-serial implementation style. The digit-serial architectures are best suited for applications for which bit-serial architectures are too slow and bit-parallel

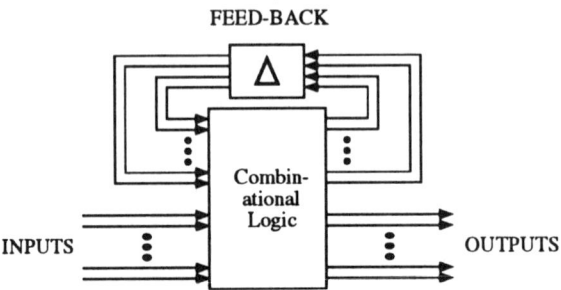

Fig. 1.1: Typical Synchronous Circuit.

architectures are faster than necessary and require more hardware.

Several high-level synthesis systems have been developed for DSP system synthesis. Early systems performed synthesis using bit-serial methodology [5][6]. These systems led to the next generation of synthesis tools such as Lager [11] and hyper [12] and later versions of the Cathedral system [13][14]. High-level synthesis allows rapid prototyping of DSP circuits optimized with respect to area and power consumption (for a certain constraint on throughput determined by the real-time application). High-level synthesis techniques have been addressed in several publications including [15][16][17][18][19][20]. High-level DSP synthesis using a library containing functional units implemented with heterogeneous implementation styles (such as bit-serial, digit-serial and bit-parallel) has been presented in [21].

This book is primarily concerned with implementation and synthesis of bit-serial and digit-serial architectures. The digit-serial architecture design methodology described in this book formed the basis of the digit-serial compiler (or synthesis tool) Parsifal developed at the General Electric Research Center [22].

1.2 Synchronous Circuits

Synchronous circuits are made up of blocks of combinational circuitry separated by latches [23]. Thus the feature that distinguishes them from arbitrary circuits is the absence of unclocked feed-back loops. A typical synchronous circuit is depicted in Fig. 1.1. As in Fig. 1.1 the symbol Δ will be used throughout this book to represent a single clock cycle delay, otherwise referred to as a clocked latch. The operation of such a circuit is controlled by the application of a *clock signal* that divides time into periods called *clock cycles*. Although many different types of clock signals are possible, in this book we will limit consideration to a two-phase clock signal as shown in Fig. 1.2. The clock consists of two signals, called *Phi_1* and *Phi_2*. The periods when the two clocks are high do not overlap.

1.2. SYNCHRONOUS CIRCUITS

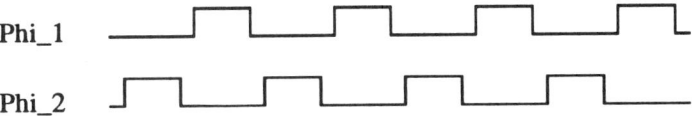

Fig. 1.2: Two Phase Clock Signal.

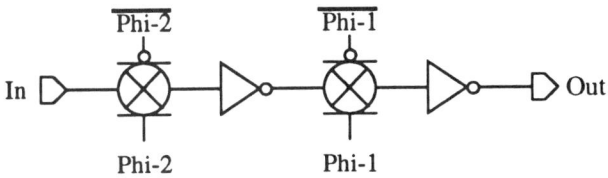

Fig. 1.3: Delay Cell.

Since the delay cell is basic in the design of synchronous circuits, we will fix a specific implementation of that cell to be used throughout this book. A suitable design for the delay cell in CMOS technology is shown in Fig. 1.3. The circles with crosses denote transmission gates. In the synchronous circuit design methodology described in this book, the *Phi_1* and *Phi_2* clock signals are used **only** inside such delay latches. Note that the *Phi_1* half-latch comes second so that the output of the delay cell transitions to its new value at the rising edge of *Phi_1*. Signals will be latched into the input half-latch (that is the *Phi_2* half-latch) of the next delay-cell at the fall of the *Phi_2* clock. For correct operation of the circuit, it is necessary that all signals be stable before the fall of the *Phi_2* clock. Since there are no feed-back loops between latches, the signals will eventually become stable as long as the clock period is long enough. The timing convention used throughout this book is shown in Fig. 1.4 which indicates that all signals change their values only at the rising edge of *Phi_1* and are latched at the falling edge of *Phi_2*.

The actual details of the delay latch are unimportant, and any appropriate

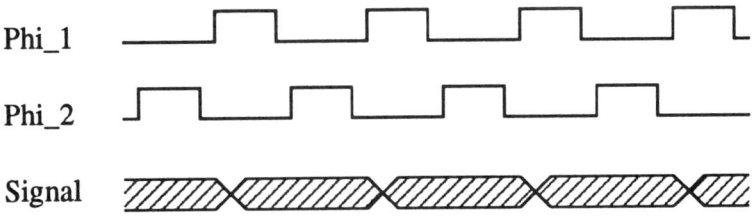

Fig. 1.4: Timing Diagram for Signals in Synchronous Circuits.

alternative technology and clocking scheme could be adopted. However, it is important in a compiler environment in which cells are to be joined together in somewhat unpredictable ways to have a well defined consistent clocking scheme.

1.3 Bit-Serial Architecture

The implementation of digital filters using bit-serial methodology was first presented in [3]. Two's complement multiplier architectures were then presented in [4]. Then the bit-serial methodology was used to develop silicon compilers for implementation of digital signal processing integrated circuits. The first such attempt was the development of the FIRST silicon compiler, which is described in detail in the the book of Denyer and Renshaw [5]. Other compilers which have exploited the bit-serial architecture have been the Cathedral I silicon compiler developed at IMEC in Belgium [6] and the bit-serial silicon compiler developed at General Electric and which has been described under various names in the literature (Inpact, BSSC, Parsifal I) [22]. The name Parsifal-I will be used in this book to describe the General Electric bit-serial silicon compiler. Bit-serial computation will not be a major theme of this book, since it has been thoroughly discussed elsewhere. However, a proper understanding of bit-serial computation is necessary to approach digit-serial computation, which will be discussed in detail in this book.

In bit-serial computation, arithmetic and logical values of all signals are referred to as words. Each word is transmitted along a single wire one bit at a time, and normally consecutive words on one wire follow each other without a break. The most usual method of bit-serial transmission of arithmetic values is least-significant-bit (LSB) first. Computations are carried out on the bits of each of the serial operands as they arrive and the result of the computation is likewise produced in bit-serial format.

The best way to understand bit-serial computation is to consider a bit-serial adder circuit. As a starting point, consider a single-bit full-adder circuit. A full-adder circuit takes three inputs, (A, B and C) and produces two outputs (X and Y). Each of the inputs is a single bit having value 1 or 0, and hence, their sum lies between 0 and 3, and consequently may be represented by two bits. The output $<YX>$ is the two-bit representation of the sum of A, B and C. Thus, the full-adder is defined by the equation

$$2Y + X = A + B + C .$$

A possible Boolean formulation of the logic necessary for the design of a full adder is

$$X = A \oplus B \oplus C$$
$$Y = (A \wedge B) \vee (A \wedge C) \vee (B \wedge C)$$

where symbols $\wedge$, $\vee$ and $\oplus$ represent Boolean AND, OR and exclusive OR operations respectively. Fig. 1.5 represents a full adder circuit.

1.3. BIT-SERIAL ARCHITECTURE

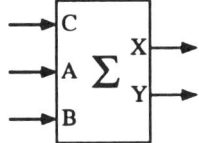

Fig. 1.5: Symbol for a Single-bit Full Adder.

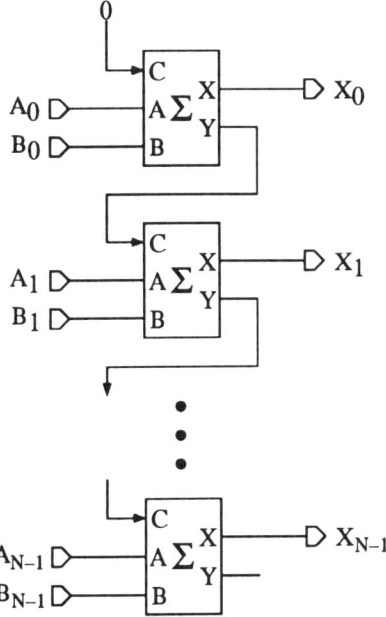

Fig. 1.6: A N-bit ripple carry adder.

A parallel N-bit adder may be constructed by connecting N single-bit adder cells together as shown in Fig. 1.6. In this adder, the C-input of each single-bit full adder is connected to the Y output of the previous bit. Although logically, all of the three inputs of the bit-slice are equivalent, the C-input as shown is sometimes called the carry-in input, and Y is called the carry-out. For the computation to be completed, it is necessary (in the worst case) for the carry to ripple through the whole length of the adder in a ripple-carry adder as shown in Fig. 1.6. The carry ripple operation imposes an upper limit on the speed of large adders. There are of course faster methods of addition, such as carry-skip [24][25], carry-look ahead [26] and carry-select [27] addition, but these will not be described here. In the adder of Fig. 1.6, all N bits arrive at one time at the $A<0:N-1>$ and $B<0:N-1>$ input ports and the computation is

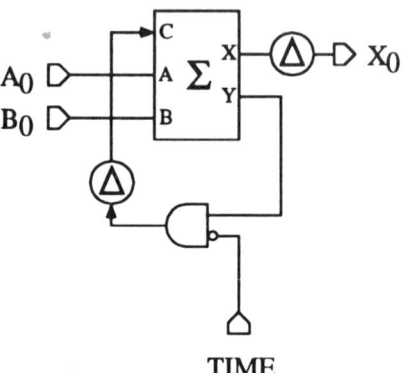

Fig. 1.7: Bit-serial adder.

carried out all at once on all bits in parallel. For this reason, such a circuit is referred to as a bit-parallel arithmetic operator.

In bit-serial arithmetic, on the other hand, the bits of each data word arrive one at a time. Consider Fig. 1.7 which shows a bit-serial adder circuit. The symbol Δ represents a single clock-cycle delay.

In the bit-serial adder the input signals A and B arrive on single input lines, one bit at a time with the least-significant bits arriving first. For this reason, this adder may be referred to as an LSB-first bit-serial adder. During any clock cycle, the current bits of A and B are added together along with the carry from the previous clock cycle. A sum and carry-bit are produced. The sum bit is output after a single-cycle delay, whereas the carry bit is delayed one cycle and fed back to the input of the adder in the next clock cycle. The AND gate shown in Fig. 1.7 is used for resetting the carry to zero at the start of each new operation. In particular, an input named $TIME$ in Fig. 1.7 is required to mark the end of each word and the beginning of the next. Since consecutive words follow each other without a gap, some indication is necessary for each operator to tell it to start a new operation. It may easily be deduced that the required form of the $TIME$ signal in Fig. 1.7 is 1 during the last bit of each word and 0 otherwise.

Latency: The number of clock cycles that elapse between the arrival of the first bit of the input words at the input of an operator and the appearance of the first bit of the output word at the output is known as the *latency* of the operator. The bit-serial adder shown in Fig. 1.7 has latency 1 (one) due to the presence of the delay at the output of the adder. The definition of latency assumes that all the inputs arrive in the same clock cycle and that there is only one output signal. This is not the case with all bit- or digit-serial operators, and for more complex operators it is not easy to define latency in a natural

1.3. BIT-SERIAL ARCHITECTURE

manner. A more general notion of timing is provided by the *timing diagram* of an operator, which will be defined later in Section 6.1.

It is part of the general philosophy of bit-serial computation [3] that individual operators should latch their outputs, so that no two consecutive bit-serial operators are cascaded together in the same clock cycle. This is the reason for the presence of the delay latch on the output of the bit-serial adder (see Fig. 1.7). Without this latch, the bit-serial adder would have latency zero.

1.3.1 Comparison of Bit-Serial and Bit-Parallel Computation

If the adders of Fig. 1.7 (bit-serial) and Fig. 1.6 (bit-parallel) are compared, various advantages and disadvantages of bit-serial computation may be perceived.

Advantages :

- The size of a bit-serial adder is considerably smaller than a bit-parallel adder, since it contains only one single-bit full adder cell. It is therefore possible to construct much more complicated networks of bit-serial than bit-parallel operators within a given constraint on circuit size.

- The number of input and output wires is much smaller for the bit-serial adder since only one wire is required for each input word. Therefore in laying out a network of bit-serial cells, the overhead due to wiring will be much smaller than with bit-parallel operators.

- Since there is no carry-ripple in a bit-serial operator the maintainable clock speed will be greater in a bit-serial circuit than in a circuit based on bit-parallel operators. Even using more sophisticated high-speed operators, bit-parallel arithmetic will be slower than bit-serial.

- If inputs to a bit-serial integrated circuit are supplied in bit-serial format, and outputs are similarly generated in bit-serial format, then any problems with pad-limited designs will be avoided. This is useful where either bit-serial chips communicate with each other or parallel-serial converters are supplied off-chip.

Disadvantages :

- Whereas a bit-parallel operator processes each word in one clock cycle, it takes a bit-serial operator W clock cycles to process a W-bit word. Hence bit-parallel arithmetic allows a much higher word rate, even though a higher clock rate may be possible with bit-serial.

- The feed-back delay and AND gate used for carry-reset in Fig. 1.7 represent an overhead that is absent in bit-parallel arithmetic.

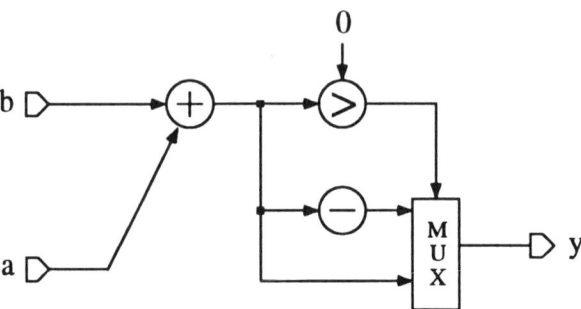

Fig. 1.8: Bit-Serial Circuit.

- The policy of inserting stage-latches after every operator in a bit-serial circuit means that the overhead due to stage latches is relatively high. In some cases these latches may be omitted, but it may be argued that the cost of doing so (in terms of decreased clock speed) is more severe in the case of bit-serial arithmetic than with bit-parallel. For instance, it may be observed that two cascaded bit-parallel ripple-carry adders will run almost as fast as one, since the carry-ripples in the two operators take place simultaneously rather than consecutively.

A reasonable conclusion is that bit-serial arithmetic has its place in the implementation of data-flow algorithms of medium complexity and medium data rate, whereas bit-parallel operators may be used for the implementation of data-flow algorithms of low complexity and high data rate. This conclusion will be reexamined when digit-serial arithmetic is discussed later.

1.3.2 Bit-Serial Circuit Architecture

Bit-serial arithmetic is generally used in so-called data-flow architectures where operators are connected together in a realization of a data-flow graph. Usually there is a minimum of control circuitry, the circuit being largely self-controlled. Such an architecture has been used in the FIRST [5]) Cathedral I [6] and Parsifal I [22] silicon compilers. Fig. 1.8 shows a typical simple bit-serial circuit. The circuit is meant to be representational only and is not meant to function just as shown. The circuit computes the absolute value of $a + b$. It may be succinctly described by the behavioural description

$$x := a + b;$$
$$y := (x > 0)?x : -x;$$

Here the ternary operator "xx ? yy : zz" is modelled on the similar operator in the C language [28] and means:

1.3. BIT-SERIAL ARCHITECTURE

"**If** xx is $TRUE$ **then** assign yy, else assign zz."

It is implemented by a multiplexor.

Suppose that the word-length in Fig. 1.8 is 16. Then a and b, being arithmetic operators, are represented by strings of 16 bits. The result of the comparison ($x > 0$), on the other hand is a Boolean quantity with value either $TRUE$ or $FALSE$. Normally, such a value may be represented with one bit. It may be noted, however that the comparison operator receives a new input only every 16 cycles. It will therefore produce only one output value every 16 cycles. Consequently, it is appropriate to represent a Boolean $TRUE$ value by a sequence of 16 consecutive 1-bits and Boolean $FALSE$ by 16 consecutive 0-bits. With this convention, the multiplexor in Fig. 1.8 will select bit-by-bit the 16 bits of x (if $x > 0$ is $TRUE$) or the 16 bits of $-x$ (if $x > 0$ is $FALSE$). This point may be noted :

- Boolean values in a bit-serial circuit have the same word-length as arithmetic values. Boolean $TRUE$ is represented by a word of 1-bits and $FALSE$ by a word of 0-bits.

- More generally, all words in a bit-serial circuit must have the same length. It is rarely feasible to change the word-length at some point in a data-flow graph, since this would mean that words of data are arriving and leaving that point at different rates.

In fact it is possible to conceive of circumstances in which the second rule above might be circumvented. For instance, it may be possible to discard every second word and simultaneously double the word-length, or alternatively have different parts of the circuit running at different clock rates. However, changing word-length is not supported by any of the bit-serial compilation systems referenced above and will not be considered here. See [10] for a discussion of multi-rate circuits.

1.3.3 Scheduling

Let us assume that the outputs of each of the operators in the circuit of Fig. 1.8 are delayed by latches. It is possible to trace the timing of signals through the circuit. Suppose that (the LSB of) a word of the signal b arrives at the input to the adder at some time $t = 0$. In order for the adder to function properly, input a must arrive simultaneously, i.e., the two inputs to the circuit must be synchronized. The adder will be said to have been scheduled at time $t = 0$. Because of the one-cycle latency of the adder, its output $x = a + b$ will be ready at time $t = 1$. The timing requirements on the adder may be represented by a diagram such as Fig. 1.9, where the parenthesized numbers on the inputs and output represent the required timing.

Now, x and its negative are passed to the multiplexor. Because of the one cycle latency of the negation circuit, these two signals will arrive at times $t = 1$

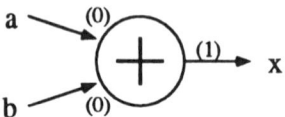

Fig. 1.9: Timing Diagram for Bit-Serial Adder.

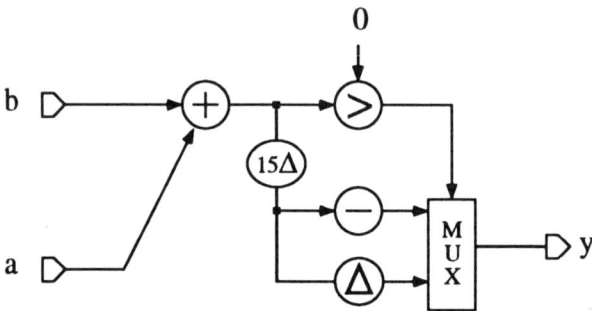

Fig. 1.10: Synchronized bit-serial circuit.

and $t = 2$ respectively. Now consider the comparator circuit. The decision as to whether $x > 0$ depends on the sign bit of x, which is the last bit of the word. For this reason, bit-serial comparators generally have a large latency, equal to the word-length of the circuit. In this case, with a word-length of 16, the comparator has latency 16, and its output will not be ready until time $t = 17$.

The three inputs to the multiplexor in Fig. 1.8 arrive at times $t = 1$, $t = 2$ and $t = 17$. Since a multiplexor will normally be designed to expect all its input signals simultaneously, the circuit of Fig. 1.8 will clearly not function correctly just as shown. In order to synchronize the inputs to the multiplexor, it will be necessary to schedule the multiplexor at time $t = 17$ and delay the inputs so that they all arrive at time $t = 17$. This may be done by the insertion of delays on some of the signals. Fig. 1.10 shows the correctly synchronized circuit. The output y is produced at time $t = 18$, and the circuit is said to have total latency 18. Algorithms for carrying out optimal scheduling of a data-flow circuit will be described in chapter 6 in a more general context.

1.3.4 Timing

The circuit of Fig. 1.10 is still not quite complete. In particular, it was seen in the bit-serial adder design of Fig. 1.7 that a *TIME* input is required to that operator to indicate the division between words. Similarly the comparator and negator in Fig. 1.8 also require such a *TIME* signal. The multiplexor carries

1.3. BIT-SERIAL ARCHITECTURE

out a simple bit-by-bit multiplexor function, treating all bits in the same way regardless of their position in a word and does not require a *TIME* signal. Because of the scheduling of the circuit, the various operators in the circuit receive their inputs at different times, and hence must receive differently timed timing signals.

The timing mechanism used in Parsifal is as follows :

- Only one type of timing signal is used. This is a periodic signal that is high for one clock cycle in each word, and is otherwise low. There are only W different timing signals, where W is the number of clock cycles in a word (i.e., the word-length in bit-serial circuits).

- The various timing signals are called $TIME_0, TIME_1, \ldots, TIME_{W-1}$ where the subscript indicates the clock cycle (within each word) in which the signal is high.

- The $TIME_0$ signal is usually an input to the complete circuit, indicating the timing of the input data signals. The other timing signals are all generated in one location by delaying the $TIME_0$ clock signal by the appropriate number of cycles.

- The *TIME* signals are routed throughout the circuit, each operator being connected to the correct one(s). Some operators may require connecting to more than one of the timing signals.

The advantage of this scheme is its simplicity. There are a maximum of W timing signals present in the circuit. On the other hand, in large circuits, the timing signals may have large fan-out, so care must be taken that the circuitry generating the timing signals has sufficient driving strength.

An alternative scheme is for each signal to have its associated timing signal and for each operator to produce in addition to its output value an output timing signal. This method has the disadvantage that the number of wires to route is doubled (for bit-serial). Furthermore, if signals are delayed, then their associated timing signals must be delayed also at the cost of additional circuitry.

Fig. 1.11 shows the circuit of Fig. 1.10 with the correct timing signals indicated. For readability, the signals $TIME_i$ are labelled T_i. It is assumed that the negation and comparator circuits, like the adder circuit of Fig. 1.7, require a timing signal high during the last bit of each word.

1.3.5 Previous Values

The circuit of Fig. 1.11 carries out computations on one "generation" of the input signals at a time. The computation is carried out over and over for each new set of inputs. In other cases it is desirable to carry out computations involving present and previous values of the inputs. A common example of

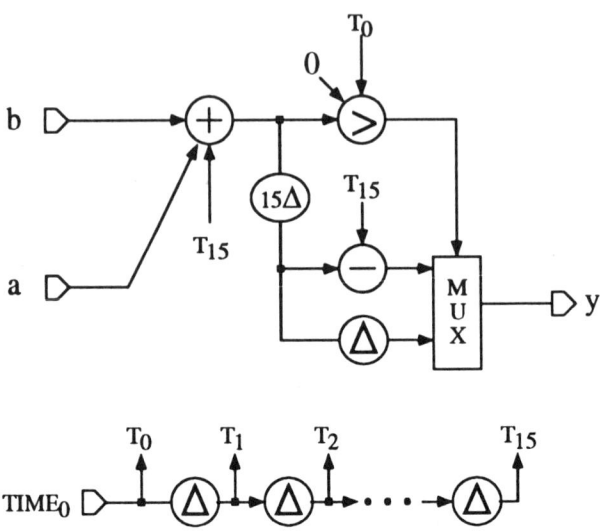

Fig. 1.11: Complete Bit-Serial Circuit.

this is in digital filter design [29]. A 4-tap FIR filter may be described by an assignment

$$y(n) = a3.x(n-3) + a2.x(n-2) + a1.x(n-1) + a0.x(n)$$

where n represents the iteration number. In the language used by Parsifal, this would be described by the formula

$$y = a3 * x[-3] + a2 * x[-2] + a1 * x[-1] + a0 * x; \qquad (1.1)$$

The symbol $x[-i]$ means the i-th previous value of x, and it is implicit that the computation is to be carried out over and over for each new input value. It is instructive to see how such a circuit is constructed. Fig. 1.12 shows the unscheduled circuit diagram for the FIR filter described by (1.1). The z^{-1} operator represents the sample delay and the triangular operator represents constant multiplication.

The correct way to handle the scheduling of the z^{-1} operator is to consider it as having a latency equal to $-W$, where W is the word-length, as shown in Fig. 1.13. Now, considering the scheduling of Fig. 1.12, it may be calculated that the input to the $a3$-multiplier is ready at time -48 (assuming 16-bit word size), and that its output is ready at time -32 (assuming the multiplier has latency 16). The output of the $a2$-multiplier is ready at time -16, however. Consequently, in order to synchronize the inputs of the first (rightmost) adder, it is necessary to insert 16 cycles of delay on its second input. With both inputs ready at time -16 now, the output will be ready at time $t = -15$. Continuing

1.3. BIT-SERIAL ARCHITECTURE

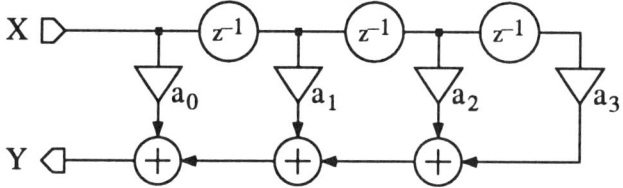

Fig. 1.12: Unscheduled FIR filter.

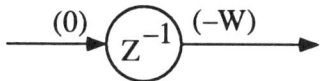

Fig. 1.13: Timing diagram for z^{-1} operator.

this analysis, the complete circuit may be scheduled. Once this is done, the z^{-1} operators may be removed from the circuit, since their only purpose was to influence the scheduling. The scheduled circuit is shown in Fig. 1.14.

Comparing this figure with the unscheduled circuit (Fig. 1.12) two points may be observed :

- The positions of the delays in Fig. 1.14 do not in any way correspond to the positions of the z^{-1} operators in Fig. 1.12.

- It is not possible to identify distinct nodes in the circuit corresponding to the values $x[-1]$, $x[-2]$ and $x[-3]$. In particular, the inputs to the four multipliers are all the same signal, x, whereas in the circuit description the inputs to the four multipliers are the signals x, $x[-1]$, $x[-2]$ and $x[-3]$ respectively.

The lack of any specific hardware corresponding to a z^{-1} operator may seem confusing at first to a DSP designer used to thinking in terms of such sample delay operators. The delay operators in Fig. 1.14 may loosely be seen as implementation of the sample delays, but in some cases, even these delays will be integrated with the latency delays of the operators in the circuit and will disappear entirely.

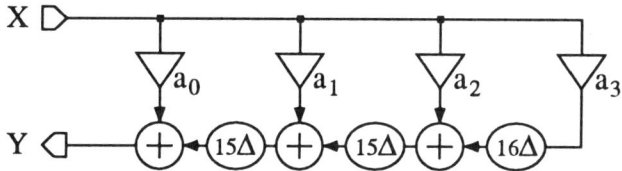

Fig. 1.14: Scheduled FIR filter.

1.4 Digit-Serial Architecture

The least-significant-digit (LSD)-first digit-serial architecture may be thought of as a generalization of the bit-serial architecture described in section 1.3. There is another form of digit-serial computation that uses most-significant digit first arithmetic (see chapter 13), but for the present, the term digit-serial will mean LSD-first. In digit-serial computation, words of data of size W are divided into digits of size N, where N is a divisor of W, and are transmitted along N separate wires, least significant digit first. A complete word is transmitted in W/N clock cycles and consecutive words follow each other without a break. The number of bits per digit or *digit-size* (N) will be referred to as the *width* of a digit-serial signal, and W/N will be referred to as its *length*. The sequence of W/N clock cycles will be called a *sample period*.

In order to understand digit-serial computation, let us consider a digit-serial adder cell, as shown in Fig. 1.15. The digit-size in this adder is 3-bits. In the first clock cycle, the three least-significant bits of A (A_0, A_1 and A_2) and of B (B_0, B_1 and B_2) are presented to the array of full adder cells. The carry-in to the least-significant position will be a zero. During the clock cycle, a ripple-carry addition of the three bits is performed and produces three bits of sum and a carry-out bit. This carry-out bit is delayed one clock cycle and fed back to the least-significant bit position. In the next clock cycle, it is combined with the next digits (each digit containing three bits) of A and B to produce the next digit of output. Thus, the operation is carried out three bits at a time. If we assume the clock rate of the digit-serial and bit-serial architectures to be the same, then the effective sample-rate of the digit-serial architecture is 3 times higher than the bit-serial since 3 bits are processed in one clock cycle in the digit-serial architecture. For a detailed analysis of area and speed performance in digit-serial architectures, see chapter 7.

As with the bit-serial adder of Fig. 1.7, a timing signal, *TIME*, controls the resetting of the carry at the end of each word. The *AND* gate and delay at the bottom of Fig. 1.15 are used for resetting and delaying the carry.

Of course, although Fig. 1.15 shows a 3-bit adder, any digit-size is possible, as long as it divides the word-length. It is also possible to design digit-serial circuits where the digit-size need not be a divisor of the word-length (at the expense of an increase in control circuitry). This is discussed in detail in chapter 8. If the digit-size equals 1, then the digit-serial adder becomes a bit-serial adder. If the digit-size equals W, then there is only one digit in each word, and the digit-serial adder is in fact a bit-parallel adder. In this case, the carry delay and reset circuitry become redundant. A digit-size between these two extremes may be chosen to make an appropriate trade-off between throughput and circuit size. In a real-time DSP application, the digit-size is chosen to match the clock speed of the hardware and the sample speed of the application.

The digit-serial adder circuit provides a model for digit-serial computation

1.4. DIGIT-SERIAL ARCHITECTURE

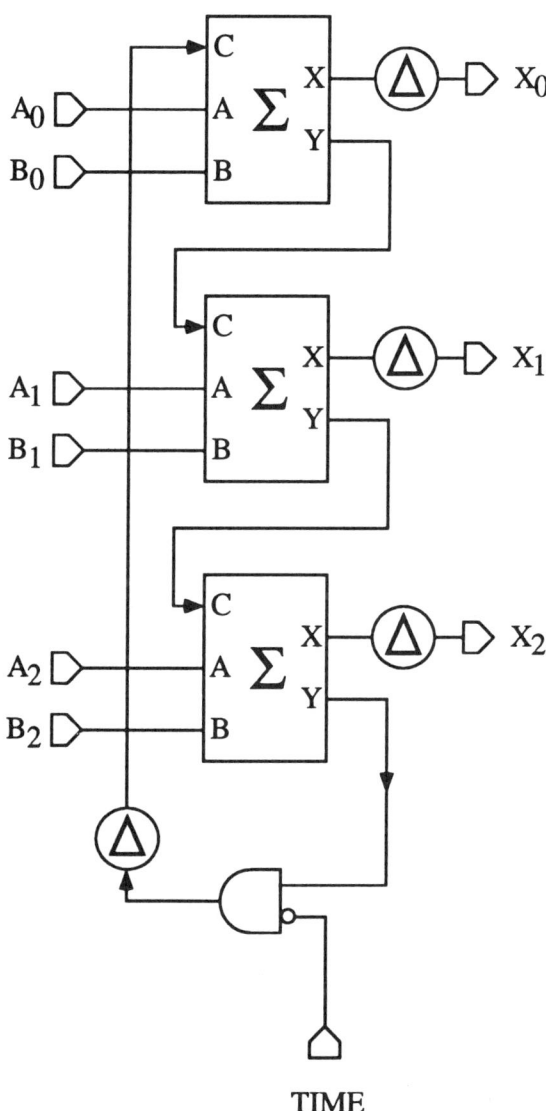

Fig. 1.15: Digit-serial adder.

in general. It is possible to design a set of arithmetic and logical operators that conform to the scheme of digit-at-a-time computation on LSD-first data. Comparing this with bit-serial computation, a number of simple observations may be made.

- Digit-serial operators will in general be larger than bit-serial operators.
- Since each operand requires N wires to transmit, the routing costs will be higher in digit-serial than in bit-serial circuits.
- Digit-serial computation will lead to a higher throughput, because each computation is carried out in only W/N clock cycles, instead of W as in the bit-serial case.
- The carry reset and delay circuitry represents a smaller overhead in digit-serial operators than in bit-serial operators. In Fig. 1.15, for instance, the cost of the carry-reset/delay circuitry is amortized over three bit-slices, whereas in bit-serial operators the same overhead is associated with a single bit-slice.

Logic values in Digit-serial: Whereas arithmetic values are transmitted over N wires in digit-serial circuits, it does not make sense to transmit logic values in the same way. In digit-serial, as in bit-serial circuits, $TRUE$ and $FALSE$ values are appropriately represented by a word of 1 or 0 values. However, it would be redundant to use N wires to transmit N identical signals representing a word of all 1's or 0's. The rule is, therefore:

- In digit-serial circuits, logic $TRUE$ or $FALSE$ values are transmitted one-bit wide.

Because of this implementation of logic values, compared with arithmetic operands they are relatively less costly in digit-serial than in bit-serial circuits.

Parallel/Digit-Serial conversion: It may (sometimes) be acceptable to have inputs and outputs to bit-serial chips in bit-serial format, since bit-serial/parallel converter chips are readily available. On the other hand, parallel/digit-serial converters are not generally available as commercial ICs. For communication between individual chips in a set using the same digit-size and word-length, it may be possible to use digit-serial IO, but in general, digit-serial chips will need to contain the necessary parallel/serial and serial/parallel converters.

1.4.1 Example Digit-Serial Circuit

The discussion of scheduling and timing for bit-serial chips applies with few restrictions to digit-serial circuits. The circuit of Fig. 1.11 is shown in Fig. 1.16

1.4. DIGIT-SERIAL ARCHITECTURE

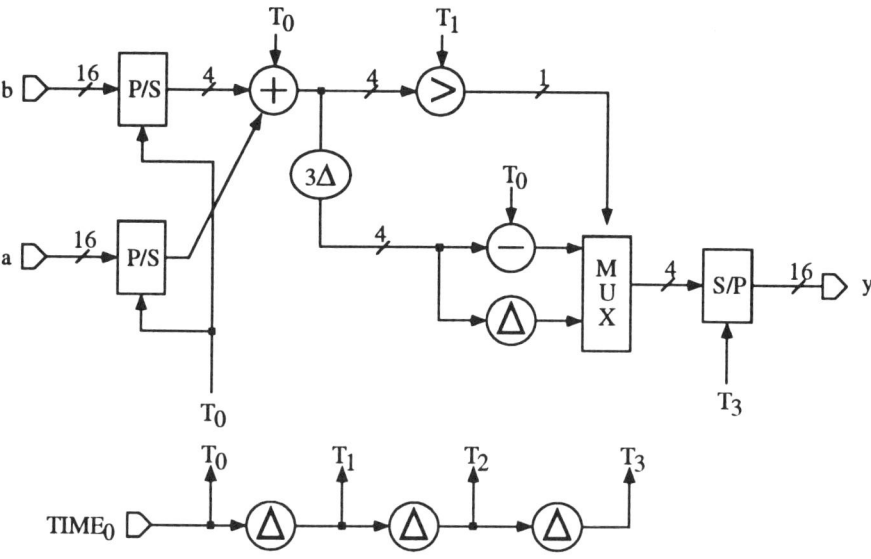

Fig. 1.16: Digit-serial circuit.

in a digit-serial implementation using a digit-size of $N = 4$ and a word-length $W = 16$.

The timing signal $TIME_0$ in this case is a periodic signal with period equal to 4 clock cycles, since there are 4 clock cycles in every sample period. The signal T_i has a high pulse at time $t = i$ and every subsequent 4 clock cycles. The two input signals a and b are each 16 bits wide, and a new value of each of a and b must be provided every 4 clock cycles. At time $t = 0$, the signals are latched into the internal parallel/serial latches and transmitted in digit-serial format in the subsequent 4 clock cycles. The first digits of the two inputs will arrive at the adder at time $t = 1$. The arithmetic operations are carried out in digit-serial format. Each of the digit-serial operators (except the multiplexor) requires a timing signal with a high pulse in the last digit of each word, as shown. It is assumed that the comparator has a latency of 4 clock cycles and that the other operators have a latency of one cycle. Note that the output of the comparator is a logic signal having width 1 bit (repeated or held for one complete sample period of clock cycles). This is used to control the multiplexor. The output of the multiplexor is a digit-serial signal representing the result of the computation.

The first digit of the result will arrive at the output of the multiplexor at time $t = 7$, and during time periods 7, 8, 9 and 10, the result will be clocked into an internal register in the serial/parallel (S/P) converter. Finally, at time $t = 11$, the timing signal T_3 will cause the assembled output word to be placed on the parallel output bus where it will remain for 4 clock cycles until replaced by the next output word. The complete operation has a latency of 11 clock cycles.

As far as the external circuitry is concerned the operation of the digit-serial circuit is carried out on parallel data, since both inputs and outputs are parallel words. It is necessary for the external circuitry to arrange for the input data to be provided at the correct rate and the output values to be read at the correct time. It is possible to arrange for the digit-serial chip itself to generate the internal signal $TIME_0$ and to provide appropriate timing signals to the external circuitry. That is, the digit-serial circuit can control the timing of the internal and external circuitry. Alternatively, all timing signals may be provided from the external circuitry.

1.5 Digit-Serial Operators

The example circuit in Fig. 1.16 shows the use of many of the important digit-serial operators. Other digit-serial operators will be discussed briefly here; a more thorough discussion of digit-serial operators are deferred to later chapters.

1.5.1 Arithmetic Operators

Addition/subtraction: The most common arithmetic operators are addition, subtraction and multiplication [30][31]. The basic design for a digit-serial adder was already shown in Fig. 1.15. It is clear that a subtractor cell could be designed in much the same way, or else a single cell could be designed to carry out either subtraction or addition. This latter course was chosen in Parsifal, as is described in chapter 2. Other useful outputs for a digit-serial subtractor or adder would be an overflow indicator and an indicator of the true sign of the result in the absence or presence of overflow.

Arithmetic negation may be accomplished either by subtraction of the operand from 0, or more economically by a special cell using half-adders rather than full adders.

Multiplication: Multiplication operations fall into two classes – multiplication of two variable numbers and multiplication of a variable by a constant. Variable-by-variable multiplication is usually carried out by an array of cells. The multiplier used in Parsifal is discussed in chapter 3. It consists of an array of bit-slice cells, one for each bit (not digit) in a word. Functionally, it is a parallel/digit-serial multiplier in which one of the values (the *multiplier*) is fed into a digit-serial/parallel register and latched, one bit in each of the bit-slice cells. Each bit-slice cell carries out a digit-by-single bit multiplication and accumulation. The other input value (the *multiplicand*) is then multiplied, one digit at a time by the parallel multiplier value. Though the cell array normally has

1.5. DIGIT-SERIAL OPERATORS

one bit-slice for each bit in the multiplier word, it is possible to use a shorter array in which the multiplicand value will be multiplied by a truncated multiplier value. Thus reduced precision will be traded for an area saving.

For constant-variable multiplication, there are two options. One possibility is that the constant multiplier design should be substantially the same as a variable-by-variable multiplication, incorporating savings possible because the constant of multiplication is known. Vaguely speaking, in some multiplier designs, if a given bit-position in the multiplier constant is zero, then a zero value (i.e., nothing) needs to be accumulated, and so the cell design can be simpler in that bit position. The specific savings possible depend on the multiplier design. The other possibility is that the constant multiplication is carried out by quite a different technique, such as shifting and addition or subtraction. This technique is discussed in chapter 12.

Division: Division is not an operator that is explicitly implemented in Parsifal. It is, however, possible to carry out a division operation by the use of addition/subtraction operators, as will be discussed in chapter 2.

1.5.2 Comparison Operators

Inequality testing: The greater-than ($>$), less-than ($<$), not-greater-than ($\leq$) and not-less-than ($\geq$) operators on arithmetic data are quite similar to the subtraction operator. In fact to compare two values, the simplest manner is to subtract them and look at the sign of the result. One pitfall is that it is not good enough simply to use the sign-bit of the output of a subtractor, since if overflow occurs, the result will have the wrong sign. Thus, overflow must be taken into account. One solution is to subtract the two values and then use *sign-bit $\oplus$ overflow* as the result of the comparison. By swapping the two inputs and selecting the initial carry-in to the subtractor to be a 1 or 0, it is possible to get all four of the above comparison operations.

In accordance with the conventions described above for digit-serial computation, the result of the comparison is a single-bit value, which must be repeated (i.e., sampled and held) for a complete sample period. Since the result of a comparison operation cannot be known until the most-significant digits of the two operands are seen, comparison will have a latency of one complete sample period in LSD-first digit-serial computation. This relatively high latency for such a simple operator is certainly a draw-back of LSD-first computation, for it leads to the necessity for inserting scheduling delay cells in the circuit, such as the 3Δ delay in Fig. 1.16. Most-significant-digit (MSD) first arithmetic (see chapter 13) avoids this difficulty.

Equality testing: The equality ($=$) comparison operator is somewhat different from the other comparison operators described above in not involving subtraction. However, in other respects it is the same, involving a digit-by-digit comparison and a carry from digit-to-digit indicating the result of the comparison on the digits already seen. It also has a latency of one sample period. However, in this case, MSD-first digit-serial equality testing must also have a latency of a full sample period. However, equality comparisons are much less common in Digital Signal Processing (DSP) algorithms than the inequality comparisons just discussed. A cell design that uses the same cell to do addition, subtraction, equality and inequality testing is described in section 2.2.

Comparisons against zero (for instance $a < 0$ or $a \leq 0$) are sufficiently common that it may be worthwhile treating them separately, since substantial saving may be possible. For instance, to implement the comparison $a < 0$ it is sufficient to latch the sign bit of a and repeat it for one sample period. The comparison $a \leq 0$ is not quite so simple, but can certainly be done with less circuitry than a comparison $a \leq b$. The capability of using special simple cells for such comparisons is designed into the Parsifal software and is used for bit-serial, but not for digit-serial designs.

1.5.3 Logic Operators

Simple cells are necessary for the implementation of Boolean (one-bit-wide) operators. These operators include AND, OR, and NOT. Less frequently used are the corresponding bit-wise AND, OR and NOT operators on arithmetic (digit-wide) data. The design of these cells is sufficiently simple not to warrant further comment.

The multiplexor operator as used in Fig. 1.16 is unique among the operators considered so far in having three inputs. One of the inputs is a Boolean input and the other two are arithmetic (digit-wide) values. The multiplexor selects one of the two arithmetic inputs depending on the Boolean value. It is assumed that the one-bit-wide Boolean input is made up of words of all 1 or all 0-bits.

1.5.4 Constants

Digit-serial constant values occur in such expressions as $y := x + 0.12345$. A digit-serial constant is not, however, represented by a set of wires hard-connected to predetermined 0 or 1 bit values, but rather by a periodic signal repeating the digit-serial constant in each sample period. Such a signal is generated by a sort of parallel/digit-serial converter with the parallel value hard-wired to predetermined 0 and 1-bits. Alternatively, digit-size (N) number of independent circular buffers can be used where each circular buffer contains sample period (W/N) number of buffer registers.

Intelligent constant generation: It is possible to be a little clever in doing this. The method used in Parsifal for generating constants can be explained

1.5. DIGIT-SERIAL OPERATORS

with an example. Suppose that the digit-size is 4, and we wish to encode the constant 0.4. This value has a binary representation $0.011, 0011, 0011, \ldots$ where the period represents the binary point and the commas indicate the division into digits. As may be seen, all the digits of this constant are identical. Considering each bit position separately, it may be seen that bits 0 and 1 in each digit (the two least significant bits) have the value 1, whereas bits 2 and 3 in each digit have the value 0. Thus the digit-serial constant may be generated by hard-wiring bits 0 and 1 to a 1 value and bits 2 and 3 to a 0 value. Thus, the constant is implemented using no active circuitry.

In general, the constant value is examined one bit-position at a time. If the bit position contains a constant value for all digits in the word, then that bit is implemented by hard wiring. If not, then an appropriate periodic signal must be generated for that bit position. For instance in the constant $0.101, 1110, 0100, 1111$ bit 2 contains a constant 1 value, whereas the other three bits have varying bit values for the different digits. Closer examination shows that bits 1 and 3 carry the same bit pattern 0101. Consequently, there is no need to duplicate the circuitry that generates this periodic pattern. A proper design of the cell also allows the Parsifal routing software to economize on interconnect by using just one wire to carry the two identical signals to their destination. Furthermore, no intercell wiring will be used to carry the constant 1 and 0-bits, for they are connected locally in the destination cell.

Generators: The decisions made in implementing constant values are carried out by Parsifal independently of the main compiler software, by a separate so-called *generator* program, the sole task of which is to generate descriptions of digit-serial constant cells. This allows the compiler to instruct the generator to provide a constant 0.4 cell without worrying about the details of the design. There are many other types of generator programs used by Parsifal. The use of separately compiled generator programs is the key to easy extendability of the capabilities of Parsifal. It also results in a clean software design by keeping the main Parsifal compilation program free from detailed operator-specific choices which would otherwise clutter the code. The generator mechanism is described in detail in chapter 4.

1.5.5 Delay Operators

Delay cells are needed for synchronization delays as shown in Fig. 1.16. Delays of Boolean (one-bit wide) signals as well as arithmetic (digit-wide) and also parallel signals of arbitrary width may be called for. The compiler calls a generator to construct suitable delay cells as required. More will be said about the construction of delay cells in section 2.1.2.

1.5.6 Shift Operators

Shift operators are required for carrying out arithmetic shifting of digit-serial arithmetic values. Shifting a value x one place right (denoted $x >> 1$) is the same as division by 2. Sign bits must be shifted in at the left hand end (most significant bits) of the arithmetic word. Shifting one place left ($x << 1$) is equivalent to multiplication by 2, zero-bits being shifted in to the right hand end of the word. Of course shifting left may cause overflow if the sign bit of the word changes.

With digit-serial representation of arithmetic values, shifting causes some bits to move across digit boundaries. Thus some bits are shifted into another digit, whereas other bits remain in the same digit, but are moved into a different position in the digit. Thus shifting involves delaying some bits in each digit (to move them into a different digit), rotating the bits in a digit and some mechanism for repeating the sign bit in shift-right operators. A design for shifters using a small number of different cells is discussed in section 2.3.

1.5.7 Converters

Converter modules for converting between digit-serial and parallel data formats are shown in Fig. 1.16. The concept of these converters is quite simple. For parallel-serial conversion, a parallel word is latched into a digit-serial shift register and is shifted out one digit at a time. A serial-parallel converter carries out the above action in reverse. Since it takes one complete sample period to shift out a digit-serial constant, a new value cannot be latched more often than once in every sample period. This means that the data rate on the parallel bus is slower than the data rate on the serial output.

In order to allow the data rate of the parallel bus to be matched to that of the serial output, parallel-serial converters are available in Parsifal having one parallel input and up to W/N serial outputs, where W is the word-length, N is the digit-size and W/N is the sample period. Words on the parallel input bus are latched as often as every clock cycle into different registers. The individual words are then shifted out on different digit-serial channels. For instance, the A and B inputs in Fig. 1.16 could be interleaved on a single parallel input bus. A parallel-serial converter could then be constructed to separate them and direct them to separate digit-serial outputs. The dual operation of interleaving separate digit-serial inputs onto a single parallel output bus may be accomplished by a suitable serial-parallel converter. Serial-parallel and parallel-serial converters are discussed in greater detail in section 4.4.

1.5.8 Timing Signal Generation

The *TIME* signal may either be generated on-chip or be provided as an input to the chip. It can be generated on-chip by using the method described for constant

1.6. LAYOUT OVERVIEW 23

generation above. The various phases of the *TIME* signal are generated by delaying it by successive clock cycles. In a large circuit, the *TIME* signals have a large fan-out, since most operators in the circuit need at least one of the phases of the *TIME* signal. Therefore, sufficient driving power must be provided to drive this fan-out.

1.5.9 Special Cells

The set of cells already described gives sufficient capability to design a large number of digit-serial circuits. It is possible to add extra capability by providing extra types of cells. Some of the extra cells available in Parsifal are described in the following paragraphs.

ROMS: A ROM capability is provided. Simple addressable ROMs of arbitrary size may be built in accordance with a specification file. The main purpose for ROMs is to provide control signals for controlling the action of the digit-serial circuit, or for the implementation of functions as look-up tables. ROMS will be described more thoroughly in section 4.4.5.

Data Registers: Addressable data registers are used to latch values from an asynchronous parallel input bus. The output of a register is a parallel data word of any specified width. If required the output may be converted to serial by a parallel-serial converter. Data registers may be used for instance to store the coefficients of a programmable filter. The syntax and use of data registers will be more thoroughly discussed in the section 4.4.4.

Finite state machines: Finite state machines would be a useful feature. Though not implemented in current version of Parsifal, the generation of simple FSMs would be a relatively easy user-addition to current capabilities. The main problem is to find a method for laying out the circuit in a manner compatible with the general layout scheme used in Parsifal, described in chapter 5. As described in section 4.2.1 a generator could then be written to generate the layout in response to specific requirements. Parsifal is designed to be user-extendable, so this could all be done by an experienced user without requiring any change to the compiler software. The use of generators to extend the capabilities of Parsifal is described in section 4.2.1.

1.6 Layout Overview

The layout style used by Parsifal is an example of an effective way of laying out digit-serial circuits. It is essentially a standard-cell approach in which the complete circuit is made up from cells of a given fixed height, arranged in rows separated by routing channels. An example of a standard cell is the digit-serial adder shown in Fig. 1.15.

1.6.1 Bit-Slicing the Standard Cells

To allow any arbitrary digit-size, the standard cells from which the circuit is constructed are made up from smaller cells stacked vertically using bit-slicing. In order to understand the principles of bit-slicing digit-serial operators, let us consider a digit-serial adder cell as shown in Fig. 1.17. In this figure, it may be seen that the adder cell may be divided up into two major types of sub-cells containing essential circuitry. First, there are the cells marked *SLICE* in Fig. 1.17. These cells must be repeated as many times as there are bits in the digit. Second, there is a single cell marked *CONTROL* in Fig. 1.17. This cell contains circuitry that must be contained just once in the digit-serial operator cell. To complete the digit-serial adder, there is a cell marked *CAP* in Fig. 1.17 which contains no active circuitry, but carries the *VSS* power bus and makes a connection of the wire holding the carry signal.

Thus, the digit-serial adder may be broken down into three types of cell, a *CAP* cell containing minor connections, several *SLICE* cells containing circuitry that is repeated for every bit in the digit, and a *CONTROL* cell that contains once-only circuitry. Many digit-serial operators have the same sort of division. In fact, most arithmetic operators may be sliced up in such a natural way. For instance, adders, subtractors, shifters, bit-slices of multipliers, comparators and format converters may easily be bit-sliced in this manner. Other cells such as Boolean operators do not naturally allow bit-slicing, since they act on one-bit wide signals. However for the sake of consistency, all operators are designed in such a way that they may be built from an extendable stack of cells to form standard cells of uniform height.

1.6.2 Complete Chip Layout

Finally, the complete chip layout of a digit-serial chip is shown in Fig. 1.18. Some details of algorithms used to achieve this layout are given in chapter 5. This completes the overview of the digit-serial style of computation.

1.6. LAYOUT OVERVIEW

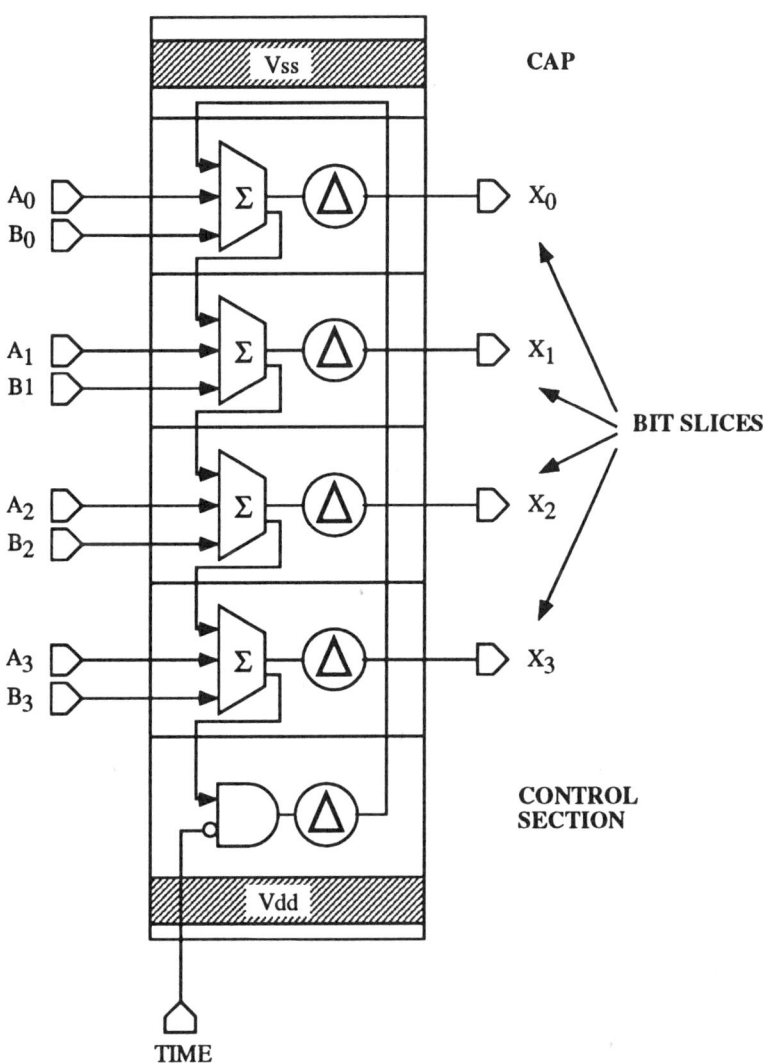

Fig. 1.17: Digit-Serial Adder Cell.

26   CHAPTER 1. DIGIT-SERIAL ARCHITECTURE

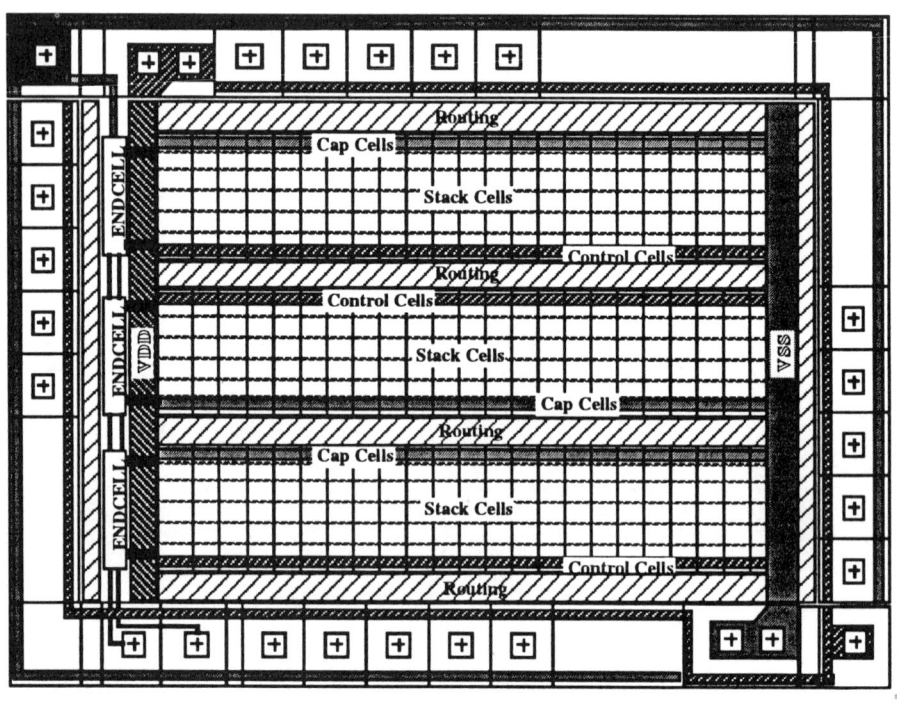

Fig. 1.18: Digit-Serial Layout.

Chapter 2

Digit-Serial Cell Design

2.1 Digit-Serial Layout

In the digit-serial layout shown in Fig. 1.18, the individual operator cells were all pre-designed to have the same height. This gives special advantages in that the whole chip may be conveniently and efficiently laid out in rows separated by routing channels. In the Parsifal digit-serial compiler, in which the user is allowed to choose an arbitrary digit-size (number of bits per digit), operator cells are constructed to correspond to the specified digit-size. For a given specified digit-size, assumed constant for the whole chip, the height of all types of cells must be the same. Clearly, making the operator cells bit-sliced is a commonly used technique for the construction of parallel operators. Digit-serial operators are more complicated than parallel operators, however, and this leads to additional complication in the construction of bit-sliced digit-serial operators. The construction of a bit-sliced digit-serial multiplier is the most involved of all the common operators.

2.1.1 The Standard Template

It was seen in section 1.6.1 that the bit-sliced design of most digit-serial arithmetic cells can be divided up into cells of three types – a *CAP* cell, several *SLICE* cells and a *CONTROL* cell. For this reason, the template shown in Fig. 2.1 was taken as a basic template for the design of digit-serial cells in the Parsifal compiler. In general, in order that given a fixed digit-size, all digit-serial cells should have the same height, the following conventions were adopted:

1. All digit-serial operators should be made up as a stack of 3 types of cells : a *CAP* cell, N *STACK* cells (where N is the digit-size), and a *CONTROL* cell.

28   CHAPTER 2. DIGIT-SERIAL CELL DESIGN

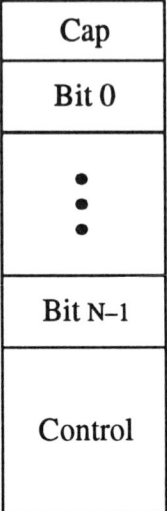

Fig. 2.1: Digit-Serial cell template.

2. The *CAP* cells for all the different cell types should have the same height. The same is true of the *SLICE* and *CONTROL* cells.

3. The *CAP*, *SLICE* and *CONTROL* cells for a given operator type should all have the same width.

Thus a digit-serial adder cell is made up of cells of three types called *ADDER_CAP*, *ADDER_SLICE* and *ADDER_CONTROL*, and the height of a complete N-bit digit-serial adder cell is

height($ADDER\_CAP$) + N * height ($ADDER\_SLICE$) + height ($ADDER\_CONTROL$).

In reality, it was found that it was not possible to hold precisely to this convention in the design of all cells, and the ways in which different cells deviated from this convention will be noted.

2.1.2 Delay Cells

The design of digit-serial delay cells will be considered in this section. As noted in section 1.4, there will be signals of several different widths present in a digit-serial circuit. The two most common signal widths will be 1-bit (as for instance in Boolean logic signals) and N-bit, where N is the digit-size. It is desirable to design delays cells for arbitrary width signals from a small number of different cell types. In an N-bit delay, it is appropriate for the delayed signals to pass

2.1. DIGIT-SERIAL LAYOUT

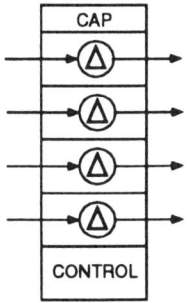

Fig. 2.2: Digit-Serial delay cell.

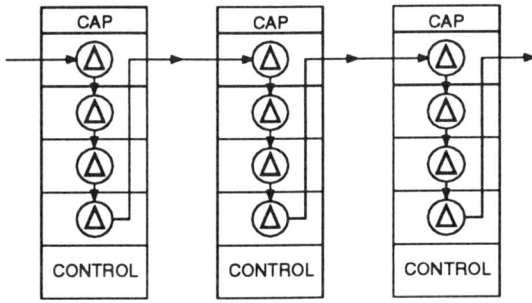

Fig. 2.3: Cascade of one-bit wide delay cells.

directly across the cell, so that it may fit conveniently between two digit-serial operators, as illustrated in Fig. 2.2. On the other hand, such a layout is not appropriate for instances where a signal of width 1 is to be delayed for several clock cycles. In such a case, a more appropriate arrangement is shown in Fig. 2.3. In order for cells of these and other types to be created conveniently a small number of cell types were designed. The essential delay circuitry is contained in the cell of Fig. 2.4. Apart from this there are two distinct sorts of connector cells as shown in Fig. 2.5 and Fig. 2.6. Not shown in any of these cells are the necessary clock signals that must pass from bottom to top of the whole *DELAY* stack. Apart from these cells, there are the usual

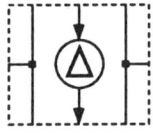

Fig. 2.4: Cell *DELAY_SLICE1*.

Fig. 2.5: Cell *DELAY_CONNECTOR1*.

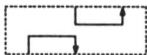

Fig. 2.6: Cell *DELAY_CONNECTOR2*.

CONTROL and *CAP* cells as well as a second slice cell *DELAY_SLICE2* having the same connections as *DELAY_SLICE1*, but no active circuitry. It acts as a filler where required. The *DELAY_CONTROL* cell contains no circuitry other than clock inverters and buffers used to drive the clock up the DELAY stack. A digit-serial delay cell consists of a stack made up of a *DELAY_CAP* cell, N *DELAY_SLICE/DELAY_CONNECTORi* pairs and a *DELAY_CONTROL* cell at the bottom. By choosing the appropriate *DELAY_CONNECTOR* cells, one of many connection topologies is constructed. For instance, the digit-serial delay cell of Fig. 2.2 is constructed using *DELAY_CONNECTOR2* cells as shown in the Fig. 2.7. On the other hand, the one-bit wide delay cell is constructed by separating the *DELAY_SLICE* cells with *DELAY_CONNECTOR1* cells as shown in Fig. 2.8.

2.1.3 Deviation From The Standard Template

Fig. 2.7 and Fig. 2.8 illustrate how the design of the delay cells deviate from the standard template used for the design of the digit-serial adder cell. In the delay cell, two cells take the place of the single *SLICE* cell. In particular, for each bit in the digit, a *DELAY_SLICE* cell and a *DELAY_CONNECTOR* cell are added to the stack of cells making a complete digit-serial delay. In order that the delay stack should be the correct height for each digit-size, a simple relationship must hold, namely

height(*DELAY_SLICE*) + height (*DELAY_CONNECTOR*)
= standard *SLICE* height.

The digit-serial delay stack differs from the standard template in another way. Since the *DELAY_CONTROL* cell has very little active circuitry, containing only clock inverters and buffers, it does not need to be as high as the *ADDER_CONTROL* cell. In fact, in the space of a standard *CONTROL* cell, it is possible to fit two extra *DELAY_SLICE*s as well as the *DELAY_CONTROL*. This is expressed by the relationship

height (*DELAY_CONTROL*) + 2 * (height (*DELAY_SLICE*) + height (*DELAY_CONNECTOR*)) = standard *CONTROL* height.

2.1. DIGIT-SERIAL LAYOUT

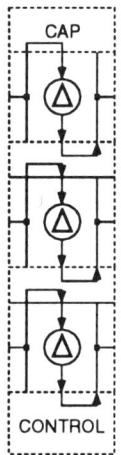

Fig. 2.7: Construction of digit-width delay cell.

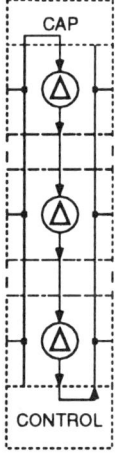

Fig. 2.8: Construction of one bit wide delay cell.

These extra slices allow two extra delay stages to be added to a cell in implementation of a one-bit-wide delay cell as in Fig. 2.8.

2.2 The Add/Subtract/Compare Cell

A digit-serial adder cell was described in Section 1.6.1 to illustrate the method of bit-slicing digit-serial operators. It clearly shows the cap, bit-slices and control sections of the standard template. The actual cell as designed and included in the Parsifal digit-serial library has more general capabilities than the simple adder cell shown there. It is also slightly larger than a specialized adder or subtractor cell, but this is made up for by its versatility. Fig. 2.9 shows a four-bit digit-serial add/subtract/compare cell. The extra output at the bottom of the adder in each bit slice carries the value of $A \oplus B$ (exclusive OR), where A and B are the two adder input bits. The adder is easily designed in such a way that this signal is available without the need for extra circuitry. The purpose of this output will be explained later. The general method of operation of this cell is an addition in which the carry ripples through 4 bits before being latched and fed back to the top of the array in the next clock cycle. The carry value is also reset to an initial value (*Reset_bit*) on the least-significant-digit of input. A *Mode* input signal controls the inversion of the B input bits. There are two timing signals *TIME_LSD* and *TIME_MSD* which are high during the least-significant-digit and most-significant-digit of the input respectively. These signals are used to control the resetting and latching of values. Correct timing of these timing signals is ensured by the Parsifal synchronization software.

To do addition, *Reset_bit* and *Mode* are set to zero causing the carry to be reset to a zero value and addition to be carried out in a straightforward manner. To do subtraction, *Reset_bit* and *Mode* are connected to 1. This causes the B input to be complemented and a 1 to be carried in to the initial carry-in position, resulting in a subtraction of B from A. In either case, addition or subtraction, the *SignBar* output gives the inverted true sign of the result. The true sign is the mathematically correct sign of the addition or subtraction operation. It will differ from the computed sign, or high order bit of the output, if overflow occurs. The true sign of the result of a computation is given by $SA \oplus SB \oplus CO$ where CO is the carry out of the last bit of the computation and SA and SB are the sign bits of the A and B inputs (after possible inversion of B). To see this, note that this is precisely the formula for the sum in an imagined higher order bit which would be produced if the words were sign extended by one bit. In the case of such a hypothetical extension, no overflow would occur, since the range of numbers is doubled. During the final digit of computation of the result, as marked by the *TIME_MSD* signal, the value of $SA \oplus SB \oplus CO$ is latched and held for one complete sample period to indicate the inverted true sign of the preceding computation. In comparing two numbers to determine which is greater, we need to subtract them and look at the true sign of the

2.2. THE ADD/SUBTRACT/COMPARE CELL

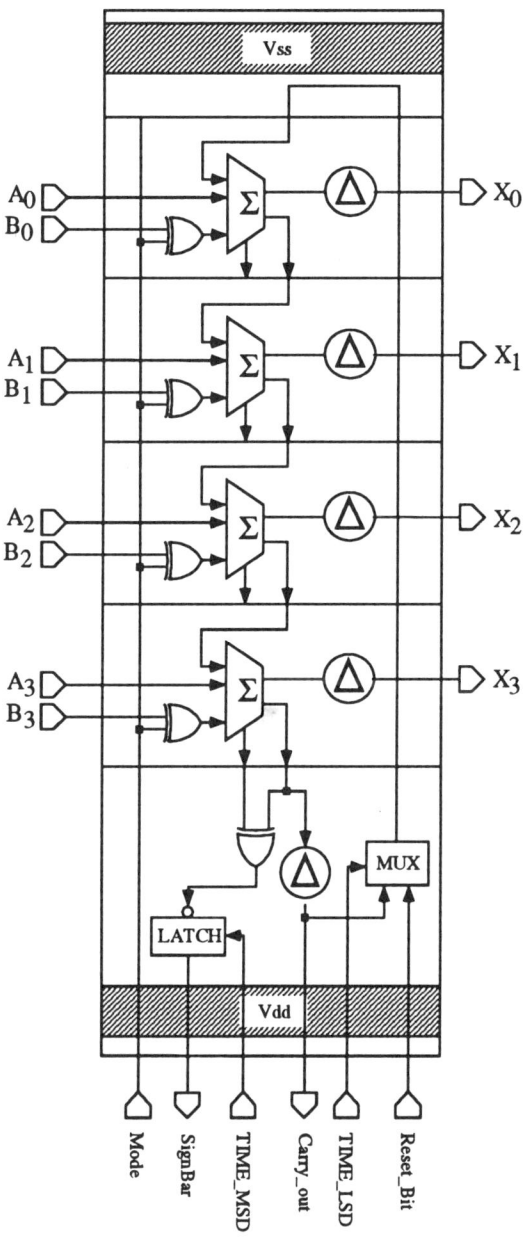

Fig. 2.9: Digit-serial add/subtract/compare cell.

result. Hence, setting $Mode = Reset\_bit = 1$, the value of $SignBar$ gives the result of the comparison $(A \geq B)$. On the other hand, setting $Mode = 1$ and $Reset\_bit = 0$ gives the result of $(A > B)$. Reversing the A and B inputs gives comparisons $A \leq B$ and $A < B$.

2.2.1 Division

Division is difficult in least-significant-bit first serial computation because it naturally takes place on a most-significant-bit first basis. On the other hand, addition and multiplication occur most naturally least-significant-bit first. One possible solution to this dilemma is to use a most-significant-digit first redundant data format (see [32]). This is not possible using least-significant-digit first digit-serial architecture. It will be shown, however, that this add/subtract/compare cell can be used to do division using the non-restoring division algorithm given below [30][31]. A is the dividend, B the divisor and Q the quotient. It is assumed that $|A| < |B|$.

```
/* Make sure B is positive */
if (B < 0)
    {
    B = -B;
    A = -A;
    }

/* Loop computing each of the W bits of the result */
for i = 1 to W
    {
    /* Add or subtract */
    if (A > 0)
        A = A - B;
    else
        A = A + B;

    /* Shift A ready for next iteration */
    A = A << 1;

    /* Bit of result is given by sign of A */
    Q_{W-i} = sign-bit of A
    }
```

The word $(Q_{W-1}, Q_{W-2}, \ldots, Q_0)$ is the quotient, with the binary point directly after the sign bit. Fig. 2.10 shows a non-restoring division circuit. For simplicity, we use the convention that the input shown on the bottom of the cell is connected to the $Mode$ and the $Reset\_bit$ inputs, and the output from the

2.3. DIGIT-SERIAL SHIFTING

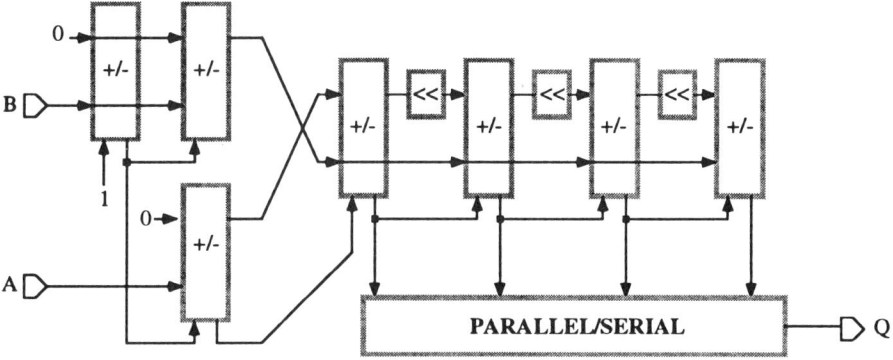

Fig. 2.10: Non-restoring division.

bottom of the cell is the *SignBar* output. Thus a 1 input means subtraction and a 0 means addition. A 1 output means a positive result and a 0 means a negative result. The cells denoted $<<$ are shift-left-1 cells. Synchronizing delays are not shown here, since they depend on the digit-size. We show here the (unrealistic) case where all the cells have zero latency. The first part of the circuit detects whether $B < 0$, and if it is, changes the sign of both A and B in order to ensure that $B \geq 0$. Then a non-restoring division algorithm carries out the division by repeatedly adding or subtracting B from A with the decision to add or subtract being determined by the sign of the previous addition/subtraction result.

2.3 Digit-Serial Shifting

Whereas shifting by fixed amounts in a fully-parallel environment is easily achieved by hard-wiring, shifting in a serial context requires the use of active circuitry. There are two separate problems in shifting digit-serial operators: shifting by complete digit amounts and shifting by less than a digit. As will be shown, a general shift, either right or left may be achieved with only two different types of bit slice. In order to understand digit-serial shifting, we consider bit-serial shifting first. Consider a bit-serial 2-bit left-shift operation with word-length of 6. The input and output streams are as follows with the leading (low order) bits to the right.

...	b5	b4	b3	b2	b1	b0	a5	a4	a3	a2	a1	a0	...
...	0	0	b3	b2	b1	b0	0	0	a3	a2	a1	a0	...

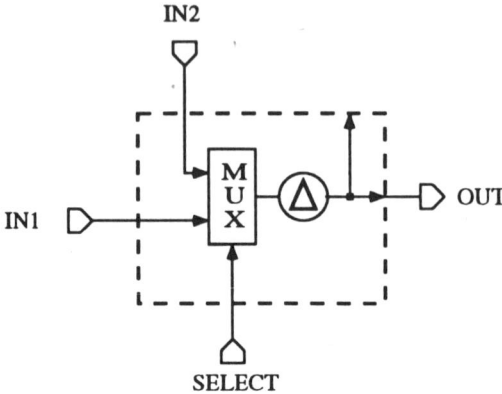

Fig. 2.11: Bit-serial shifter cell.

Here the vertical bar denotes the division between words. Notice that overwriting the two high order bits of the previous word with 0 gives an effective left shift by two places. Thus a left shifter is simply a cell which passes its input straight through or replaces it with a zero. Now consider a right shift by two places with repetition of the sign bit. The input and output streams are:

...	b4	b3	b2	b1	b0	a5	a4	a3	a2	a1	a0	c5	...
...	b4	b3	b2	a5	a5	a5	a4	a3	a2	c5	c5	c5	...

Right shifting is achieved by overwriting the low order bits of each input word by the previous output value. A cell which will accomplish either bit-serial right or left shifting is shown in Fig. 2.11. The input *IN2* is connected either to a zero value (for left shifting) or to the output (for right shifting). The multiplexor selects one of the two inputs on the basis of a properly timed alternating select signal which controls the amount of shift.

Now let us turn to digit-serial shifting. For illustration, we choose a word of 12 bits divided into 4-bit digits and consider a 6-bit left shift. The complete operation is illustrated in Fig. 2.12. The input stream appears in Fig. 2.12(a), where a heavy bar denotes the division between words. The shift operation can be conceived as taking place in three stages : zero-substitution, differential shifting and cyclic bit-rotation. In the first stage (Fig. 2.12(b)) zeros are substituted in place of the trailing (high order) bits of each of the individual bit streams, just as in the bit-serial left shift. Notice that, in each word, in the high order bit positions of the digits, two zeros are substituted whereas in the low order positions, one zero is substituted. In the second stage (Fig. 2.12(c)), the high order bit positions are delayed one position (one clock cycle). Finally,

2.4. CORDIC OPERATORS

in the cyclic rotation stage, the bit streams are rotated cyclically, producing the correct output shown in Fig. 2.12(d). Right shifting is accomplished in a similar manner to left shifting except that instead of zero-substitution, sign-bit repetition is used, as with bit-serial shifting.

Right and left shifts of all degrees may be accomplished using only the two different types of bit-slice as shown in the example Fig. 2.13 which shows the configuration necessary to do the 6-bit left shift considered above. Since two different amounts of shift are involved (either 1 or 2 places in the example above), two different control lines are required to select the multiplexors in the two types of cells. Normally, these two control signals are generated in the control section of the cell stack but the circuitry to accomplish this is not shown here. Note that in this design each bit stream is subjected to an additional one-cycle delay not considered in Fig. 2.12. This is done to preserve the systolic, or pipelined nature of data transmission. The final cyclic permutation of the bit streams implemented in Fig. 2.13 by bit-sliced braiding cells may alternatively be left to be handled by the routing software.

2.4 Cordic Operators

The design of the Add/Subtract/Compare cell makes it particularly suitable also for the implementation of Cordic operators, such as in the computation of the arc-tangent function, or equivalently, the conversion from cartesian to polar coordinates in the plane [33] [34]. A brief description of the Cordic algorithm follows.

Given a pair of coordinates (x, y) in the plane, representing a vector, v, the task is to compute the polar coordinates of the point. A sequence of preliminary steps may reduce the problem to consideration of a point in the first octant of the plane (that is, one may assume that $x \geq y \geq 0$). The polar coordinates of a general point may easily be obtained by appropriate addition or subtraction of some multiple of $\pi/2$ to the argument of the vector.

Next, one proceeds by rotating the vector v by a sequence of positive or negative angles θ_i until it lines up with the positive x axis. The sum of the rotations will then be the argument (i.e., θ) of the original vector. Rotating a vector through an angle θ_i may be achieved by multiplying it by the matrix $\begin{pmatrix} \cos \theta_i & -\sin \theta_i \\ \sin \theta_i & \cos \theta_i \end{pmatrix}$. In general, this will require four multiplications. Instead of doing this, one multiplies by the matrix $\begin{pmatrix} 1 & -\tan \theta_i \\ \tan \theta_i & 1 \end{pmatrix}$ which will require only two multiplications. The length of the vector will be multiplied by a factor $1/\cos \theta_i$, and this will be accounted for at the end. After a sequence of rotations through angles $\pm \theta_i$ the vector will be multiplied by a factor $1/\prod \cos \theta_i$.

The sequence of rotation angles are chosen such that each successive rotation rotates the vector towards the x-axis. In other words, if the vector lies in the

CHAPTER 2. DIGIT-SERIAL CELL DESIGN

c8	c4	c0	b8	b4	b0	a8	a4	a0
c9	c5	c1	b9	b5	b1	a9	a5	a1
c10	c6	c2	b10	b6	b2	a10	a6	a2
c11	c7	c3	b11	b7	b3	a11	a7	a3

(a) Input Stream

0	c4	c0	0	b4	b0	0	a4	a0
0	c5	c1	0	b5	b1	0	a5	a1
0	0	c2	0	0	b2	0	0	a2
0	0	c3	0	0	b3	0	0	a3

(b) After zero substitution

0	c4	c0	0	b4	b0	0	a4	a0
0	c5	c1	0	b5	b1	0	a5	a1
0	c2	0	0	b2	0	0	a2	0
0	c3	0	0	b3	0	0	a3	0

(c) After differential delays

0	c2	0	0	b2	0	0	a2	0
0	c3	0	0	b3	0	0	a3	0
0	c4	c0	0	b4	b0	0	a4	a0
0	c5	c1	0	b5	b1	0	a5	a1

(d) Output after cyclic bit rotation

Fig. 2.12: 6-bit Digit-Serial Left Shift.

2.4. CORDIC OPERATORS

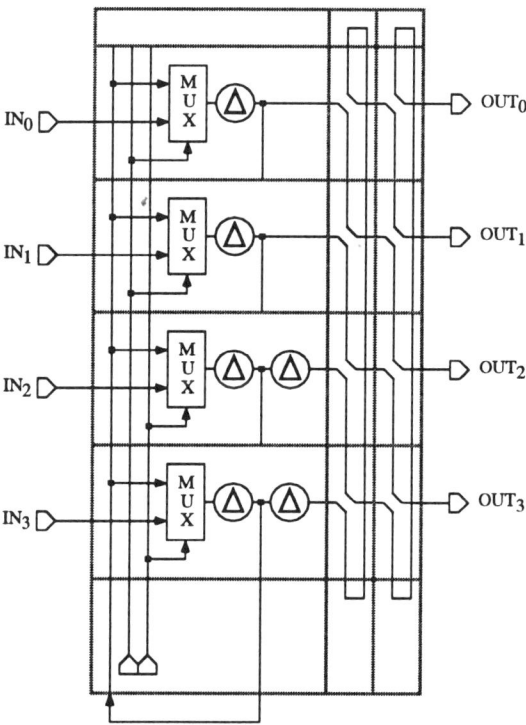

Fig. 2.13: Digit-serial shifter.

first quadrant, then the next rotation will be negative. If it lies in the third quadrant, then the rotation will be positive. The magnitudes of the angles θ_i are chosen so as to avoid doing any actual multiplication at all. This is done by choosing $\theta_i = \arctan(2^{-(i+1)})$. With this choice, successive rotations involve only a multiplication by the matrix $\begin{pmatrix} 1 & -2^{-(i+1)} \\ 2^{-(i+1)} & 1 \end{pmatrix}$ or its inverse. Thus the rotations may be done using only shifts and adds.

The algorithm makes use of an operator *addsubcomp* with syntax

$(c, sign) := $ addsubcomp (a, b, sel) ;

where c is assigned the value $a + b$ if $sel == 0$, and $a - b$ if $sel == 1$. In either case, *sign* is assigned the value 1 if the result c is positive, and 0 if it is negative. Such an operation is implemented using the add/subtract/compare cell.

At the start, the vector $(x, y) = (x_0, y_0)$ is in the first quadrant. This is represented by the variable $sign_0 = 1$. The first rotation is negative and the first stage of the cordic operation is

$x_1 \quad\quad := x_0 + y_0 >> 1;$
$(y_1, sign_1) := $ addsubcomp $(y_0, x_0 >> 1, sign_0);$
$\theta_1 \quad\quad := -atan_0;$

At subsequent stages, the operation carried out is

$dx_i \quad\quad\quad\quad := y_i >> (i+1);$
$dy_i \quad\quad\quad\quad := x_i >> (i+1);$
$x_{i+1} \quad\quad\quad := $ addsub $(x_i, dx_i, !\, sign_i);$
$(y_{i+1}, sign_{i+1}) := $ addsubcomp $(y_i, dy_i, sign_i);$
$\theta_{i+1} \quad\quad\quad := $ addsub$(\theta_i, atan_i, sign_i);$

where $atan_i$ is a stored constant value equal to $\arctan(2^{-(i+1)})$ and the *addsub* function is same as *addsubcomp* except that it does not evaluate the *sign*. Thus, each stage may be implemented using only three add/subtract/compare cells, plus two shifters.

One bit of accuracy of the result is gained at each stage. After a number of repetitions equal to about half the number of bits of working accuracy the value of $y_i >> (i+1)$ is effectively zero, and the value of x_i will not change any further. Consequently, subsequent steps of the algorithm may be simplified to

$dy_i \quad\quad\quad\quad := x_{i-1} >> 1;$
$(y_{i+1}, sign_{i+1}) := $ addsubcomp $(y_i, dy_i, sign_i);$
$\theta_{i+1} \quad\quad\quad := $ addsub$(\theta_i, atan_i, sign_i);$

In the last stages of this algorithm, one is effectively carrying out a division to evaluate $\Delta\theta = \arctan(y_i/x_i) \approx y_i/x_i$.

The value of θ_{i+1} found after the final step is equal to the argument of the original vector, whereas the length of the original vector is equal to $x_i * \prod \cos(\theta_j)$.

2.4. CORDIC OPERATORS

The value of $\prod \cos(\theta_j)$ is a constant equal to approximately 0.60725911. Note that if only the length of the vector, but not its argument are required, then the final iteration steps of the algorithm are not required and the algorithm converges in about half as many steps.

Chapter 3

Multipliers

3.1 Multipliers

In this chapter, we will examine serial multipliers, both bit-serial and digit-serial. Bit-serial multipliers have received a lot of attention ([4], [8]) in the literature, and the reason for including a discussion of them here is to allow for a better understanding of the digit-serial multipliers that will be discussed later. In discussing digit-serial multipliers, attention will be focussed on those multipliers that may be easily generalized to arbitrary digit-width.

This chapter will deal only with LSD-first multipliers, the description of MSD-first multipliers being left to chapter 13.

3.2 Parallel Array Multiplier

In order to understand serial multipliers it is appropriate first to study the design of parallel-by-parallel multipliers. Consequently, we begin by considering a simple array multiplier design and restrict attention to multiplication of unsigned numbers. Hand multiplication of unsigned or positive binary numbers is done in much the same way as the familiar way in which ordinary decimal numbers are multiplied. Individual $1 \times W$-bit products are formed and are added together to obtain the complete $W \times W$-bit result. The following example shows the product of 1001×1011 (binary) $= 9 \times 11$ (decimal).

$$
\begin{array}{cccccccc}
 &   & 1 & 0 & 1 & 1 &   & 1 \\
 & 0 & 0 & 0 & 0 &   &   & 0 \\
 0 & 0 & 0 & 0 &   &   &   & 0 \\
 1 & 0 & 1 & 1 &   &   &   & 1 \\
\hline
 0 & 1 & 1 & 0 & 0 & 0 & 1 & 1 \\
\end{array}
\qquad (3.1)
$$

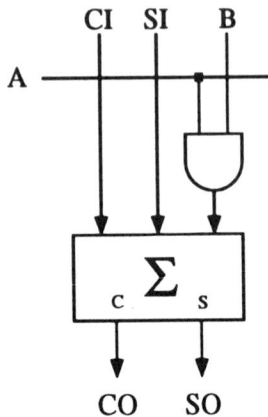

Fig. 3.1: Basic multiplier cell.

The method of carrying out this computation in a hardware array differs from the usual method of hand computation in that carry-save arithmetic is used (see chapter 12) to minimize propagation paths. Furthermore, the array is made square by shifting each successive row one place to the right. The basic cell in a multiplier is a single-bit multiply accumulate cell as shown in Fig. 3.1. Fig. 3.2 shows a 4 × 4 array of such basic cells designed to multiply unsigned binary numbers. The bits of the multiplier, B, are passed down the columns of the array and the bits of the multiplicand, A, are passed across the rows. At each node in the array, the bits A_i and B_j are multiplied together (a one-bit multiplication is done using a single AND gate) and added to the carry and sum bits from the previous rows of the array. Sum-out and carry-out bits are produced. The carry-out bit is passed directly downwards and the sum-out bit is passed diagonally down to the right. The W (in the figure $W = 4$) least-significant bits of the product are produced at the right hand side of the array. At the bottom of the array, the remaining sum and carry words are combined using a full W-bit adder to produce the high order bits of the product.

3.3 Bit-Serial multiplication

The multiplier of Fig. 3.2 may be turned into a parallel/serial bit-serial multiplier by folding the rows of the array into one row and inserting delay latches. The design produced is essentially a parallel/serial multiplier as considered by [4]. Fig. 3.3 shows the essentials of the design. In this multiplier, the bits of the multiplier, B, are presented in parallel, the least-significant-bit on the right. The bits of the multiplicand, A, are presented serially, least-significant-bit first and the result, X, is produced on the right, least-significant-bit first, with la-

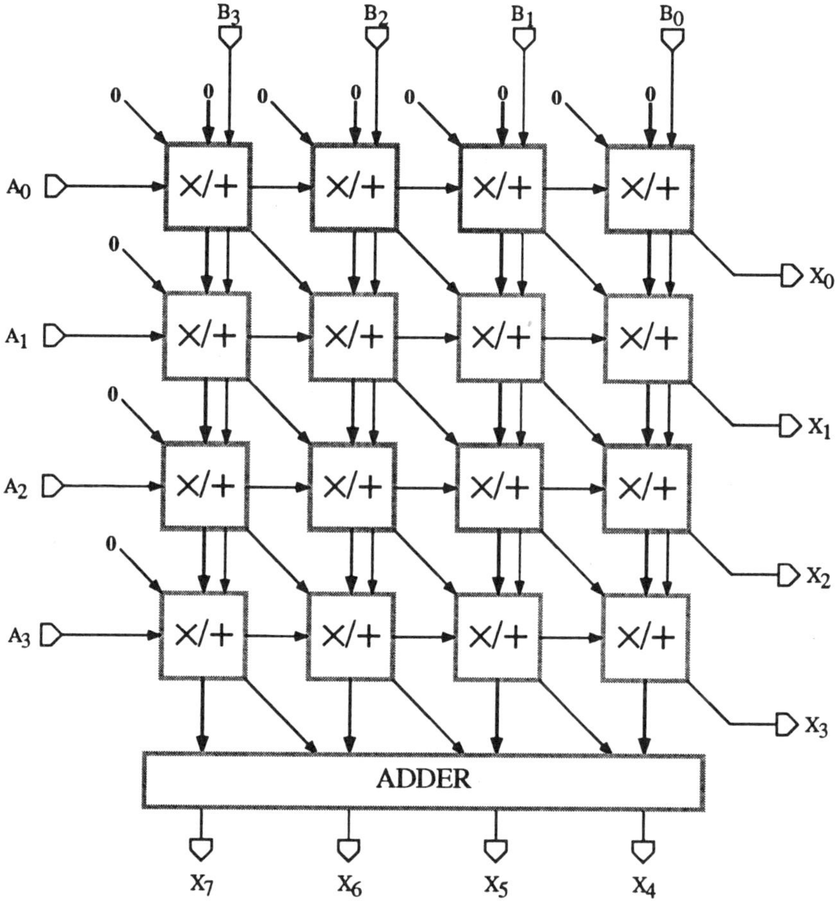

3.3. BIT-SERIAL MULTIPLICATION

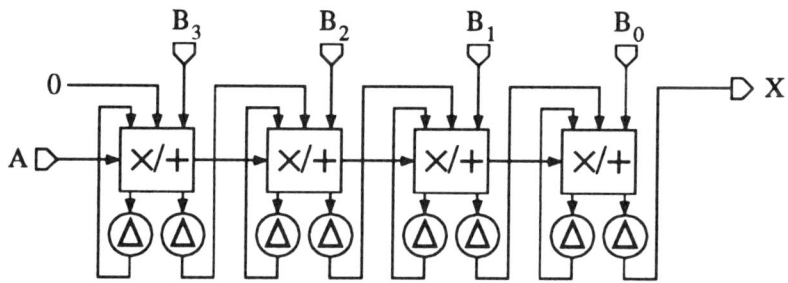

Fig. 3.2: Parallel unsigned array multiplier.

Fig. 3.3: Simplified bit-serial P/S multiplier.

tency one with respect to the arrival of the first bit of A. At each clock cycle, the present bit of A is multiplied bit-by-bit by the multiplier value represented by bits $B_3 \ldots B_0$. The multiplication is carried out by the array of AND gates. Each bit of the resultant product is summed with the previous sum and carry bits by the array of adders. The resultant sum and carry output bits are delayed for one cycle. The sum bit is passed one place to the right and the carry bit is returned to the same cell for next operation. At each cycle, the rightmost sum output bit is input to the right cell and is used for computation of the next bit of the product. For simplicity, the circuitry necessary to reset the sum and carry to zero for the beginning of a new operation is omitted from Fig. 3.3.

If A has W bits and B has W_1 bits, then during the W cycles that A is presented at the multiplier, the W least-significant bits of the result are output at X. In general, however, the most-significant bits of the product are more interesting than the least-significant bits. Fig. 3.3 does not show how to produce the high order bits of the product. One way to obtain the complete $W \times W_1$-bit product is to extend A by the addition of W_1 extra zero bits. (Remember we are considering unsigned addition.) The complete $(W + W_1)$-bit product will be produced at the X output.

Unfortunately, extending A by the addition of W_1 extra zero or sign bits is inadmissible in the bit-serial architecture described in chapter 1. If W is the word-length and $W = W_1$, then A must be extended to length $2W$ to obtain the correct $2W$-bit product. This means that the multiplier can carry out only one operation every $2W$ cycles. This conflicts with the requirement that in a bit-serial architecture with word-length W, the words of length W should follow each other without a break. The extension of A to double the length disrupts the natural flow of data. If the flow of data is to be maintained at the rate of one new word every W clock cycles, then either the $2W$-bits of the product must be produced on separate output wires or the product must be truncated to W bits. Computation of the $2W$-bit product using two output wires is described next. Computation of the truncated W bit product is described in section 3.3.2 in the context of signed multiplication.

Taking the hint from the parallel array multiplier, we see that the high-order part of the product is obtained by summing the final *sum* and *carry* words. Naturally, this should be done with a bit-serial adder. Consequently, after partial products corresponding to the W bits of the multiplicand, A, have been summed by the multiplier array, the remaining *sum* and *carry* words are latched into serial-out shift registers and shifted out through a final bit-serial adder as shown in Fig. 3.4. A timing signal is necessary (but not shown) for marking the beginning of each new computation. This signal which should be coincident with the first bit of A is used to latch the *carry* and *sum* words into their shift registers, and at the same time to reset them to zero at the input of the adder array cells. The low order word of the product (X_L) will appear serially at the XL output, and the high order word of the product (X_H) will be

3.3. BIT-SERIAL MULTIPLICATION

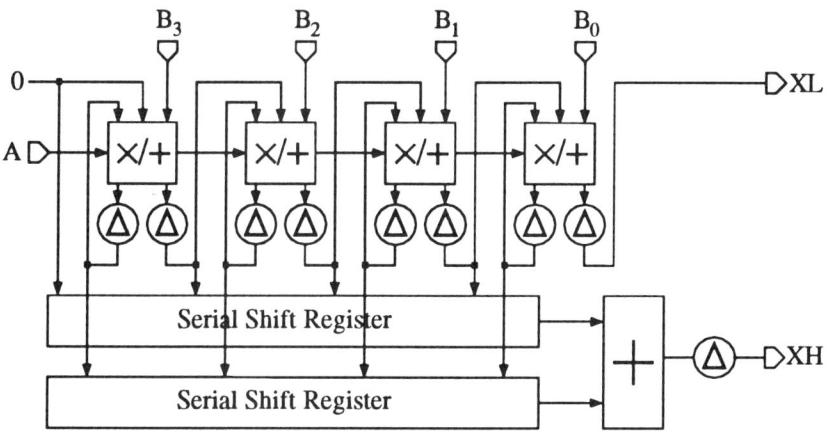

Fig. 3.4: Bit-serial multiplier.

produced at XH, exactly one word delayed with respect to X_L. The latency of X_L will be one cycle with respect to A.

3.3.1 Signed Multiplication

If the numbers 1001 and 1011 in example 3.1 are interpreted as two's complement numbers they are equal to -3 and -5 respectively. It is not surprising that if all values in (3.1) are interpreted as two's complement quantities, then the result would be incorrect. This contrasts with the situation with addition where unsigned and two's complement values may be added in the same way. Correct signed multiplication may be carried out in various ways. First, consider the case of multiplication of a positive multiplier by a negative multiplicand as in (3.2). Consider the example -5×5, or in binary, 1011×0101. It is appropriate to sign-extend the multiplicand a sufficient number of bits to the left as shown here:

$$
\begin{array}{cccccccccc}
1 & 1 & 1 & 1 & 1 & 0 & 1 & 1 & & 1 \\
0 & 0 & 0 & 0 & 0 & 0 & 0 & & & 0 \\
1 & 1 & 1 & 0 & 1 & 1 & & & & 1 \\
0 & 0 & 0 & 0 & 0 & & & & & 0 \\
\hline
1 & 1 & 1 & 0 & 0 & 1 & 1 & 1 & &
\end{array}
\qquad (3.2)
$$

Extension of the words any further to the left will not change the result, since this will result only in sign extension of the result. Now write this example again with an "s" indicating the sign bit in each word :

$$
\begin{array}{ccccccccc}
1 & 1 & 1 & 1 & s & 0 & 1 & 1 & 1 \\
0 & 0 & 0 & s & 0 & 0 & 0 & & 0 \\
1 & 1 & s & 0 & 1 & 1 & & & 1 \\
0 & s & 0 & 0 & 0 & & & & 0 \\
\hline
1 & 1 & 1 & 0 & 0 & 1 & 1 & 1 &
\end{array} \tag{3.3}
$$

It may be observed in carrying out the sum of partial products that the sum-in to each sign bit (except the one in the first row) is equal to the sum-out of the sign-bit in the previous row. This means that the sign extension bits are unnecessary and the multiplication may be carried out as follows :

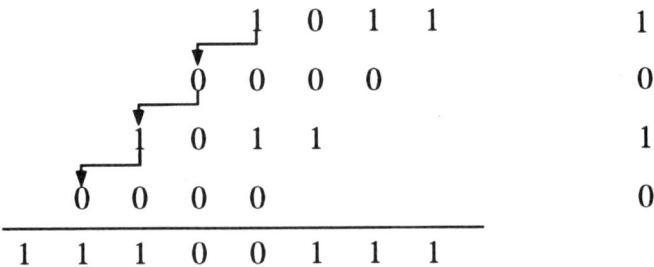

Unfortunately, this is not enough to handle multiplications by negative multipliers as may be easily verified. The reason is that the sign bit of a two's complement number has negative weight (see section 12.1). The correct action therefore, if the multiplier is negative, is to negate the partial product word corresponding to the sign bit of the multiplier. This is illustrated in the next example where the numbers $1011 = -5$ and $1001 = -7$ are multiplied :

$$
\begin{array}{ccccccccc}
 & & & 1 & 0 & 1 & 1 & & 1 \\
 & & 0 & 0 & 0 & 0 & & & 0 \\
 & 0 & 0 & 0 & 0 & & & & 0 \\
0 & 1 & 0 & 1 & & & & & 1 \\
\hline
0 & 0 & 1 & 0 & 0 & 0 & 1 & 1 &
\end{array} \tag{3.4}
$$

The last partial product row in this example is obtained by negating the multiplicand value 1011. In adding the partial products, the method of (3.3) is used. Fig. 3.5 shows the modification to the parallel array multiplier necessary to allow it to multiply two's complement numbers. Note that for the A_3 partial product term the B word is negated as required. This is done by inverting the bits of B and adding A_3 in at the least-significant-bit position. If $A_3 = 1$, then the resultant term is $-B$, whereas if $A_3 = 0$, then it is 0.

3.3. BIT-SERIAL MULTIPLICATION

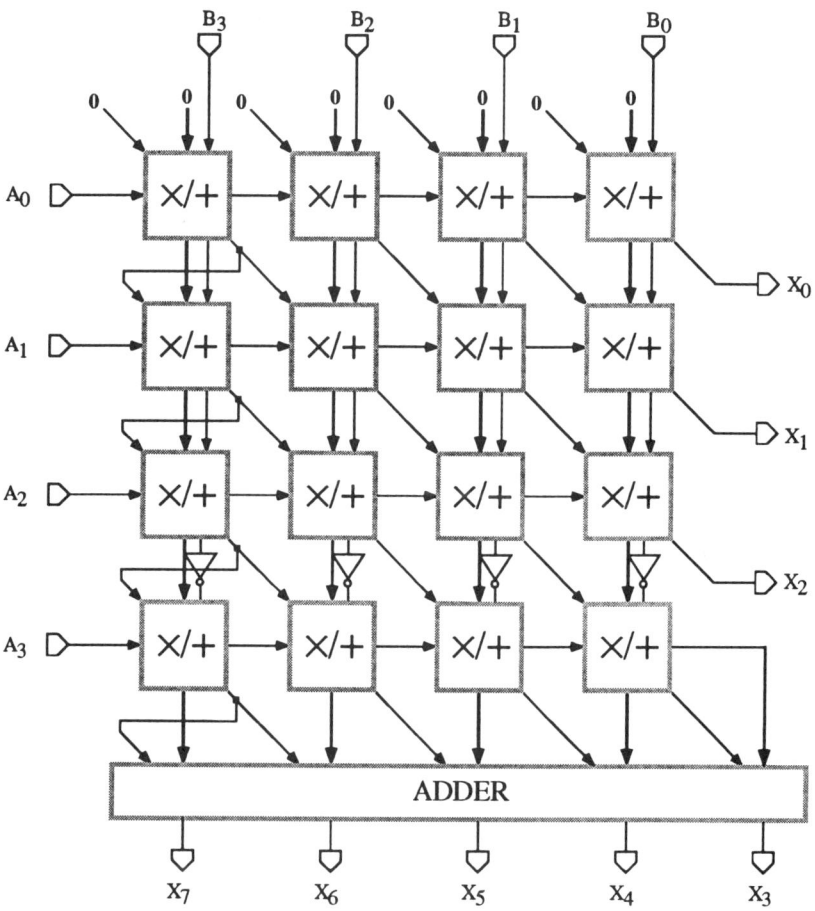

Fig. 3.5: Two's complement array multiplier.

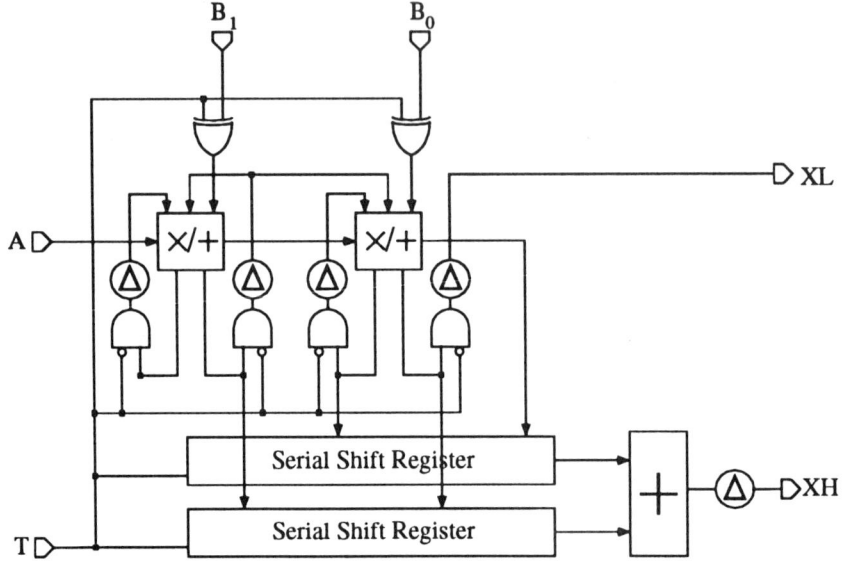

Fig. 3.6: Two's complement bit-serial multiplier.

For a $W \times W_1$-bit two's complement product the result may be represented in $W+W_1-1$ bits, except for the product of the two maximum negative values. In particular, except for the product of values 1000 and 1000, the output bits X_7 and X_6 of the multiplier in Fig. 3.5 will be the same.

From the parallel two's complement array multiplier we can derive a bit-serial two's complement multiplier by folding the rows of the array into one row as before. Fig. 3.6 shows such a bit-serial multiplier. For simplicity, only a two-bit multiplier is shown. Note that the sum-out bit in the left-most position is recycled as sum-in to the same bit in order to handle negative multipliers (B). The timing signal, T, must be high during the last bit (sign bit) of A. It serves to invert the bits of B, to reset the carry-out and sum-out bits ready for the next computation, and simultaneously to latch them into the serial shift register. The shift registers should be designed such that they repeat the left-most bit.

If the inputs A and B are interpreted as integers, this multiplier actually computes $2 \times A \times B$ by supplying a 0 in the least significant bit position and ignoring the most-significant redundant bit. The latency of X_L is 0 with respect to A, and the latency of X_H is W. Normally when using multiplication in a serial circuit, it is not necessary to keep the full precision result of a computation, and some of the $2W$ bits resulting from a $W \times W$ product are discarded. In fact, in order to keep a steady flow of serial data, exactly W bits must be kept, unless the product is to be represented in double precision (that is, as two

3.3. BIT-SERIAL MULTIPLICATION

words).

The format notation $< W.W_1 >$ is used here to represent a data word of W bits of which W_1 lie to the right (fractional side) of the binary point. Thus, $< W.0 >$ format means that the numbers represent integers, whereas $< W.W - 1 >$ means that the numbers are purely fractional, lying in the range $[-1, 1)$.

Which W bits of a $W \times W$-bit product should be kept depends on the formats of the input words and the desired format of the output word. It may be verified that if the multiplier of Fig. 3.6 is used to multiply two words of format $< W.W-1 >$, then the output word X_H will be the $< W.W-1 >$ format product. This is the most common case in serial designs – the low order part of the product may be ignored. However, there are instances in which it will be desirable to extract a different set of W bits from the double-precision product. (In such cases it is necessary to beware of overflow.) Because the high-order word immediately follows the low-order word, it is easy to extract an arbitrary contiguous set of W bits from the two words using a simple multiplexor to switch periodically between the high and low order words.

3.3.2 Serial Multiplication with Constant Word Length

The implementation of bit-serial parallel-serial multipliers with full precision multiplication result computation was described in sections 3.3 and 3.3.1. To maintain a constant word-length throughout the architecture, only W most-significant bits of the $W \times W$-bit multiplication result need to be computed. This can be achieved without computing the least-significant W bits of the result. In this multiplier, a constant word-length is maintained at all signals in the architecture. Consider the two's complement multiplication $A \times B$ where B is 4-bit signal represented by $B_3.B_2B_1B_0$. The product $X = A \times B$ is described by

$$\begin{aligned} X &= -A \times B_3 + A \times B_2 2^{-1} + A \times B_1 2^{-2} + A \times B_0 2^{-3} \\ &= -A \times B_3 + [A \times B_2 + [A \times B_1 + A \times B_0 2^{-1}] 2^{-1}] 2^{-1} \ . \end{aligned} \quad (3.5)$$

The multiplication operation described by (3.5) is shown in tabular manner in Fig. 3.7 and the corresponding implementation is shown in Fig. 3.8. The form of the equation (3.5) is obtained by the use of Horner's rule [6]. Note that the multiplication of a signal by 2^{-1} represents right shift by one bit and is also referred to as a *scaling* operation. This operation performs two tasks simultaneously. First it truncates the least-significant bit of the word and second it sign extends the most-significant bit of the word. Thus a constant word-length is maintained for all signals. Note also the use of the inverted A_i bits in the last row of Fig. 3.7 which is required for two's complement multiplication. Further note that the most-significant bit B_3 is added in the lsb result position to generate the two's complement of AB_3. In Fig. 3.7, two rows are added at a time and the carry-ripple operation is shown in dashed lines. The solid lines with arrows to the left represent sign extension of partial sums. The multiplexor to reset the carry-input of the last adder on the right to B_3 has not been shown in Fig. 3.8.

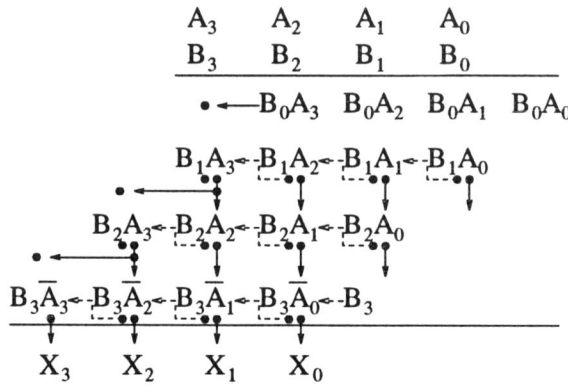

Fig. 3.7: Carry ripple multiplication table.

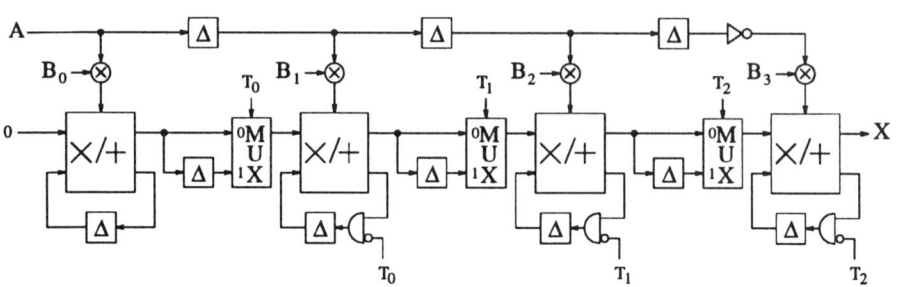

Fig. 3.8: Bit-serial multiplier with constant word-length.

3.3.3 Size of the Multiplier

It should be evident that the multiplier of Fig. 3.6, except for the final bit-serial addition, is easily divided up into identical bit-slices which may be stacked side-by-side to make multipliers of any word-length. Although the size of the multiplicand (A) is fixed by the word-length, W, of the architecture, the size of the multiplier (B) may be arbitrary. If the number of bits of precision of B is smaller than the word-length, then important space saving is possible by using fewer bit slices in the multiplier. The design as shown in Fig. 3.6 does not depend on the size of the multiplier and multiplicand being equal. Depending

3.3. BIT-SERIAL MULTIPLICATION

on which bits of the multiplier are used, care must be taken to select the correct bits from the dual output streams using appropriate multiplexing scheme.

3.3.4 Constant Multiplication

For constant multiplication, in which the multiplier value is known in advance, significant advantages may be had by carefully examining the form of the multiplier constant. In a compiler context, such as Parsifal, this form of optimization may be routinely carried out. In general, the total number of bit-slices required in a multiplier of type Fig. 3.6 is equal to the number of places between the first and last significant bits of the multiplier constant. This may be done by selecting the right bits from the dual streams of the double-precision multiplier output, and appropriate post-shifting. Minimizing the number of bit-slices used in the multiplier has obvious advantages in terms of the hardware cost. In addition, it will in general also minimize the latency of the multiplication operator. This is important for many reasons. It will tend to minimize the total latency of the complete circuit, it will usually lead to a lower cost in the number of required shimming delays, and it may be important to enable tight feedback loops to be made feasible. These issues are discussed in Chapter 6.

If the bits of the multiplier value are known in advance, then the design of the individual bit-slices of the multiplier may be simplified. This is particularly true in the case of a zero bit, since the bit-product $A_i B_j$ is zero in this case, and hence does not need to be generated or added. However, the gains produced by this optimization are smaller than those resulting from minimizing the multiplier length.

An alternative method of implementation of constant multiplication using shift and add (or subtract) operators is discussed in Chapter 12. This is particularly useful in the case where the coefficients have a simple form, or where term sharing may be possible as in the filter designs discussed in Chapter 12. However, it is not clear in what circumstances it is preferable to the design of Fig. 3.6.

Since the complexity of the multiplier depends on the number of significant bits of the multiplier coefficients. It makes sense, therefore to simplify the coefficients as much as possible. In filter designs, it may often be possible to simplify the form of the filter coefficients without significant effect on the form of the filter response. This observation is used in Cathedral I ([6]) to optimize filter designs.

3.3.5 Serial-Serial Multiplication

If two serial quantities, A and B are to be multiplied together using the parallel/serial multiplier of Fig. 3.6, then before the multiplication can start, B must be clocked into a parallel register and latched. The parallel register is most conveniently designed built into each bit-slice of the multiplier. Consequently the word B must arrive serially at the multiplier in advance of A. Consider a complete $W \times W$ bit multiplication. It proceeds in three stages. During the first stage of W cycles, the word B is serially clocked in to the array and is latched. The signal T in Fig. 3.6 can be used to activate the latches. During the second stage of W cycles, the partial products are generated and summed and the low order product word is produced. Meanwhile, the next B word is being serially read into the array. At the end of the second stage, the residual sum and product words are latched into the shift registers in response to T. During the third stage, they are serially scanned out and summed to produce the high order product word. Meanwhile the next multiplication is proceeding in stage 2.

3.3.6 Software Support

In a compiler environment a software array generator must be used to handle the generation of different varieties of multipliers. For instance, the Parsifal silicon compiler has a generator to assemble multipliers from basic cells. It takes into account the format of the two input words and the desired format of the output word, plus a user indication of how many and which bits of the multiplier B are to be used in the multiplication. This is essential for the efficient handling of low precision values, such as limited accuracy inputs, say from an A/D converter. For instance, it may be desired to do 16-bit computations even though some of the input data has only 10 bits of accuracy. Remember that in a serial architecture all data must have the same word-length.

The multiplier generator begins by computing the relative timing of the two inputs, A and B. They will not always be separated by a complete sample period of W cycles. The generator will also compute the necessary timing for the latch signal T. It will then compute which bits of the two words of output need to be extracted in order to produce the correct format output and if necessary will include a multiplexor cell to extract them. It will also compute the correct timing of the signals used to control the multiplexor. Finally, it will compute the latency of the output signal. The generator returns to the compiler a complete net-list description for the multiplier as well as the correct timing for all its input and output signals. The timing information is used by the compiler for the correct scheduling of the complete data flow graph of the circuit containing the multiplier, as described in chapter 6.

3.4 Digit-Serial Multiplication

The bit-serial multiplier design has been described in detail in order to simplify the description of the digit-serial parallel/serial multiplier.

Essentially, a digit-serial multiplier is a $W \times N$ parallel-serial array multiplier. Sum and carry bits ripple through N stages in the multiplier array before

3.4. DIGIT-SERIAL MULTIPLICATION

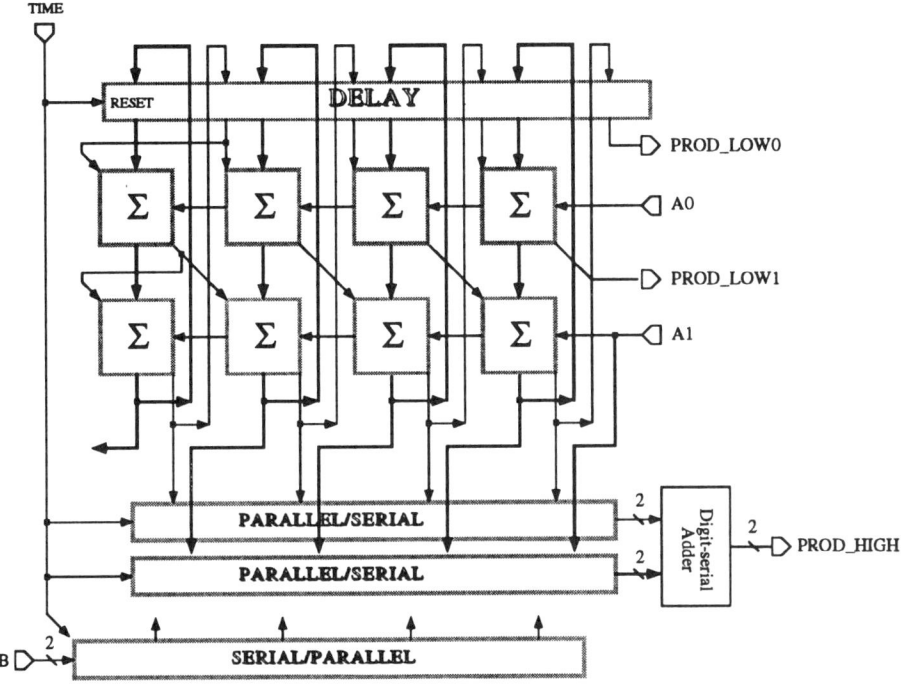

Fig. 3.9: Digit-serial Multiplier.

being latched and passed back to the top of the array to continue when the next digit arrives in the next clock cycle. Booth multiplication [35] is also feasible, as was shown for radix-4 multiplication by Smith and Denyer [8]. This scheme was not implemented for digit-serial multiplication in Parsifal, partly because of the difficulties with odd digit sizes, and partly because the extra speed afforded by Booth multiplication was not necessary to match the speed of other operators.

In adapting the array multiplier of Fig. 3.5 for digit-serial operation, the downward ripple of the sum and carry bits is interrupted by a delay and then folded back on itself to give an $N \times W$ array of cells. Fig. 3.9 shows this array for the case of a 2-bit digit-size. The bits of B are latched into a serial-in/parallel-out register and supplied one bit per column of the array. The bits of A, on the other hand are supplied one digit at a time and fed across the rows. Meanwhile the next word of B is being loaded into the serial/parallel register. A signal, $TIME$, high during the last digit of A, causes the next value of B to be latched and output in parallel in the subsequent cycle. The same signal causes the sum and carry values to be reset to zero. The low order part of the product appears in digit-serial format at the right hand end of the array. In order to obtain the high order word of the product, the final sum and carry words are latched into a parallel-in/serial-out register and passed out to the right, through a further digit-serial adder. Since the carry word has double the weight of the sum word it is shifted one place left before being latched into the serial-out register. As with parallel multiplication, the sign bit of A takes the low order place left vacant by the shift of the carry word. In this way, both low and high order words of the product are produced in digit-serial format, the high order word following the low order word one sample period later on a different set of wires. In other words, the double precision product is in standard double precision format ([8]). Note that the $2W-1$ bit result of the multiplication is filled out to $2W$ bits by the insertion of a 0 in the least-significant-bit place of the low order word. Consequently, the high order word contains the W most significant bits of the product.

Fig. 3.10 shows a more detailed diagram of a stack of three cells indicating how the multiplier stack conforms to the basic template of cap, multiple bit-slices and control section as explained in section 1.6.1. In particular, this figure shows how the inversion of the B input is done in the most-significant-bit place. The B input to the bottom cell in the stack passes through an XOR gate and is inverted during the last clock cycle of the multiplication as marked by the $TIME$ signal. The parallel/serial and serial/parallel converters for the B input and sum and carry outputs are contained partly in the control section of the cell stack. Details are omitted from Fig. 3.10.

The concept of the parallel/serial and serial/parallel converters is quite simple, but their layout is complicated by the need to bit-slice them to accommodate all digit sizes. How this is done is shown in Fig. 3.11. The complete layout for the multiplier cell departs from the standard template by having two separate bit-sliced sections – one for the bit-sliced adder array and one for the combined serial/parallel and parallel/serial converters shown in Fig. 3.10. The multiplexor and delay elements shown in Fig. 3.11 are contained in the $CONTROL$ section of the multiplier stack. In Fig. 3.11, if the parallel bits (denoted as PAR) on the top are processed, then these bits which were input to the the converter at the same time are delayed properly and output in digit-serial form. Otherwise, this converter outputs a delayed version of the digit-serial input (denoted as SER_INi in left). In order to maintain compatibility with cells constructed according to the standard template, the sum of the heights of the two types of bit-slice used in the multiplier equals the height of a standard-template bit SLICE cell.

3.5 Low Latency Multiplication

The multipliers considered so far share the disadvantage that one of the inputs must be converted to parallel form before the multiplication can commence. This means that the latency of the multiplier with respect to this operand is at least one sample period before even the low-order part of the result is output.

3.5. LOW LATENCY MULTIPLICATION

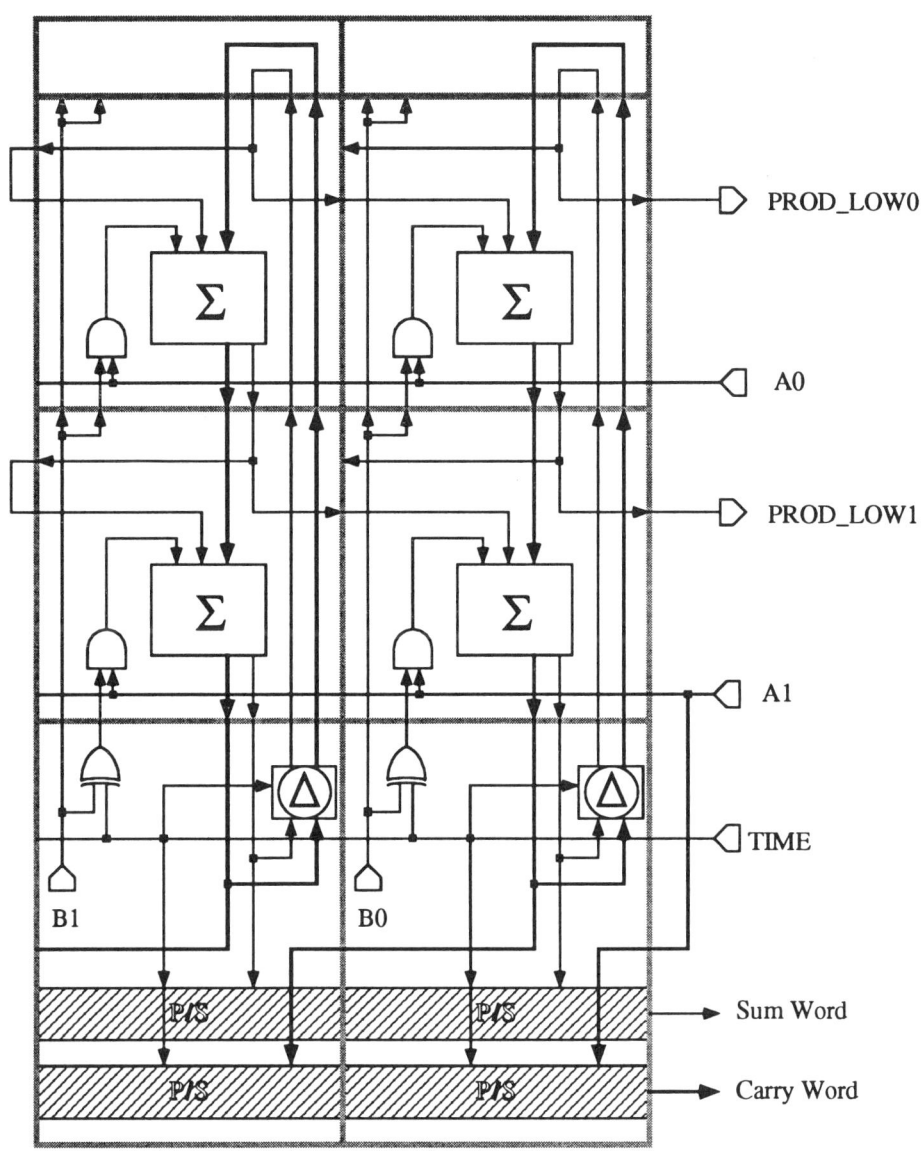

Fig. 3.10: Digit-serial multiplier layout.

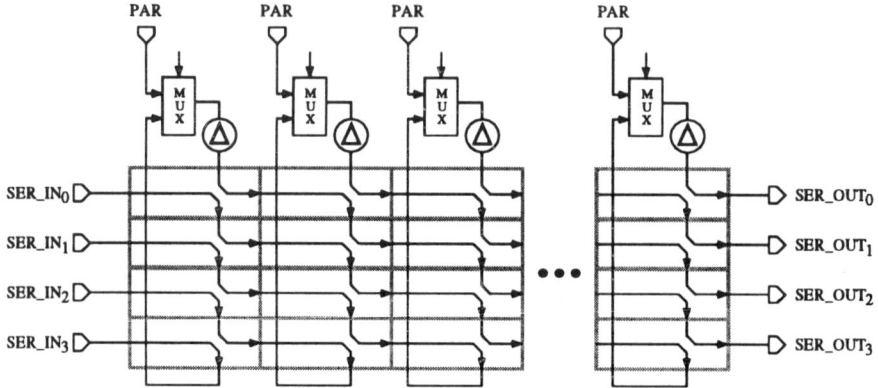

Fig. 3.11: Bit-sliced parallel/serial converter.

Two complete sample periods will elapse before the first digit of the high-order word of the result is output. This delay is theoretically unnecessary, since once the first k digits of both operands are known, it is possible to compute the first k digits of the resulting product. In fact it is possible to design a digit-serial multiplier having low latency, such that both input operands arrive at the same time and the first digit of the product is output immediately (or in the following clock cycle). Supposing that the output is subject to a one-clock cycle delay, this multiplier has latency 1 clock cycle before the output of the low-order part of the product.

Low latency multipliers for bit-serial computation were introduced by Chen and Willoner ([36]). Expansions and modifications to their idea were given by other authors in [37],[38] and [39]. Smith and Denyer discuss low-latency bit-serial multipliers in [8] and conclude that because of their greater size their utility in bit-serial design is limited. Although this conclusion is probably valid in most cases, there are instances where the use of low-latency multipliers may be necessary to meet feedback timing constraints. In addition, the extra latency of the parallel-serial type of multiplier does lead to extra hardware in the form of scheduling delays which will need to be inserted to compensate for the high multiplier latency. The same considerations apply equally to digit-serial multiplication. Consequently, the design of low-latency digit-serial multipliers will now be discussed. The description of the multipliers will be given at a conceptual level, details being omitted for the sake of conciseness.

Let A_k and B_k be the arithmetic values represented by the first k digits of each of the two operand values A and B. Suppose that the product $A_k B_k$ has been computed, to the extent that appropriate partial products have been accumulated resulting in the output of k output digits and the computation of sum and carry words which if added would provide the remaining digits of

3.5. LOW LATENCY MULTIPLICATION

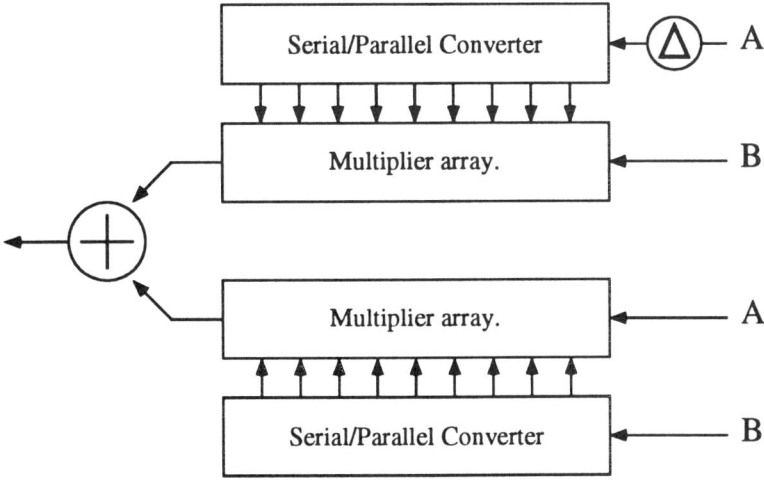

Fig. 3.12: Low latency digit-serial multiplier.

the product. Let a_{k+1} and b_{k+1} be the arithmetic values represented by the $(k+1)$-st digits of A and B, each appropriately weighted by the appropriate digit power. Thus $A_{k+1} = A_k + a_{k+1}$ and $B_{k+1} = B_k + b_{k+1}$. In the $(k+1)$-st clock cycle it is necessary to compute the product $A_{k+1}B_{k+1}$. Simple algebra shows that

$$\begin{aligned} A_{k+1}.B_{k+1} &= (A_k + a_{k+1}).(B_k + b_{k+1}) \\ &= A_k.B_k + A_k.b_{k+1} + a_{k+1}.(B_k + b_{k+1}) \\ &= A_k.B_k + A_k.b_{k+1} + a_{k+1}.B_{k+1} \ . \end{aligned}$$

Since the product $A_k.B_k$ has been evaluated already, it is necessary in computing the new result to update the previous result by adding $A_k.b_{k+1}+a_{k+1}.B_{k+1}$. Note that this formula indicates that the new digit, b_{k+1} must be multiplied by the previous value of the operand A_k, whereas the digit a_{k+1} must be multiplied by the current value B_{k+1} of the other operand. This leads to the design (Fig. 3.12) of a low latency multiplier. Each of the two multiplier arrays is of the parallel/serial type. Initially the two parallel-serial converters contain a value of zero and they are filled up digit-by-digit as the digits of the two operands arrive. At any time, the bits of the serial/parallel converter not yet filled must contain zero values. In order for the multiplier to be able to handle a steady stream of input values, each new input operand word following directly after the previous one, it is necessary to provide for the resetting of the serial/parallel register and also for resetting the sum and carry words in the multiplier array. In addition, if both the high and low order parts of the product are required, then it is necessary before the start of a new operation to latch the remaining sum and carry words (two of each) and to sum them using separate

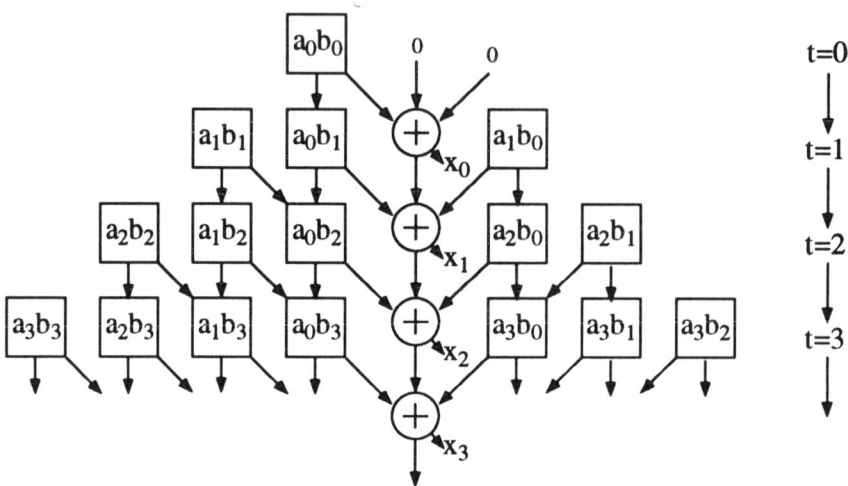

Fig. 3.13: Array for carrying out low latency multiplication.

digit-serial adders. The detail of these operations are quite analogous to those described for the parallel-serial digit-serial multiplier described in section 3.4.

The operation of a low-latency bit-serial multiplier may be illustrated clearly by considering an array multiplier as shown in Fig. 3.13. This figure shows two arrays, one on the right and one on the left. In each side, the carry-bits from each cell ripple straight downwards, whereas the sum bits ripple towards the center. It can be seen that in the left-hand array the bits b_k are fed along the rows and the bits a_k are fed down the columns. In the right-hand array, on the other hand, bits a_k are fed along the rows and bits b_k are fed down the columns. Time increases from top to bottom as shown on the right side of the figure. It can be seen that no bit b_k or a_k is used before time $t = k$. By folding the arrays after each row, as was done with the array multiplier of Fig. 3.2 a bit-serial low-latency multiplier is formed. In this folding, the absent cells in the upper rows of the array become cells producing sum and carry values of zero. The adaptation of this array to signed bit multiplication may be accomplished using the techniques discussed for bit-serial signed multiplication as in Fig. 3.6. The high-order part of the product may also be produced using similar techniques. The multiplication takes place in two stages. In the first stage the low order word of the result is produced by accumulating partial products in the array. At the end of the first stage, all remaining sum and carry words are latched and summed serially during the second stage to produce the high order part of the product. Meanwhile, stage one of the next multiplication is being carried out in the array.

It is possible to modify this general design in many different ways, such as

3.5. LOW LATENCY MULTIPLICATION

interleaving the two arrays into one array. This can be done in such a way as to reduce the number of sum and carry words to three instead of two. Another modification reduces the number of such words which need to be latched and summed serially during stage two of the multiplication from three to two by making use of unused parts of the array during the first stage of the next multiplication. Various other techniques and tricks to save hardware are also possible. One idea is based on the observation that much of the array is unused during part of the computation, being occupied accumulating zeros. It is therefore possible at the cost of some complexity to interleave two multiplications on the same hardware array whereby one computation ripples to the right and the other to the left.

The array Fig. 3.13 is easily adapted to digit-serial operation simply by interpreting each cell as being a digit-by-digit multiplier cell similar to the array part of Fig. 3.2 (for a 4-bit digit). There is no conceptual difficulty in connecting up the sum and carry output digits correctly between cells.

Chapter 4

Digit-Serial Input Language

In this chapter, the input language used by the Parsifal compiler will be described. Many different languages have been used for synthesis systems. For instance, the Cathedral system uses Silage ([40]) as its input language. The FIRST compiler uses its own language which is described in detail in [5]. It is not the purpose of this chapter to discuss the various merits of various hardware description languages, but rather to give an example of a simple language that has proven suitable for describing digit-serial designs.

There is a basic choice in selecting an input language in choosing between an existing language and inventing a new language. In the first approach, advantage can be taken of existing tools that understand the language in question. Furthermore, inputs can be shared between different design or simulation systems. The VHDL language though perhaps more suitable for simulation and documentation has been used in synthesis systems ([41]). The choice of Silage as the input language for the Cathedral compilers was no doubt motivated by these considerations. On the other hand, the advantage of developing a new language for a synthesis system is that the designer has complete control over the constructs of the language and may add new features to cover specific requirements. This was the choice made in developing the input language for Parsifal. The resulting language is called BSL (Bit-Serial Language), which reflects its origins. It is used for the description of digit-serial as well as bit-serial designs. It has also been extended to the description of multi-chip designs in the *DIODES* rapid prototyping system ([42]), but the language extensions required for this will not be described here.

4.1 Cells

The complete design of a chip is made up from *cells*, and in fact the complete design is a cell. There are several types of cells, leaf cells, stack cells, composite cells and symbolic cells. Each of these cell types will be described in its turn.

In general a cell will begin with its header

CELL <*cellname*> (<*IO-description*>);

and will end with the words

END <*optional cellname*>;

One of the cells in the input description represents the complete design; in other words, that circuitry made up of other smaller sub-cells constitutes the complete chip to be created. This cell is termed the *processor* and starts with the keyword **PROCESSOR** instead of **CELL** .

The *IO-description* or *interface* is common to all cell types. It lists the names of the input and output signals to the cell, as well as their width (in bits) and a possible timing description. The complete header for a 4-bit digit-serial adder similar to the one shown in Fig. 1.15 is as follows.

CELL ADDER_4 (**IN** : A[4](0), B[4](0);
                          OUT : X[4](1);
                          IN : TIME (0));

This description indicates that the two input signals A and B are each 4 bits wide, as is the output X. The adder has a further input signal called *TIME* as shown in Fig. 1.15. The *TIME* signal has no width specified, so it is assumed to have the default width, namely one bit.

The numbers in parentheses indicate the timing of the IO signals as shown in a timing diagram such as Fig. 1.9. They indicate that the (first digits of) the two input signals arrive together at time $t = 0$ and the first digit of the output will be produced one cycle later at time $t = 1$. The *TIME* signal is similarly to arrive at time $t = 0$. This signal is different from the others in that it is a periodic signal of the timing type described in section 1.3.4. As stated there, such a signal is to be connected to the correct phase of a global *TIME* signal which has a high pulse in the last digit of the word[1]. Such a signal is given special treatment during the scheduling of the circuit and in fact is ignored until after all operators have been scheduled (see section 6.1). It is necessary for the compiler to understand the special nature of this signal, and the following convention is adopted for this purpose.

---

[1] By convention, the high pulse of the time signal is the last cycle of the *TIME* word. Hence, if the *TIME* word is timed to arrive at time $t = 0$ then its high pulse arrives at time $t = -1$.

4.1. CELLS

Any signal with name starting with the letters *TIME* is a timing signal, and is to be connected to correct phase of the global *TIME* signal.

A different example of a cell header is that of a comparator.

CELL GREATER_THAN (**IN** : A[4](0), B[4](0),
TIME(0);
OUT : GT(W));

The *GREATER_THAN* cell has the unusual property that the latency of its output cannot be predicted out of context – it depends on the word-length. In fact, as shown in Fig. 1.16, the latency of a comparator is equal to the sample period, since the result of a comparison cannot be known until the last digit of the inputs has been seen. This explains the timing specification W for the output GT. The letter W means one sample period. General linear expressions such as $5W + 7$ or $3W - 1$ are acceptable timing specifiers.

4.1.1 Leaf Cells

At the bottom of the design hierarchy is the leaf cell. There are three aspects to a leaf cell, namely its interface, its layout and its simulation description, though not all of these need be present. The interface description of the leaf cell is of the form described in the last paragraph.

The Layout Description

The layout description of the chip gives all the information as to the size of the cell and the locations of the ports. A given IO port to the cell may have pins (that is connection points) on the top, left, right or bottom edges of the cell, and multiple ports for the same signal are supported. Here is a description of the *ADDER* control cell :

LAYOUT
  HEIGHT 10000;
  WIDTH 5000;
  A_1 = **TOP** @ 500;
  A_2 = **LEFT** @ 700;
  VDD = **LEFT** @ 500, **RIGHT** @ 500;
  CLOCK1 = **LEFT** @ 700, **RIGHT** @ 500;

All measurements are given in 100-ths of microns. In cases where more than one location for a pin is given, connection may be made to either one of the alternative locations. The pins must be connected internally.

Since the correctness of the layout is dependent on the correctness of the information provided in the layout description of the leaf cells, it is important

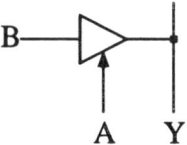

Fig. 4.1: Semantics of the NEXT keyword.

Fig. 4.2: Semantics of the tristate operator.

that this information be correct. In order to ensure this the layout description is extracted automatically from a GDS-II Stream[2] description of the laid-out cells.

The Simulation Description

In order that a circuit may be simulated it is necessary to specify a description of the function of each individual leaf cell. In Parsifal, this description is given in terms of a small number of simple logical operators :

+ : Logical OR of two Boolean (bit) values.

* : Logical AND of two Boolean values.

~ : Logical negation of Boolean value.

− > : Tristate driver.

In addition, the keyword **NEXT** is used to represent a one-cycle delay (Fig. 1.3). For instance, the expression

  NEXT A := B;

represents the circuitry shown in Fig. 4.1. The tristate operator "− >" has the following syntax.

  Y := A − > B;

where A and B are any expressions and represents the circuitry shown in Fig. 4.2. The tristate operator is not used frequently in bit-serial or digit-serial chips. A common example, however is in a serial/parallel converter, which has a circuit of the type shown in Fig. 4.3.

As an example, the bit-serial adder shown in Fig. 1.7 is described as follows:

---

[2]Stream is a layout description language analogous to CIF [43], developed by Calma corporation and still widely used for chip layout.

4.1. CELLS 67

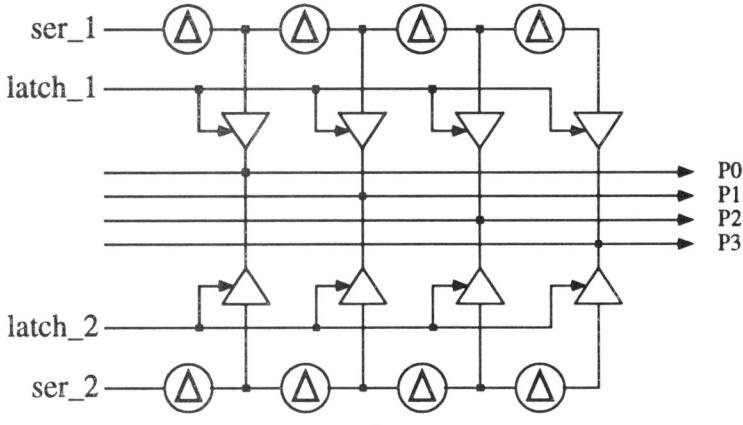

Fig. 4.3: Simple Serial/Parallel Converter.

EMULATION
  CARRY_OUT := A*B + C*(A+B);
  NEXT SUM := A*B*~ C + ~ A*B*C + A*~ B*C;
  NEXT CARRY := ~ TIME * CARRY_OUT;
  . . .

Any circuit made up of AND and OR gates, inverters, tristate drivers and delay cells can be described precisely by logic descriptions of this type. It is not suggested that all bit- or digit-serial cells are actually of this simple form, but rather that all cells can be approximated by circuits of this type, indeed approximated closely enough to provide a bit-exact simulation. Normally, when designing a digit-serial circuit using Parsifal the main concern is the logical correctness of the design. The low level electrical correctness and questions of signal timing are built into the cell designs and compiler software and need not concern the user. If more accurate simulation is required interfaces have been provided to allow simulation by the Hilo–trademark and Mimic–trademark simulators.

4.1.2 Stack Cells

Digit-serial operators are constructed from leaf cells according to the template shown in Fig. 2.1 or a similar template. A stack of cells is built simply by placing the cells on top of each other. This is described by a simple syntax simply listing the cells making up the stack and the connections between them. For instance, a 3-bit digit-serial adder is described as follows.

  CELL ADDER (**IN** : A[3](0), B[3](0);
            OUT : X[3](1);

                    IN : TIME(0));
STACK
    C **of** ADDER_CAP;
    S1 **of** ADDER_SLICE;
    S2 **of** ADDER_SLICE;
    S3 **of** ADDER_SLICE;
    CONT **of** ADDER_CONT;
CONNECT
    A_1 **to** S1.A;
    A_2 **to** S2.A;
    CONT.CLK1 **to** . . .

The specified cells are placed in the specified order from top to bottom. The connections between the cells must be explicitly given, even though the correct corrections are made by abutment. Note that the individual pins of a signal such as $A[3]$ having width greater than one are denoted as A_1, A_2 etc. Usually stack cells are generated automatically.

4.1.3 Composite Cells

Larger, so called *composite*, cells may be constructed from stack cells or from smaller composite cells. The syntax is similar to that of a STACK cell, except that the keyword **INSTANCE** is used instead of *STACK*. The difference is that no relative placement of the cells is implied, the placement being left to the layout software to decide. Common examples of composite cells are

1. Multipliers made up of several similar bit-slices and other slices as described in Chapter 3.

2. Banks of asynchronous registers.

3. The complete top level description of a chip after compilation.

As with stack cells, it is often not necessary for the user to create composite cell descriptions, since these either exist in a library, or are produced automatically by *generator* programs provided with the Parsifal system. Although the relative placement of the slices in a composite cell is not specified, in general the bit-slices making up major blocks of circuitry such as a multiplier or register bank will always be placed together automatically. This is because cells for which the required connections may be achieved efficiently by cell abutment will preferentially be placed together. The way this is done is described in Chapter 5.

4.1.4 Symbolic Cells

The main design description is usually entered by the designer in the form of a *symbolic cell*. The syntax for a symbolic cell is somewhat more complicated

4.1. CELLS

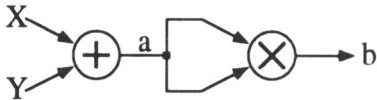

Fig. 4.4: Circuit.

that that of the former cell types.

Applicative Languages

A symbolic cell is an *applicative* description of the desired function of the cell. The symbolic language is a so-called *single assignment* language, which means that each variable in the description receives only one assignment. The order of the assignment statements is not important. Single assignment applicative languages are popular for the description of digital signal processing circuits. For instance, the Silage language ([40]) used in Cathedral is applicative.

The idea behind an applicative language is that it represents a set of equations that are satisfied by the variables in the circuit, rather than as with a prescriptive language which represents a sequence of actions (assignments) to be carried out. Consider the following simple example in which the variables X and Y are inputs to the circuit.

    b := a * a;
    a := X + Y;

In this example, variable b is specified to be equal to $a * a$. On the other hand, a is specified as being equal to $X + Y$. Consequently, b must be equal to $(X + Y) * (X + Y)$. Note that the result of interpreting the two statements as assignments carried out in the order written is quite different. In general, an applicative description may be realized physically by interpreting each statement as a block of combinational circuitry carrying out the arithmetic or logical operations specified. The output of the block is the left side of the statement. All the blocks are connected together to form a complete circuit. The circuit resulting from the above example is shown in Fig. 4.4. It is important that there should not be loops in the applicative description such as in the following example.

    b := a + X;
    a := b + Y;

Consider what happens if these two statements are interpreted as equations. Substituting the first equation into the second gives

    a := (a + X) + Y

from which it may be deduced that $X + Y = 0$. This is a statement about the input values and makes no sense in the context of circuit designs.

Previous values: As described so far it seems that applicative languages are able to describe simply connected circuits only (that is, those without feedback loops). In order to extend the range of circuits, the notion of previous values is used. The operation carried out by an applicative description is interpreted as being carried out over and over again forever in an infinite loop :

> **while** (1)
> {
>   read input values.
>   compute the values of all variables.
>   write output values.
> }

Each execution of the loop is called an iteration and a new sample is processed in each iteration. The time to execute an iteration is referred to as sample period. Given a variable X, the notation $X[-1]$ denotes the value of X in the previous sample period, and in general $X[-nn]$ where nn is a positive integer represents the value of X in the nn-th previous sample period. Such previous values may be used in expressions just like other variables. The only difference is that it is not permissible to make an assignment to the previous value of a variable. Previous values are used essentially in the description of digital filters. For instance, a 3-tap FIR filter may be described by the single statement

$$y := a0 * x + a1 * x[-1] + a2 * x[-2];$$

Similarly, an IIR filter may be described by a statement of the type

$$y := a0 * x + a1 * x[-1] + a2 * x[-2] + b1 * y[-1] + b2 * y[-2];$$

Because of the feedback in this IIR filter example it is possible that such a circuit cannot be implemented because of the impossibility of scheduling. It depends on the implementation of the multiplication operator. This point will be discussed in Chapter 6.

Symbolic cell syntax: Although the main part of a symbolic cell description is the set of applicative assignment statements, there are other necessary parts necessary to make a complete circuit description. For instance, the following is the complete description of the FIR filter implemented in Digit-Serial Arithmetic with a digit-size of 4.

> **SYMBOLIC CELL** FIR (**IN** : X[4]; **OUT** : Y[4]);
>
> **WordSize** 16;
> **FixedSize** 15;
> **DigitSize** 4;

4.1. CELLS

    FOR addition USE ADDER;
    FOR multiplication USE MULTIPLIER CALL "multgen";
    FOR bitdelay USE DELAY CALL delaygen;
    FOR controller USE CONT CALL contgen;

    CONST
        a0[4] = 0.345;
        a1[4] = 0.456;
        a2[4] = -0.045;

    SPECIFICATION

    $Y := a0 * X + a1 * X[-1] + a2 * X[-2]$;

    END FIR;

Arithmetic format description: The first three lines of the chip description after the the cell header specify the default arithmetic format to be used in the design. This example specifies a word-length of 16 bits divided into digits of 4-bits each and the word-length is denoted by the keyword **WordSize**. The **FixedSize** keyword is used to specify the number of bits to the right of the binary place, in this case 15. Thus, consider a word $a_{15}a_{14}\ldots a_0$. This word will be transmitted four bits at a time, beginning with the low order bits $a_3\ldots a_0$. Interpreted as an arithmetic value it is equal to $2^{-15}\sum_{i=0}^{15} a_i 2^i$, since a **FixedSize** of 15 is specified. Note that the position of the binary point is not important in adding, subtracting or comparing two arithmetic values, as long as the position of the binary point is the same in both the words. In multiplication, on the other hand, the position of the binary point is important. For instance in the multiplication of two format $< 16.15 >$ operands[3] to produce a $< 16.15 >$ format result, a different set of the full precision 32 output bits are selected than if two $< 16.14 >$ operands are multiplied to get a $< 16.14 >$ result.

    The possibility of allowing the format of each individual variable to be specified independently was considered at various times during the development of Parsifal, however it was never implemented, since it did not seem to be an important feature in terms of user demand. If necessary, it is possible for the designer to obtain the same effect by explicitly shifting arithmetic values.

    There are problems with allowing different fixed-point formats in that decisions must be made as to how fixed-point values with different formats are to be combined. For instance, if two numbers are to be added together then it is necessary that the two binary points line up. If they do not, then one of the operands must be shifted so as to line up the two binary points. One of them

---

[3]The format $< nn.mm >$ denotes a word of nn bits of which mm lie to the right of the binary point, that is, a format with **WordSize** = nn and **FixedSize** = mm.

must be shifted right, resulting in loss of low order bits, or the other must be shifted left, risking overflow and the loss of the most significant digits. Since the latter is the more undesirable outcome, a sensible choice could be to line up the binary points by shifting the appropriate operator right. However, for intermediate results in calculations, such as the value of $a + b$ in the computation of an expression $(a + b) + c$ the choice is not so clear. For instance, the values of a and b may always be such that the most significant bits cancel when added. Then it will be safe to shift $a + b$ left (if necessary) before adding to c. In general, there is no fool-proof way of doing arithmetic involving fixed-point numbers. It may be best to leave the explicit manner of combining them up to the designer.

Declarations: Passing for the moment over the *FOR* statements in the symbolic cell example we come to constant declarations. Note that the width of the constant value must be specified. Constant arithmetic values are implemented by circuitry that periodically repeats the desired arithmetic constant in digit-serial format.

The FIR example, consisting as it does of a one-line specification, does not require the use of any subsidiary variables. If such variables are used, then they must be declared. The appropriate syntax is as shown in the next example.

VAR
    avar[4], bvar[4]; /* 4-bit wide variables */
    bool1, bool2; /* 1-bit wide values */

In general variables of any width may be declared. The *length* of each variable (that is the number of clock cycles between successive instances of the variable) must be equal to (word-length / digit-size), the number of digits in each word. The necessity of this is discussed in Chapter 1. The only restriction on signal width is that arithmetic values that are to be operated on by digit-serial operators must all have width equal to the digit-size. As these examples show, Parsifal uses a very simple typing mechanism, different types of signals being differentiated only by their width. This has proven sufficient for the many bit- and digit-serial chips designed. Other languages (for instance Silage [40]) have a more extensive typing syntax.

Specification: The semantics of the specification section of the symbolic cell description have been described in the foregoing brief description of applicative languages. The syntax is simple, resembling the syntax of the C programming languages, but choosing PASCAL's ':=' assignment operator. Expressions may be built up from a set of primitive operators. Here is a full list of the available operators.

Arithmetic operators $*, +, -, <<$ and $>>$.

4.1. CELLS

Relational operators $<, >, <=, >=, =, !=$

Logic Operators $\&, |$ and $!$

Choice operator x?y:z

Each of these operators has a name known to the compiler. For instance, the addition operator is called "add", multiplication is called "multiply", and so forth. The intended meaning of these operators should be self-evident, except perhaps for the multiplexor operator. This is identical with the operator having the same syntax in the C language ([28]). The operators have a fixed built-in precedence which is the same as that for the equivalent operators in C, as specified in [28]. Although each of the operators has an intended meaning, as reflected by the operator name, these meanings are not enforced.

It was taken as a guiding principle in the design of the Parsifal software that no knowledge should be built in about properties of any operators. In other words, there should be no built-in rules specifying algebraic properties, such as commutativity or distributivity[4]. Further, the possible manners of implementation of an operation are not built in. Experimental and test versions of the software have occasionally violated this principle, for instance in the work on tree-height minimization of associative and commutative operators reported in [44]. Nevertheless, the adherence to this principle facilitates the development of a flexible and extensible system. In the ideal case, the set of basic operators would be extensible, and their precedence and right- or left-associativity would be specified by the user. However, this would considerably complicate the parsing of the language.

It is possible, however, for the user to implement each of the operators in any way desired. For instance, the user could perversely decide to implement the addition operator '+' using a subtractor. How the *binding* of operators is done is specified in the next paragraph.

Operator Binding

When the compiler sees an addition operator (for instance) in the chip specification, a decision must be made as to how this is to be implemented.

Any information that the software needs about any particular operator must be provided to it. The method used by Parsifal to specify operator binding is through the FOR statements given in the circuit description. For instance, the statement

FOR addition **USE** DS_ADDER;

is an instruction to the compiler that when an addition operator (i.e., '+') is encountered in the circuit description, then a cell named DS_ADDER is to be

---
[4] Assumptions of associativity are necessary for simple parsing.

used to implement it[5]. It is necessary that the cell *DS_ADDER* be already known, that is, has appeared already in the input circuit description. The minimum that the compilation software needs to know about the *DS_ADDER* is its interface description as provided by its cell header, as described in section 4.1. Descriptions of the basic library cells, including interface, layout and simulation descriptions are kept in standard libraries, which must be included in the input description before the symbolic cell is parsed.

Certain operators cannot be implemented in such a simple manner. For instance, bit- and digit-serial multipliers are built up from smaller bit-slices. The number of slices will in general be equal to the number of bits (not digits) in the word, but may sometimes be different. Since it is not reasonable to hold multipliers for all possible word lengths in the library, multipliers must be built once the word-length is known. The statement

> **FOR** multiplication **USE** DS_MULT **CALL** "multgen";

means

> To implement the multiplication ('*') look for a cell with name starting with *DS_MULT*[6] and if one is not found, then call the program called *multgen* to produce one.

The program *multgen* referred to here is an example of a *generator* program. A generator program is a separately compiled executable program which may be run independently, but in this case is launched directly by the compiler itself. The compiler intercepts the output of the generator and interpolates it into its input stream. It interrupts the parsing of the primary input to parse the output of the generator program and when that is finished, returns to the primary input. The output of a generator must be a description of the required cell, as either a stack cell, a leaf cell, a composite cell or even another symbolic cell description. It may refer to any of the cells already parsed, or it may define any number of levels of subcells before defining the desired cell itself.

The reason for implementing the generators in this manner, rather than as a simple subroutine call, is that it is then independent of the main compiler executable. It is possible for a user to add new generators to implement new operations or replace existing generators without needing to recompile the Parsifal compiler. The generator programs are thought of as not being part of the Parsifal compiler itself, but rather as being a set of utility programs available to the user. This gives the user a great degree of control over the implementation of all operators, and is in accord with the principle of keeping these details separate from the compiler.

---

[5] In retrospect, it may be argued that a better syntax for this statement may have been FOR '+' USE DS_ADDER;

[6] To be precise, the cell will be called *DS_MULT_nn* where *nn* is an integer representing the word-length.

4.2 Function Calls

As in most computer languages, it is possible to call functions in the BSL language. There are several special features, however, which make BSL suitable for describing Digital Signal Processing (DSP) designs.

Parameters

In the use of functions in the specification of DSP designs, it is useful to distinguish between the *signal* and *non-signal* arguments to the subroutine call. For instance, suppose that we are writing a shift-left operation[7]. In a particular instance, we may wish to shift a signal x by 5 places. The *shift_left* function will be called with arguments x and 5. However, there is a difference between x which is an actual signal carried by a set of wires in the circuit and 5 which is not a signal in the circuit, but is simply a specification of how many places to shift x. It is appropriate to separate the arguments to the subroutine call into *signals* and *parameters*. The required syntax for the call to the shift-left subroutine is

y := shift_left (x ; 5);

The signals come first and are separated from the parameters by a semi-colon. the general syntax will be

y := some_function $(s_1, s_2, \ldots, s_n; p_1, p_2, \ldots, p_m)$;

Of course, functions may be used at any point in an expression, just as in a standard software language. For instance,

y := shift_left(a+b; 5) + shift_left(a−b; 4);

Multiple output functions: In electronic circuits, it is common that a block of circuitry has more than one output. For instance, when two arithmetic values are added together, their arithmetic sum is produced. It is possible, however that overflow may occur, in which case the result will be invalid. It is possible to design a digit-serial adder in such a way that it outputs both the sum and an overflow indicator. This is the case with the digit-serial adder used in Parsifal. Such an adder will be referred to by a statement of the form

(y, overflow) := sum (a, b);

The general form of a function call is then

$(y_1, y_2, \ldots, y_k)$ = some_function $(s_1, s_2, \ldots, s_n; p_1, p_2, \ldots, p_m)$;

---

[7]Of course, a built-in operation for shift-left exists, so the user will normally not need to supply another.

Built-in operators : The built-in operators in the BSL language are all equivalent to function calls. There is no difference in the manner Parsifal handles expressions of the form $a + b$ and $add(a, b)$. Each of the built-in operators corresponds to a particular named function. For instance the '+' operation corresponds to the function *add* as previously indicated. Some of the built-in operators correspond to functions with parameters. For instance, the expression

$$a << 6$$

is equivalent to a function call *shift_left*$(a; 6)$. Similarly, a constant multiplication $x * 0.1234$ or $0.1234 * x$ is equivalent to the function call *const_multiply*$(x; 0.1234)$.

Bound cell names: Every function call contained in the specification of a symbolic cell must be bound to a particular cell. The cell binding is specified by a FOR statement as already indicated. Suppose, for example, that a binding is specified by a statement

    FOR somefunction **USE** basename;

Now, when a reference to the function, *somefunction* is encountered, the compiler will look for a cell called *basename_xxxx* to implement this function. The *basename* is specified in the *FOR* statement, whereas the string *xxxx* is obtained by suitably encoding list of parameters of the function call, separated by underscores. This convention is necessary, since the user does not want function calls such as *somefunction (x; 5)* and *somefunction (x; 6)* to be implemented by identical cells. Instead, *somefunction (x; 5)* should be implemented by a cell called *basename_5* and *somefunction (x; 6)* by cell *basename_6*.

Here is an example program segment:

    SYMBOLIC PROCESSOR fred (. . .)
        WordSize 16;
        FixedSize 15;
        DigitSize 4;

    FOR myfunction **USE** myf **CALL** "myfgen";
    . . .

    SPECIFICATION
        . . .
        (y1, y2) := myfunction (a+b, c; 5, −0.13);
        . . .
    END ;

In this case, the compiler will search the library for a cell named *myf_5_n0d13* to implement the function call.

4.2. FUNCTION CALLS

Calling generators: If the compiler does not find a cell in the library to implement a function call, it will call a generator program if one is specified. Generators will be called with a command line syntax as follows

generator_name cellname wordsize fixedsize
digitsize [parameters]

where *generator_name* is the name of the generator program specified in the FOR . . . USE . . . CALL statement, *cellname* is the name of the cell to be generated and *wordsize, fixedsize* and *digitsize* specify the data format of the current symbolic cell. The line ends with any parameters that may have been specified when the function was used. Thus, in the example of the previous section, the generator program *myfgen* will be called with the following command line :

myfgen   myf_5_n0d13   16   15   4   5   −0.13

This style of invoking commands is based on the standard conventions used in the UNIX–trademark operating system, under which the compiler runs. Other UNIX–trademark conventions are used. For instance, it is possible to specify additional flags to pass to the generator program. For instance the statement

FOR myfunction **USE** myf **CALL** "myfgen -p";

would cause the *myfgen* program to be called with the '-p' flag, the meaning of which will be interpreted by the generator program. A further UNIX–trademark convention is the specification of directories to search when looking for generator programs by use of the '-I' option when the compiler is invoked.

Bound and unbound cells: Many digit-serial cells (whether described by leaf cell, stack, composite or symbolic cell descriptions) are intended to be bound to function calls from a symbolic cell. Examples are complete digit-serial adders or multiplier cells. Such cells will be called *bound cells.* Other cells are not intended to be bound directly to function calls. They usually are building blocks used to build larger cells as specified by a stack cell or composite cell description. Examples of such cells are the *ASUB_SLICE*, *ASUB_CONT* and *ASUB_CAP* cells used to build a digit-serial adder/comparator cell. Another example is provided by the bit-slice cells, used to build up a digit-serial multiplier. These bit-slices may themselves be made up from a stack of smaller leaf cells.

We now explain how the signal arguments of a function call are identified with the IO ports of the cell that it is bound to. The inputs of a bound cell are divided into two classes : *TIME*-type inputs and non-*TIME*-type inputs, otherwise known as data inputs. The *TIME*-type inputs always have names that start with the prefix *TIME* (for example *TIME*, *TIME_2*, *TIME_LSB*). The two types of inputs may be mixed together in the cell header. The rules for correspondence of function arguments and IO ports are as follows:

- There must be exactly as many non-*TIME*-type inputs to a cell as there are inputs to the function to which it is bound. The non-*TIME*-type inputs of the cell are bound to the signal arguments of the function call in the order in which they occur in the cell header.

- There must be exactly as many outputs to a bound cell as there are outputs to the function to which it is bound. The outputs of the cell correspond to the outputs of the function in the order they are written.

- The *TIME*-type inputs are connected to the appropriate phase of the *TIME* signal after the circuit has been scheduled.

- The width of the IO signals of the function must match the width of the ports with which they are identified.

For example, if the function call

(y1, y2) := myfunction (a+b, c ; 5, −0.13);

is bound to the cell *myf_5_n0d13* which has the cell header

CELL myf_5_n0d13 (**IN** : TIME, x(1), y[4], TIME_2(3);
OUT : xout[7](5), yout(3));

then the result of the expression $a + b$ is connected to the cell input x and c is connected to y. The *TIME*-type inputs *TIME* and *TIME_2* are connected to timing signals. Similarly, the output *xout* is connected to the signal *y1* and *yout* is connected to *y2*. The width of signals $a + b$ and c must be 1 and 4 respectively (the widths of x and y), and the widths of *y1* and *y2* must be 7 and 1 (the widths of *xout* and *yout*). If the '+' operation is bound to a digit-serial adder with a digit-size of 4, then $a + b$ has width 4, not 1 as required and the specification is in error.

Timing information: Bound cells must contain timing specifications on all their input and output ports in order for the circuit to be properly scheduled. If a timing specification is absent, then the default value of 0 is assumed. The syntax of timing specifications was given in section 1.3.4 and the way they are used for scheduling is described in Chapter 6. Non-bound cells need not have a timing specification.

4.2.1 Adding New Functions

It will now be shown how a designer may add new functionality without recompiling the compiler.

Suppose that a user wants to design and use a new adder to replace the standard adder. These are the steps that should be taken

4.2. FUNCTION CALLS

- For digit-serial designs, this new adder should preferably be laid out as three cells, *NEWADD_CAP*, *NEWADD_SLICE* and *NEWADD_CONT* according to the standard template. The layout descriptions of the three cells may then be extracted automatically from the STREAM–trademark description of the cells.

- The next step is to create a stack cell description for a complete adder of the desired digit-size and include it in the input before the symbolic stack description in which it is used. Supposing that the digit-size is 4, such a description will be something like :

    CELL NEWADD (**IN** : A[3](0), B[3](0), TIME(0);
                          OUT : X[3](1));

    STACK
        C **of** NEWADD_CAP;
        S1 **of** NEWADD_SLICE;
        S2 **of** NEWADD_SLICE;
        S3 **of** NEWADD_SLICE;
        CONT **of** NEWADD_CONT;

    CONNECT
        A_1 **to** S1.A;
        .
        .
        .

    END ;

- Finally, this cell can be accessed by including a statement of the form

    FOR add **USE** NEWADD;

The method shown here requires the creation of a stack cell description suited to the digit-size being used in the circuit. If the digit-size is changed, then the stack cell description for the cell *NEWADD* must be changed. It is possible to handle general digit sizes by the use of a generator program. Instead of writing the description of the cell *NEWADD* as given above, a program can be written to generate this description. Using this method, the user creates a program called *newaddgen* which is called with a command of the form

    newaddgen <wordsize> <fixedsize> <digitsize>

and writes the cell description given above to its standard output. Such a generator can be written in such a way that it generates an adder with the digit-size specified in the command line. The binding statement

FOR add **USE** NEWADD **CALL** "newaddgen";

is then used to specify the name of the cell to use and the generator to call. When the generator is called, its output is parsed by the Parsifal compiler and the generated cell is added to the data base of known cells. Thus, the next time an adder is required, the cell *NEWADD* will be known and the generator need not be called again.

4.3 Control Structures

So far, we have not described any mechanism similar to *while*-loops, *for*-loops or *if* statements of the type found in most programming languages. Rather than include such statements directly in the BSL language parsed by Parsifal, it was decided that a macro package would be used instead. This provides the desired functionality with minimal programming effort. The standard C-preprocessor ([45]) macro expander commonly used in UNIX–trademark does not have the desired capability. Instead, the *m4* macro preprocessor was chosen. This is a macro preprocessor program distributed with UNIX–trademark, and hence widely available.

When a text is processed my *m4* it is simply copied from input to output, except for certain key words that are recognized as *macros*. A macro may be just a single word, or it may have arguments which appear in brackets following the macro name. When a macro is encountered in the input, certain text is substituted in its place. Text is made immune from macro expansion by including it in quotes, such as 'not-to-be-expanded-text'.

Apart from a small number of macros which are built in, it is possible for the user to define new macros. The most important built-in macros are

define This allows the user to define a new macro. The syntax is *define('macro-name', 'replacement-text')*. Whenever the word *macro-name* is encountered in the subsequent text, the text *replacement-text* is substituted for it. The replacement text may contain symbols $1, $2, ... which refer to the arguments of the macro.

eval Syntax is *eval(expression)* where *expression* is text which may be interpreted as an arithmetic expression. The macro causes the expression to be evaluated and replaced by its numerical value. For instance, *eval(5+5)* will be replaced by the string 10.

ifelse This allows selection between two optional pieces of text. The syntax is *ifelse(exp1, exp2, text1, text2)*. If *exp1* and *exp2* are the same then text *text1* is included, otherwise *text2* is included.

When text is generated by a macro expansion, it is itself subject to further expansion if it contains any macros.

4.3. CONTROL STRUCTURES

It is clear that the *ifelse* macro can be used in BSL descriptions to specify conditional compilation of digit-serial specification code. However $m4$ does not contain any built-in macros for constructs such as *for* or *while* loops. Therefore, it was necessary to define a small number of new $m4$ macros which provide the required capability. The most important ones are the following :

while: The meaning is the same as the *while* loop in, for instance, C or PASCAL. The syntax is

　_while('condition','action')

The _while macro itself is defined in $m4$ by the definition

　define('_while','ifelse(eval($1),1,'$2','_while('$1','$2')')')

_for: The _for macro creates a loop with index that is incremented from a starting value to an ending value. The syntax is

　_for('index',startvalue,endvalue,'text')

As an example, the macro text

　_for('i',1,10,'eval(i*i) ')

generates a list of the first 10 squares.

The _for macro itself is defined in $m4$ by the definition

　define(_for,'define('$1',eval($2))_while('($1)<=($3)',
　　　'$4''define('$1',eval($1+1))')undefine('$1')')

which is left here without explanation for the reader to puzzle out.

A separate macro is provided for loops which count downwards. Other macros _succ and _prev generate the successor or predecessor of an integer argument. Any BSL circuit description is automatically run through the $m4$ macro preprocessor, and the macros just described are provided for use in the circuit description. As an example, the *NEWADD* cell described in section 4.2.1 may be written using $m4$ macros as follows :

　define(nsize,3)

　CELL NEWADD (**IN** : A[nsize](0), B[nsize](0), TIME(0);
　　　　　　　　OUT : X[nsize](1));
　STACK
　　　C of NEWADD_CAP;
　　　_for('i',1,nsize,'
　　　　　'S'i of NEWADD_SLICE;
　　　')

```
        CONT of NEWADD_CONT;

    CONNECT
        _for('i',1,nsize,'
            'A_'i to 'S'i.A;
        ')
            .
            .
            .
    END ;
```

In this way, an adder of any digit-size may be generated by defining *nsize* to the current digit-size. Of course, the *m4* macro set may be used in any type of cell, for instance in a symbolic cell description.

4.4 Standard Libraries

To simplify the task of the designer, standard libraries are provided for all the built-in operators. These include

1. *m4* macro definitions defining the macros described above.

2. interface, layout and simulation descriptions for all the leaf cells,

3. *m4* descriptions of stack cells for all the stacks usually built from the leaf cells,

4. A file containing *FOR* statements specifying the standard bindings of the built-in operators, as well as generators for those functions (e.g., multiplication) which need them.

5. Executable programs for a variety of useful generators.

6. A STREAM–trademark file containing the layouts of all the leaf cells.

The first four of these are to be included in the input description. The generators are automatically called as required by Parsifal. The STREAM–trademark file for the leaf cells is to be merged with the output of the compiler to produce a complete description of the desired chip. Some of the important library functions made available in the standard function library will now be described.

4.4.1 Data Conversion

We describe the dual functions of parallel/serial and serial/parallel conversion.

Parallel data can be transmitted on word in each clock cycle, whereas digit-serial data requires several clock cycles to transmit a word. For instance, suppose that the word-length is 12 and the digit-size is 4. Then it takes three clock

4.4. STANDARD LIBRARIES

cycles to transmit a digit-serial word. Consider a parallel data stream which transmits a new word in each clock cycle. If this data stream is being converted to a single digit-serial stream, then the digit-serial data stream cannot handle the data rate of the parallel stream and so two out of three of the parallel data words must be discarded. In order to match the data rate of the parallel stream three separate digit-serial data streams will be required. On the other hand, it is possible that the data in the parallel stream does not change in every clock cycle, so it may be possible for a single digit-serial stream to keep up with the data rate of a parallel stream. The opposite considerations apply in the case of serial/parallel conversion. In designing parallel/serial and serial/parallel converters we require the flexibility to handle different cases as just described.

As has been frequently remarked, digit-serial data is divided up into words which last several cycles called a *sample period*. Similarly, it is convenient to divide parallel data up into groups, each group lasting one digit-serial sample period. Since it is inappropriate to refer to such a group of parallel data as a word (for it actually consists of several words), we will refer to it as a *frame* of parallel data. Thus both digit-serial and parallel data are divided up into frames of length one sample period. Just as with digit-serial data we can number the digits of a sample, so with parallel data the individual words in a frame are numbered from first to last.

4.4.2 Parallel/Serial Conversion

The function of a parallel/serial converter will be illustrated with an example. We assume that the word-length is 16 and the digit-size is 4, and hence that the sample period (or frame length) is 4 cycles. The code

(s0, s1, s2, s3) := ps(x ; 0, 1, 2, 3);

describes a parallel/serial converter that splits a parallel data stream into four serial streams, $s0$, $s1$, $s2$ and $s3$. As indicated by the four parameters in the call $ps(x; 0, 1, 2, 3)$, the four outputs of the converter consist of the parallel words which are present on the input during the clock cycles numbered 0, 1, 2, and 3 of each frame. The word present at the input at time $t = 0$ of each frame becomes the serial word $s0$, the word at time $t = 1$ becomes $s1$, etc. The cell produced by the corresponding generator will be of the form

CELL PS_0_1_2_3 (
    IN : X[16](0),
        TIME0(1), TIME1(2), TIME2(3), TIME3(4);
    OUT : A0[4](1), A1[4](2), A2[4](3), A3[4](4));

Notice that 4 *TIME* signals are used to latch the signals from the parallel bus into serial-out registers. The four output signals are staggered in time, but this is of course taken care of by the scheduling algorithm.

An alternative parallel/serial converter may be described by the code (s0, s1, s3) := ps(x; 0, 1, 3);

In this case, words are read from the parallel input during clock cycles 0, 1 and 3 of each frame. The value on the input during clock-cycle 2 of the frame is ignored. A further example shows how the I and Q components of a complex signal may be interleaved on a parallel input bus. They will be separated into two serial streams by the statement

(I, Q) := ps(x; 0, 2);

Assuming a frame of 4 clock cycles, the value on the input during times 1 and 3 will be ignored. This means that the parallel input values need to change only once in 2 clock cycles. In other words, the internal clock rate will be double the external rate.

4.4.3 Serial/Parallel Conversion

Serial parallel conversion is the reverse of parallel/serial conversion. For instance, in the opposite operation to the $I - Q$ example given above, serial I and Q signals can be interleaved into one parallel output using a statement

X := sp(I, Q ; 0, 2);

This indicates that the I signal is to be placed on the output during clock cycle 0 of every frame, and Q is to be output during clock cycle 2. This is implemented in such a way that a signal placed on the output will remain there until replaced by a new signal. For instance, since no output is specified for time $t = 1$, the value I will be maintained on the output during this clock cycle. Thus, the output will contain value I during clock cycles 0 and 1, and value Q during clock cycles 2 and 3. In the next clock cycle Q will be replaced by the next value of I.

The cell produced by the serial/parallel converter in this example will have the form

CELL SP_0_2 (**IN** : A0[4](0), A2[4](2), TIME0(0), TIME2(2);
     OUT : X[16](5));

The long latency of the serial/parallel converter is due to the fact that a parallel word cannot be output until all the digits of an input word have been assembled into one word.

4.4.4 Latches

There is a function that provides a set of loadable registers. The syntax is of the following general form

4.4. STANDARD LIBRARIES

(x1, x2, x3, x4) := latch (Xbus, lat1, lat2, lat3, lat4; 16, 4);

This provides a set of four sixteen-bit registers connected to the input bus Xbus. Values are latched into the registers in response to the latch signals *lat1*, *lat2*, *lat3* and *lat4*. Any number of registers may be attached to the bus. A *decoder* function is also provided to decode the latch signals from an address. The output of a register is a parallel word, remaining fixed as long as a new value is not latched into the register.

4.4.5 ROMs

It is possible to include a ROM in a circuit by means of the ROM generator. The ROM functions simply as a look-up table, generating an output word of specified width in response to an input address. The address may be provided from off the chip or from an internal source. The specification of a ROM is kept in a separate file. Here is a description of a very simple ROM

```
#
# This file is the description of a ROM
# containing 12 8-bit words.
#

# First line contains the number of words and
# the number of bits per word

12 8

# Subsequent lines contain the values in the ROM, LSB first

0 0 1 0 0 0 1 0
0 0 0 1 0 0 1 0
1 0 0 1 0 0 0 0
0 1 0 1 1 0 0 0
0 0 1 1 0 1 0 0
0 1 0 0 1 0 0 0
1 0 0 0 0 1 0 0
0 0 0 0 0 1 1 0
0 0 0 1 0 0 1 1
0 0 0 1 0 1 1 0
1 0 0 1 0 1 0 1
0 1 0 1 1 0 0 1
```

In a symbolic cell description, the ROM just described can be accessed by means of the FOR...USE...CALL device. The statement

FOR myrom **USE** rom1 **CALL** "make_rom rom12.pers";

causes the compiler to search for the function myrom, and then to create myrom by calling the make_rom program, passing the file name rom12.pers to the makerom program. Thus, the function call myrom will become valid for use in the SPECIFICATION section of a symbolic processor description. The statements

SIGNAL y[8], address[4];
...
SPECIFICATION
y := myrom(address);

will cause the variable *y* to receive data from the ROM referred to by the function call myrom at the address specified by *address*.

4.4.6 Static and Periodic Signals

The *static* and *periodic* functions indicate to the compiler that signals carry constant or periodic values. For instance, suppose that one of the primary inputs to the chip is an asynchronous signal intended to be used as an *enable* signal for an on-chip latch. It is important that the latch signal not be passed through any synchronization delays, but rather must be fed directly to the *enable* pin of the latch cell. Similarly, if the signal to be latched is provided from off the chip, then it should not pass through synchronization latches before being latched. The *static* function is an indication to the compiler that the signal in question should not be subject to synchronization delays. Thus, if we write

y := static (x; 4);

then no synchronization delays will be applied to the signal *y*. The parameter 4 is the width of the signal *x*. The correct way of latching the signal on a bus using an external latch signal is

y := latch (static(x_bus; 16), static(lat; 1) ; 16);

In this case a 16-bit wide latch is created. The 16-bit wide bus *x_bus* is latched when the 1-bit wide signal *lat* transitions from high to low.

Notice that it is necessary to pass the widths of the input signals as explicit functions to the function calls unless a default value, is to be used. This is because the widths of the input signals are not automatically passed in the call to the generator which will create the cell to implement the function. The size of a latch generated by the *latch* function defaults to the word-length if no parameter is specified. Similarly, the size of the *static* function is by default equal to 1. In the above example, if the word-length is 16, then the default sizes for the *latch* and *static* functions can be used to simplify the above expression to

4.4. STANDARD LIBRARIES

  y := latch (static(x_bus; 16), static(lat));

It would no doubt be preferable if the widths of all the inputs were passed automatically to the generator, but this is not the case in the present implementation of Parsifal. Notice that such a convention would allow the above latch command to be simplified to

  y := latch (static(x_bus), static(lat));

The *periodic* function is used to label a signal as periodic. For instance a digit-serial signal carrying an arithmetic constant is a periodic signal, since the digit-serial constant will be repeated with a period equal to the sample period. The fact that this is a periodic signal means that it is not necessary to pass it through long synchronization delays longer than the period of the signal. A declared constant value is understood implicitly to be periodic with period equal to the sample period. For other periodic signals, the period must be explicitly declared. This would be indicated by a statement similar to

  y := periodic(x; 4, 3);

which indicates that x is a signal of width 4 with period 3. This would be the case if x were a constant digit-serial value with word-length 12 and digit-size 4.

For instance, the output of a parallel-serial converter is periodic as long as the input parallel signal remains unchanged (maybe a static signal from off-chip). In this case, it would be appropriate to use the statement

  y := periodic(ps(x) ; 4, 3);

in the case where word-length and digit-size are 12 and 4, respectively.

The common case where parallel values are latched into an on-chip register and converted to digit-serial format for on-chip use can be described by the statement

  define(wsize,16)
  define(nsize, 4)
  define(speriod, eval(wsize/nsize))

  y := periodic (ps (latch (static (x;wsize), static (lat))),
        nsize, speriod);

Both the *static* and *periodic* functions are implemented by a null-cell containing no hardware, but simply connecting their input to their output. This operation has no ultimate effect on the generated hardware but simply serves to insert a node in the signal flow graph. In addition, these functions are understood by the compiler[8] during the scheduling and delay-insertion stages of

---

[8]This is unfortunately a violation of the principle that the compiler should not have any built-in understanding of the semantics of any operator or function.

compilation. During scheduling, they ensure that the output of the function is not included in the cost function during scheduling, and during delay-insertion they inhibit the insertion of delays.

The library contains many other functions other than those described above. These include functions for extracting individual bits, assembling parallel words from individual bits, sign extending words and many others.

4.5 Examples

Various examples of chip descriptions are given next to illustrate the concepts described in this chapter.

4.5.1 Square Root

The first example is a simple circuit for performing a square root operation [30][31]. The square root is computed using a minimax polynomial expansion. By doing pre-normalization of the number to lie in the range the range [0.5, 1.0) it is possible to achieve good accuracy with only a quadratic expansion.

```
/*
 * Does a square root of a number using a quadratic approximation.
 * This will be accurate to within 5.0e-4.
 * For negative number, it takes the square root of the absolute value.
 */

define(nsize, 4);
define(wsize, 16);

include(BSDIR/m4operators.lib);
#include "interface.lib"
```

SYMBOLIC PROCESSOR sqroot (in : xin[wsize]; out : sqrootout[wsize])

WordSize wsize;

#include "operators.h"

CONST
    a0[nsize] = -0.315138;
    a1[nsize] = -0.885683;
    a2[nsize] = -0.201354;

SIGNAL

4.5. EXAMPLES

x[nsize], sqrootx[nsize],
x0[nsize], x1[nsize], x2[nsize], x3[nsize], x4[nsize],
shift0[nsize], shift1[nsize], shift2[nsize], shift3[nsize],
y0[nsize], y1[nsize], y2[nsize], y3[nsize], y4[nsize];

SPECIFICATION

/*Serialize x */

x := ps(xin);

/* Take the absolute value of x */

x0 := x<0.0 ? -x : x;

/* Shift the number into the range 0.5, 1.0 */

shift0 := x0 < 0.00390625;
x1 := shift0 ? x0<<8 : x0;

shift1 := x1 < 0.0625;
x2 := shift1 ? x1<<4 : x1;

shift2 := x2 < 0.25;
x3 := shift2 ? x2<<2 : x2;

shift3 := x3 < 0.5;
x4 := shift3 ? x2<<1 : x3;

/* Now carry out the square root operation */
y4 := (a2*x4 - a1)*x4 - a0;

/* Now shift back */
y3 := shift0 ? y4>>4 : y4;
y2 := shift1 ? y3>>2 : y3;
y1 := shift2 ? y2>>1 : y2;
y0 := shift3 ? y1 * 0.7071068 : y1;

sqrootx := (x4 = 0.0)? 0.0 : y0;

/*Restore result to parallel format */

```
        sqrootout := sp(sqrootx);
    END sqroot;
```

In this example, the number is shifted into the desired range by a sequence of four shifts, the polynomial expansion is done and then the result is shifted back half as many places to compensate for the normalization of the input. At the final stage a multiplication by $\sqrt{2}/2$ is used to compensate for the one-place shift of the input.

Three libraries are included : the library "m4operators.lib" containing the descriptions of how to build up stack cells for the designated word-length; library "interface.lib" describing the layout of each of the leaf cells and the library "operators.h" describing the standard operator bindings. The library "m4operators.lib" contains descriptions using m4 macros for all the stack cells parametrized in terms of the digit-size. The m4 processor expands the stack cell descriptions to generate stacks of the chosen digit-size.

4.5.2 Complex Arithmetic

The next example shows how the basic arithmetic operations can be redefined to implement complex arithmetic. This does not constitute a true overlay capability such as exists in Ada or C++, since the arithmetic operators (such as * and +) cannot be used to mean both real and complex operators according to context.

This example also shows the use of m4 macros. The first part of the description defines the word-length. All operations are done on words of length 16 using a digit-size of 4. A complex number is conceived as being a word of width twice the digit-size, being made up of two digit-serial words. After the word-length is defined the library "m4operators.lib", containing stack cell descriptions, is included.

```
    /* Example of complex operations */

    define(wsize, 16)
    define(nsize, 4)
    define(csize, eval(2 * nsize))
    define(BSDIR, /usr/src/digitserial)
    include(BSDIR/lib/m4operators.lib)
```

The next code describes how complex words are to be assembled from two real quantities and vice-versa.

```
    CELL complex (in : re[nsize], im[nsize]; out : z[csize]);
        /* Makes a complex number from its real */
```

4.5. EXAMPLES

```
                    and imaginary parts */
              CONNECT
                  _for('i', 0, nsize-1, '
                        're_'i TO 'z_'i;
                        'im_'i TO 'z_'eval(i+nsize);
                  ')
        END complex;

        CELL real (in : z[csize]; out : re[nsize]);
              /* Extract real part of a complex number */
              CONNECT
                  _for('i', 0, nsize-1, '
                        're_'i to 'z_'i;
                  ')
        END real;

        CELL imaginary (in : z[csize]; out : im[nsize]);
              /* Extract imaginary part of a complex number */
              CONNECT
                  _for('i', 0, nsize-1, '
                        'im_'i TO 'z_'eval(i+nsize);
                  ')
        END imaginary;
```

Next, the implementation of the complex operators is defined in terms of real arithmetic operations.

```
        SYMBOLIC CELL complexadd (in : z1[csize], z2[csize];
                                   out : zout[csize]);

              /* Complex addition */
              WordSize wsize;
              DigitSize nsize;

              #include "operators.h"
              FOR complex USE complex;
              FOR re USE real;
              FOR im USE imaginary;

              SPECIFICATION
                  zout := complex ( re(z1) + re(z2), im(z1) + im(z2));
        END complexadd;
```

SYMBOLIC CELL complexdiff
    (in : z1[csize], z2[csize];
    out : zout[csize]);

    /* Complex subtraction */
    WordSize wsize;
    DigitSize nsize;

    #include "operators.h"
    FOR complex **USE** complex;
    FOR re **USE** real;
    FOR im **USE** imaginary;

    SPECIFICATION
        zout := complex (re(z1) - re(z2), im(z1) - im(z2));
END complexdiff;

SYMBOLIC CELL 'complexmult_'eval(wsize/nsize)
    (in : z1[csize], z2[csize]; out : zout[csize]);

    /* Complex multiplication */
    WordSize wsize;
    DigitSize nsize;

    #include "operators.h"
    FOR complex **USE** complex;
    FOR re **USE** real;
    FOR im **USE** imaginary;
    SIGNAL t1[nsize], t2[nsize];
    SPECIFICATION
        t1 := re(z1) * re(z2);
        t2 := im(z1) * im(z2);
        zout := complex (t1-t2, (re(z1)+im(z1))*
                      (re(z2)+im(z2)) - (t1+t2));

END ;

Finally a small program tests the implementation of complex arithmetic. Note how the basic arithmetic operators *add*, *subtract* and *multiply* are redefined to mean complex arithmetic. The two complex values *z1* and *z2* are input on the same input bus in the order re(*z1*), im(*z1*), re(*z2*), im(*z2*). One complex value

4.5. EXAMPLES

is output, the real and imaginary words are interleaved on the output bus.

SYMBOLIC PROCESSOR testcomplex
(in : Xin[wsize]; out: Xout[wsize]);

WordSize wsize;
DigitSize nsize;

#include "operators.h"
FOR add **USE** complexadd;
FOR subtract **USE** complexdiff;
FOR multiply **USE** complexmult;
FOR re **USE** real;
FOR im **USE** imaginary;
FOR complex **USE** complex;

SIGNAL
re1[nsize], im1[nsize], re2[nsize], im2[nsize]; z1[csize], z2[csize], zout[csize];

SPECIFICATION
/* Separate the input word into separate parts */
(re1, im1, re2, im2) := ps (X; 0, 1, 2, 3);

/* Compose the real and imaginary parts
   into complex numbers */
z1 := complex (re1, im1);
z2 := complex (re2, im2);

/* Do a calculation */
zout := z1 + z2 - z1 * z2;

/* Output the real and imaginary parts of zout
   on the same bus */
Xout := sp (re(zout), im(zout); 0, 2);

END testcomplex;

Chapter 5

Layout of Digit-Serial circuits

This chapter describes the methods used for generating a layout of a digit-serial circuit. The layout process starts with a description of all the functional units necessary for the accomplishment of the desired function and ends with the output of a complete geometry of the finished circuit. The layout method described here is that used in the Parsifal silicon compiler. Many other methods of layout could be used, including the use of a generic placement and route package. However, digit-serial circuits have special features which make it advantageous to use layout methods specifically applicable to digit-serial circuits. The methods described here have been shown to generate efficient digit-serial layouts, and that is the reason for including their description in this book. The layout methodology used for digit-serial designs grew out of that used for bit-serial designs. Various other authors have discussed the layout of bit-serial designs ([5][6]).

5.1 Digit-Serial Cell Layout Conventions

Digit-serial circuits are made up of a number of individual operators joined together in a data-flow graph by wiring. The approach taken in all bit- and digit-serial compilers has been to design a set of standard cells implementing each of the operators. The layout problem is then a standard-cell layout problem. In the Parsifal digit-serial compiler, each of the standard cells conforms to a given template as shown in the Fig. 5.1. All the cells are of a fixed height (depending on the digit-size) but may vary in width according to the complexity of the design. As may be seen in the diagram, there is a *VDD* bus passing across the bottom of the cell and *VSS* across the top. In addition to the power buses,

CHAPTER 5. LAYOUT OF DIGIT-SERIAL CIRCUITS

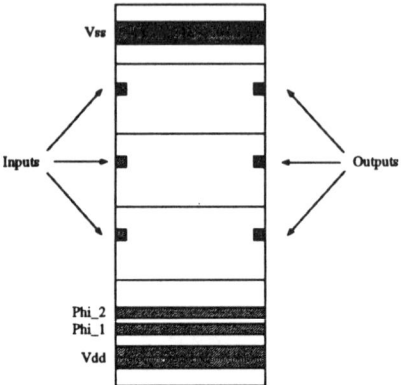

Fig. 5.1: Digit-serial standard cell

there are also *Phi_1* and *Phi_2* clock buses passing across the bottom of the cell. The width of these wires was chosen sufficient for the current-carrying capacity estimated necessary for the maximum length of a row of cells envisaged. Signal inputs and outputs to each cell are arranged in such a way that inputs are on the right and outputs are on the left, and as far as possible, the inputs and outputs are made to line up. In addition, each input or output may have alternative locations on the top or bottom edge of the cell. This placement of pins on the cells is intended to maximize the number of connections that can be made by simple abutment while allowing maximum flexibility in choosing which pin to make a connection to. In order to facilitate connections by abutment other rules are necessary. These include the following :

1. All pins on the right or left edge of a cell are in a given fixed layer (metal 1).

2. Two cells can always abut without design rule error, except in the case where a pin on one cell is not matched by a pin in the abutting cell. This is ensured by maintaining a half-design rule distance of any geometry from the edge of the cell, except where the geometry extends to the edge of the cell and forms a pin.

Certain operations, most specifically multiplication, in bit-serial are made up of many cells placed next to each other in a given specific arrangement. In these cells, the pins are (of course) designed to line up with each other left and right. It is nevertheless necessary when designing these cells to observe the above rules, since during the layout, it may be necessary to break up the array between different rows of the layout, or to insert special cells between cells of the array, as will be described further below.

5.1. DIGIT-SERIAL CELL LAYOUT CONVENTIONS

5.1.1 Layout of the Complete Chip

The basic layout of a digit-serial chip is according to the template shown in Fig. 1.18. In this template, the cells are arranged in horizontal rows separated by routing channels. Pads are placed around the periphery of the chip. The input clock signals (2 phases) are fed up the side of the chip. They are buffered in special clock driver cells (ENDCELL) at the end of the rows of cells, and the Phi_1, Phi_2 clock signals are driven along the rows.

5.1.2 Linear Layout

Since the hardware requirement for the cells used in the design is substantially fixed, the major consideration in layout will be to reduce the routing area required in the layout of the chip. The first step in the generation of the layout of the chip is to pretend that the cells can be laid out in one long line of cells. The problem of cell placement is then reduced to one of finding the best linear ordering of the cells such that the length of interconnect is minimized. Later on, the linearly arranged cells are folded into several rows to generate a two-dimensional placement. The rationale behind starting with a linear placement of cells is as follows. A good layout of the cells will have a high proportion of abutted and near-by local connections with few connections between widely separated cells. If this is true of a linear layout, then the property of local interconnection will be maintained when the line of cells is folded into several rows, so a good linear layout will lead to a good 2-dimensional layout. Furthermore, it is far easier to deal with a linear arrangement of cells than a two-dimensional one, so that is how we start.

Growing islands: The first task is to identify blocks of cells that should be placed together, for instance the cells that make up a multiplier. The layout software assumes no inherent knowledge of the desirable placement of cells, other than what can be derived from the required network connections and knowledge of the pin placements on each of the cells. The first task is to find pairs of cells that unquestionably should be placed side-by-side. Such a placement is mandated if the following conditions are satisfied:

1. All pins on the right side of one cell are to be connected to all the pins on the left side of the other cell.

2. The desired connection can be made by simple abutment.

3. None of the pins in question is required to be connected to a third pin elsewhere.

If these conditions are met, then the cells clearly should be abutted. The two cells are amalgamated together into a single block and the process continues until no further cases of this sort exist. We continue to find other cases in which cells should be abutted.

Placement of blocks: At the end of the island-growing stage, a number of blocks of cells remain that must now be placed linearly. During the next stage of block placement, the blocks of cells are treated as inseparable units. The pin locations on each of the blocks, and the required connections are known. The placement of blocks takes place in two phases – an initial placement stage and an iterative refinement stage.

Initial placement: The algorithm used for the initial placement of the blocks is probably not particularly important, since it will be changed substantially by the later refinement step. The method that is used in the compiler is a simple topological sort. An ordering relation is defined in which a block A is defined to precede a block B (written $A \geq B$) if there is a net attached to an output pin in block B and to an input pin in block A. An attempt is then made to find a topological ordering of the blocks. Specifically, at each stage a block A is placed in the ordered output list if the set of all blocks which precede A in the ordering have already been output. If there are loops in the ordering, then there will come a point at which there exists no block eligible to be output, but the ordering is not yet complete. In this case, that block is selected for output which depends on the minimum number of other blocks not yet output. The idea behind this placement technique is to let the structure of the data-flow graph be mirrored in the layout. However, since most data-flow graphs of interest are not linear, the technique itself does not lead to adequate results, and a refinement stage is necessary.

Iterative refinement: The initial layout of the blocks is improved by swapping blocks. The method used is to continue swapping blocks while such swapping leads to a better layout. Thus, no swaps that lead to a worse layout are allowed. An alternative approach would be to allow temporary moves that led to worse placements using some sort of simulated annealing scheme ([46]). Such methods may lead to a slight improvement in the placement at the cost of greater coding complexity and run times. The methods actually used, and described below, led however to satisfactory placements. As mentioned above, the cost function used as a criterion for evaluating placements is based on routing length. In particular, the cost function used is of the form

$$\sum_{n_i} (length(n_i))^k$$

where the sum is over all nets n_i. If k equals 1, then this cost function represents the sum of net length. If $k = \infty$, then it represents maximum net length. After experimenting, it was found that a value $k = 2$ gave the most satisfactory

5.1. DIGIT-SERIAL CELL LAYOUT CONVENTIONS

results. The net length of a given net is simply the distance between the first and last pins belonging to that net. In changing the placement of blocks, we proceed by swapping adjacent blocks. In swapping two adjacent blocks, the only net lengths that change are those belonging to the two blocks in question. Consequently, it is an easy task to determine the effect on net length that would result from two such blocks being swapped.

Suppose the blocks are numbered $B_1, B_2, \ldots, B_N$. Starting with $i = 1$, we evaluate the effect of swapping blocks B_i and B_{i+1}. If the swap would result in a decrease in the routing cost, then the swap is carried out and i is set equal to $i - 1$. Otherwise, the swap is not carried out, and i is set equal to $i + 1$. This sequence of operations is continued until $i = N$, at which point the algorithm stops. It is easy to see that no further improvements are possible by swapping adjacent blocks.

Although this simple algorithm does result in very great improvements in the layout, it is possible to decrease the cost function further by allowing two or more blocks to be moved together. Therefore further passes of the algorithm are attempted in which larger groups of adjacent blocks are moved together.

Insertion of routing cells: When a linear ordering of cells has been determined, certain specialized cells are inserted in the layout to preserve routability and avoid design rule errors.

Spacer cells: Wherever two cells abut in the layout and there are pins on the adjoining edges that are not intended to connect to each other, or pins that are not matched by another pin in the adjoining cell, then a spacer cell must be added to prevent unwanted connections or to avoid design rule errors. This will be a cell with one design rule spacing width. Connections must be carried across the spacer cell where abutting connections are intended and must be omitted where no connection is required. In addition, the power and clock connections must be connected across the spacer cell.

Bringout cells: In certain cases, pins on the left or right edge of some cell or block of cells must be connected to some other pins in a non-adjacent cell. These pins will be hidden when the cell is abutted to the next cell. If alternative pin locations on the top or bottom edge of a cell are provided then no special action is required. However, when such alternative pins are not provided, it is necessary to bring out the hidden pins to the top and bottom edge of the array of cells so that the connection can be made in the routing channel.

No actual geometric layout is generated for these special cells at this stage, but an entry describing the cell is entered into the cell data-base. The output of geometric layout information is generated later.

5.1.3 Two-Dimensional Placement

The next stage is the division of the linear array of cells into rows. The number of rows of cells in the layout is either computed or provided by the user (for instance in order to force a non-square aspect). The linear array of cells is then divided into equal length rows which are then zigzagged back and forth in a serpentine arrangement of the original linear cell array. In dividing up the array of cells into rows, some of the original blocks of cells constructed during the island-growing phase of the layout may be split into two rows. This is necessary in order that the rows have approximately equal length.

Pre-routing - insertion of feed-throughs: The layout is not ready to route yet, since it is not yet possible to make all the connections in the routing channels lying between the rows of cells. Although all the pins to be routed together face onto a routing channel, they do not necessarily face onto the same routing channel. It will be necessary to make feed-throughs from one routing channel to the next so that connections may be made. A feed-through is a cell that may be placed between any two cells in the row to allow a single wire to pass from top to bottom. Each row of cells is considered separately and it is determined if it is necessary to insert a feed-through between the channels above and below in order to route the chip. If a feed-through for a given net already exists in a row then no new one need be added. This can occur if some cell contains a pin belonging to the net in question with alternative pin locations on the top and bottom edges of the cell, indicating a connection through that cell from top to bottom edge.

Flipping the rows: Counting the number of feed-through cells that will be required in each of the rows is a fairly straight-forward matter, as described above. A check is made at this stage for each row whether fewer feed-throughs will be needed if the complete row of cells is flipped top-to-bottom. The decision may be made independently for each row of cells. If flipping would result in an improvement, then all the rows in the cell are flipped.

Redivision into rows: At this stage, a count has been made on the number of feed-through cells that must be inserted in each row of the layout. Adding several feed-through cells to a row will increase the length of the row significantly. In general, there tend to be more feed-throughs in the middle rows of the layout than the first and last rows. The addition of feed-throughs therefore makes the length of the rows in the layout uneven. This makes the layout look unpleasant and wastes space at the end of the rows. The solution is to repeat the division of the cells into rows using the computed increase in the length of the rows due to addition of feed-through cells during the first pass as an estimate of the extra allowance of space that must be made in each row. After the

5.1. DIGIT-SERIAL CELL LAYOUT CONVENTIONS

second pass of division into rows and addition of feed-through cells, the rows are close to being of equal length, and further iteration is not necessary.

Two-dimensional optimization: After the second and final division of the cells into rows, a pass of rearrangement of cells within the rows takes place. This two-dimensional optimization is similar in concept to the one-dimensional optimization described previously in that cells are grouped into blocks which belong together as before. These blocks are then moved about within their own row. The goal is the minimization of routing length as before. In moving blocks in one row, account must be taken of the length of routing connections to the other rows in the layout. It was found in practice that if the division of the cells into rows was based on a good linear cell arrangement, then the gains achieved through this pass of two-dimensional optimization were not spectacular. However, they were sufficient to make this optimization pass necessary.

Routing: Once the order of the cells in each row has been determined, the feed-through cells may be placed in their final position. In doing this placement, the goal is the minimization of net length for each net individually. The complete chip may be routed. The routing has been reduced to a problem of channel routing. Each pin to be connected lies on the top or bottom edge of a row of cells, and connecting these pins together correctly will ensure that all cells in the chip are correctly connected together. It is appropriate to use an ungridded router, since it is not easy to ensure that the pins on the top and bottom of the rows of cells will be on any appropriate routing grid.

Construction of the pad ring: For many applications, the compilation of a cell without pads is useful. Such a cell may be used in combination with other cells, such as analog circuitry to constitute a complete chip. Fig. 5.3 shows a layout of such a chip. In other instances, it is appropriate to complete the chip by automatic insertion of the ring of pads. This is a fairly straight-forward task, involving a certain amount of tedious programming. Several different special cells must be constructed, such as cells to carry the power buses for the cells around the corners of the chip, and cells to carry the power bus over the empty spaces between pads. The general method used, as in the complete layout was to construct special cells which are then tiled together to make the complete chip. The top level of the geometrical hierarchy of the chip is a set of non-overlapping cells which cover the complete chip area.

The determination of the placement of the pads is done either automatically or at the direction of the user. A file specifying the locations of each of the pads is prepared in which the location of the pads is specified. There are various options. In one option, the locations of the pads may be determined. Alternatively, the side of the chip that each pad must go on is given and an optional order of the pads along that side of the chip is also specified.

5.2 Chip Examples

This section shows the layouts of various chips produced by the Parsifal compiler. Specific details of the applications will not be discussed. For particulars of various chips produced by Parsifal or the BSSC compiler (Parsifal I), the reader is referred to the following papers which describe compiled chips : [47],[48],[49], [50],[51],[52], [53],[54] and [55].

5.2. CHIP EXAMPLES

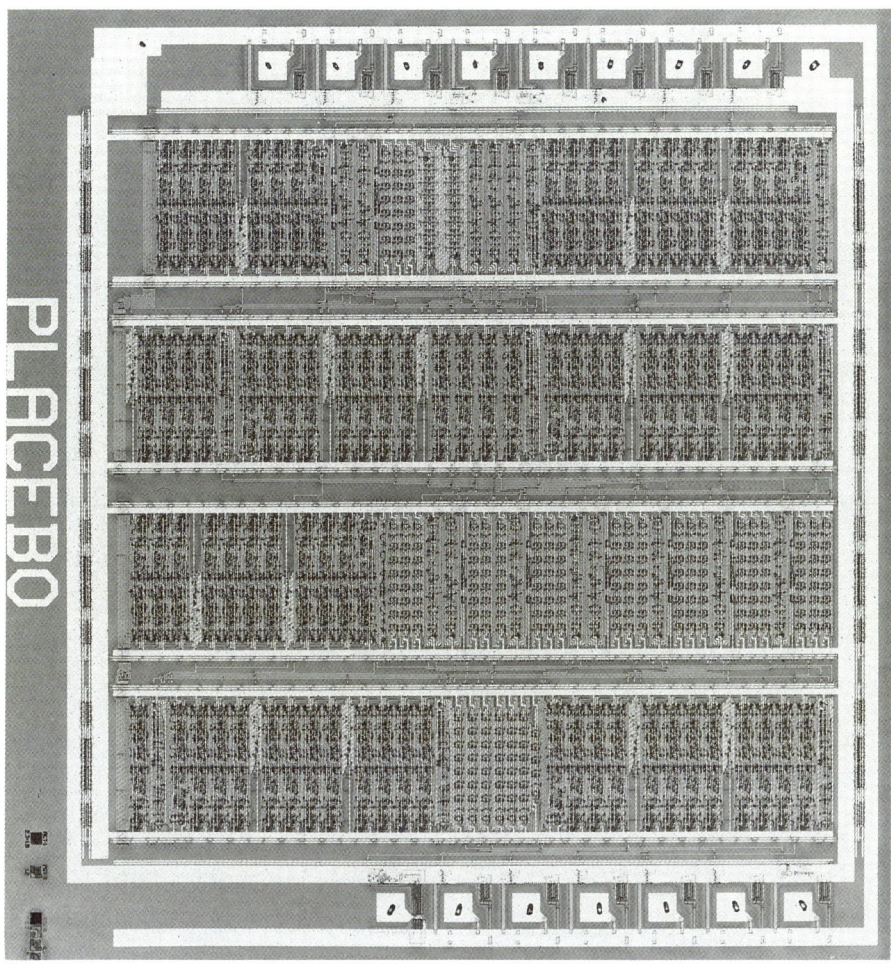

Fig. 5.2: This first layout is a small bit-serial chip. The chip is laid out in four rows of cells. The pads ring and the pads were placed automatically. In fact, the chip was laid out directly from the algorithmic description without need for user intervention. In this early version of the BSSC compiler (Parsifal I), pads were laid out on two sides of the chip only. Because the chip used serial input/output, the total number of pads is small. The disadvantage of having pads on two sides only is the increased difficulty in bonding the chip. Since this is a bit-serial chip, the width of the routing channels is relatively small.

Fig. 5.3: This die photograph shows the layout of the chip described in [48]. This chip was used by GE's power distribution business in a power management application. According to [48] the design of the digital processor including simulation took 2 weeks to complete. The three large blocks at the right side of the chip are analog circuitry (11-bit Delta-Sigma A/D converters). In this case, the chip was compiled without pads and the analog circuitry and pad ring were added by hand. Another chip that used analog circuitry is described in [51] and [52]. In that case, the layout of the chip was more automated. The analog circuitry consisted of standard cells that were placed in the row of cells, and connected automatically to the digital circuitry and the analog pads.

5.2. CHIP EXAMPLES

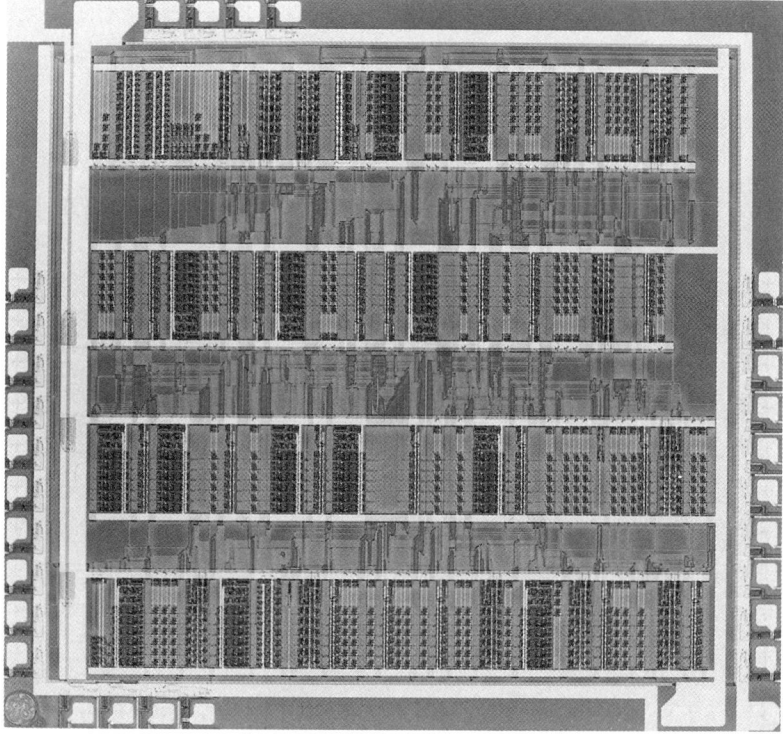

Fig. 5.4: This shows a small digit-serial chip. It is possible to perceive how the cells in the row are divided into bit-slices. The digit-size for this chip is 6 bits. The amount of the chip area used for routing is significantly greater than in the bit-serial chips. In fact, it is unusually large even for digit-serial chips. Pads are placed automatically on all four sides of the chip, in an order specified by the user, or as chosen by the compiler.

Fig. 5.5: This shows a larger digit-serial chip – the filter chip described in [53]. The digit-size is once more equal to 6 bits. The amount of area used for routing in this example is typical of a 6-bit digit chip. This chip contains a total of 8 separate digit-serial multipliers with loadable coefficients, and uses interleaves two separate digit-serial streams as described later in section 7.4.

Chapter 6

Scheduling

6.1 Scheduling

Scheduling of serial operator was treated informally in section 1.3.3. A precise algorithm for optimal scheduling will be described now. It is the task of the scheduler to ensure the correct synchronization or timing of operator inputs and outputs. Usually, correct synchronization is achieved by the insertion of delays into the circuit. As a simple example, the circuit shown in Fig. 6.1 to compute $(A+B)+A$ using the bit-serial adder of Fig. 1.7 is not properly synchronized, since the A input will arrive too early at the second adder. Inserting a one clock-cycle delay as shown in Fig. 6.2 will produce a correctly synchronized circuit. The task of synchronization is to insert delay operators in such a way as to produce a correct circuit. Since each clock cycle of delay is implemented by a hardware shift register, a secondary goal is to minimize the hardware cost by doing the scheduling using the minimum possible total number of delays. A simple metric used in circuit synchronization is that the cost of a delay is equal to the product of the width of the signal being delayed and the number of cycles of delay.

An algorithm will be described for the correct optimal insertion of delays. This algorithm is optimal in that it will produce the scheduling with minimum cost within the constraints of the problem as formulated.

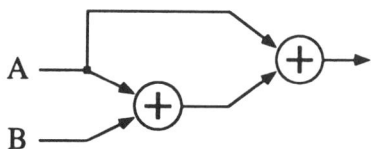

Fig. 6.1: Incorrectly synchronized circuit.

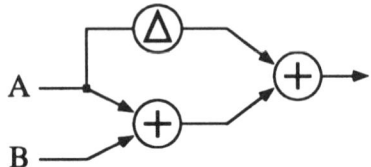

Fig. 6.2: Correctly synchronized circuit.

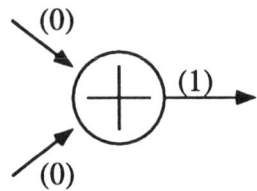

Fig. 6.3: Timing of bit-serial adder.

Operator timing: We begin with a general formulation of operator timing and latency. Consider the bit-serial adder of Fig. 1.7, which simultaneously receives two one-bit wide serial input data streams and computes their sum, also in bit-serial format. The output of the bit-serial adder is delayed by one clock cycle as shown and hence is produced one clock cycle later than the arrival of the inputs. Thus the latency of the bit-serial adder is one clock cycle. This timing relationship is represented in the *timing diagram* of Fig. 6.3. As shown, the two data inputs arrive at time $t = 0$, and the output is produced at time $t = 1$. All the timing values are related to some arbitrary point in time. In this case, we have chosen to set $t = 0$ to be the time of arrival of the first bits of the input data. The absolute time values in the timing diagram are not important, however – only the relative values are important.

The adder also receives a periodic input, denoted by *TIME*, that indicates the timing of the MSB (most-significant bit) of the two input streams. As indicated in Section 1.3.4 this timing signal is connected to the correct phase of a global timing signal, and so is treated differently from the primary adder inputs. Usually such timing signals are not shown in the timing diagram. Any bit- or digit-serial operator input port with a name starting with "*TIME*" is assumed to be connected to some phase of the global *TIME* signal. For this reason, such signals should not be treated in the same way as normal data signal in the circuit, for they will not pass through synchronizing delays in the same way as data signals. This leads to the rule :

- Timing signals are ignored in scheduling the digit-serial circuit.

It will be seen later that certain other types of signals are treated specially in

6.1. SCHEDULING

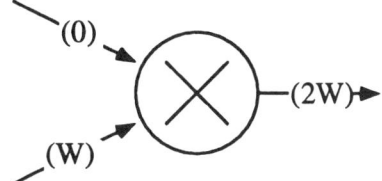

Fig. 6.4: Timing of parallel/serial bit-serial multiplier.

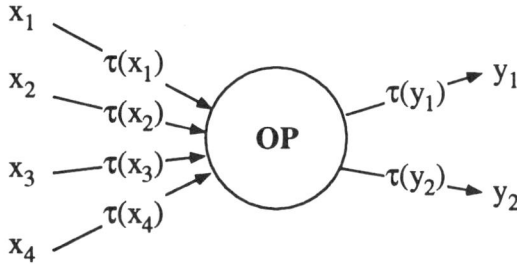

Fig. 6.5: Timing diagram of a general operator.

the scheduling process. Timing diagrams may be more complicated than that of the bit-serial adder. For instance, consider the serial/parallel or S/P multiplier, described in Chapter 3. In such a multiplier, the multiplicand input, B, must arrive W clock cycles before the multiplier A, say at time $t = -W$. During W clock cycles it is clocked into a serial shift register and is then latched at time $t = -1$ in response to a *TIME* signal. The multiplier input, A, arrives at time $t = 0$ and the partial products are accumulated during next W clock cycles. The less-significant part of the output is produced with its first bit appearing at time $t = 0$. The more-significant word of output is produced at a different output port, its first bit being produced at time $t = W$. Fig. 6.4 shows the timing diagram for such a bit-serial multiplier. In general, a digit-serial or bit-serial circuit may be described in terms of a timing diagram of the sort shown in Fig. 6.5. The timing diagram expresses a necessary relative timing of the inputs and outputs of a serial operator. In general, the operator will not operate as intended if the inputs do not arrive at the correct relative times.

Though it may seem strange, it is possible for arithmetic serial operators to have negative latency, which means that the first bit of the output is produced before the first bit of input is seen. An example of this is bit-serial left shifting. For instance, given a binary word $a_{W-1}a_{W-2}\ldots a_2a_1a_0$, the result of shifting three places to the left is $a_{W-4}a_{W-5}\ldots a_1a_0 000$. Thus it is possible to predict the first three bits of output (all zeros) before the first bit of the input is seen, and these three zero bits may be output in advance, which suggests that a

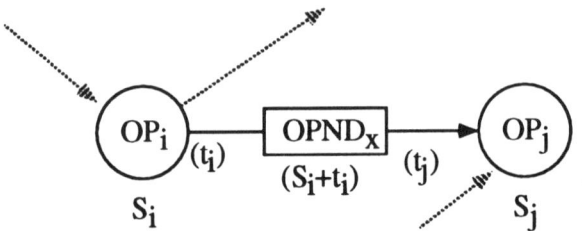

Fig. 6.6: Operator pair showing timing information.

latency of -3 cycles is possible. However, if the output of the shifter is to be latched, then the latency will be increased by one to -2 cycles, and in fact it is easy to design a circuit with latency -2 to carry out this operation.

This example may be used to demonstrate the fact that it is impossible to deduce the latency of a serial operation from any sort of low-level description of the cell, for instance in terms of gates. In order to compute the latency it is necessary to know the intended operation of the cell. To see this, consider a hypothetical operator called *ZERO_3MSB* which replaces the three high-order bits of a word with zeros (not a common operator admittedly, but suitable for illustrative purposes). Thus, given an input word $a_{W-1}a_{W-2}\ldots a_2a_1a_0$ the output is $000a_{W-4}a_{W-5}\ldots a_1a_0$. The latency of such an operation must be 0 (or 1 if the output is latched). However, it is readily seen that *ZERO_3MSB* may be carried out by exactly the same circuitry as the latency -3 shift-left circuit described above. The only difference is that in the case of the *ZERO_3MSB* circuit the three zero bits are considered to belong to the end of the current output word, whereas in the case of the shift-left operator they are viewed as belonging to the start of next output word. Thus the exact same circuitry has either latency -3 or 0 depending on the interpretation of the output.

Outline of the scheduling algorithm: The scheduling algorithm consists in assigning to each operator in the circuit a "scheduled time", which is the time at which the operation will nominally take place. Informally, each operator must be scheduled at a time after all its inputs are "ready", that is, have been output by the previous operator. We will formulate the scheduling problem as a linear programming problem. Since all scheduled times of operators must be integers, it is required to find the optimal integer solution to this problem. We will see, however, that an optimal real solution to the linear programming problem can easily be transformed into an integer solution, and so we can apply any of the known algorithms to the solution of the linear programming problem to find an optimal scheduling.

Consider a pair of operators joined by an edge as shown in Fig. 6.6. The operator OP_i has as one of its outputs the operand $OPND_x$. This operand in

6.1. SCHEDULING

turn is input to (perhaps among others) the operator OP_j. The scheduled times of operators OP_i and OP_j are denoted S_i and S_j, respectively. Further, the timing specifications of the relevant output and input ports of OP_i and OP_j are denoted by t_i and t_j respectively.

Now, since OP_i is scheduled at time S_i, the output $OPND_x$ will be ready at time $S_i + t_i$ as shown. Further, this same operand will be required as an input to the operator OP_j at time $S_j + t_j$. By the requirement that the operand cannot be used by OP_j before it is produced by operator OP_i we derive an inequality

$$S_j + t_j \geq S_i + t_i ,$$

or otherwise stated,

$$S_j - S_i \geq t_i - t_j . \tag{6.1}$$

Such an inequality holds for each pair of operands joined by an edge in the flow graph. In these inequalities the S_i and S_j are unknowns, whereas the value $t_i - t_j$ is a known constant value.

In order to handle input and output operands of the circuit properly, one may introduce special input and output pseudo-operators. An input operator is an operator with no inputs and a single output acting as the source of an input operand of the circuit. An output operator on the other hand has a single input which is attached to an output of the circuit.

Once a solution is found to this set of inequalities, the circuit may be correctly synchronized by inserting a delay equal to $S_j - S_i - (t_i - t_j)$ clock cycles between the operators OP_i and OP_j [22].

The minimum cost solution: There will in general be many solutions to the equations (6.1) and hence many ways of synchronizing the circuit. Our goal is to find the solution with the minimum cost, where our cost function is the total number of shift-register delays that need to be inserted into the circuit. Once a linear cost function is found, the problem of finding a minimum cost solution is simply a linear programming problem.

We next make an observation about delay insertion in the case of multiple fan-out. We remark that delays applied to a single signal should be applied sequentially and not in parallel. Consider a case where an output of some operator OP_O is used as an input to several other operators. Fig. 6.7 show such a situation, depicting a case where the output of OP_O is an input to three operators, OP_A, OP_B and OP_C. As shown in this example, delays of 10, 12 and 25 clock cycles need to be applied to the operand before it is input to the three operators. The total number of delays is equal to $10+12+25 = 47$. Fig. 6.8 shows an alternative arrangement of delays, where the arrays are arranged in sequence. The total number of delays is equal to 25, the length of the longest delay. This leads to an observation:

Observation 6.1. *The total number of delays that need to be applied to any node in the circuit is equal to the maximum delay that must be applied.*

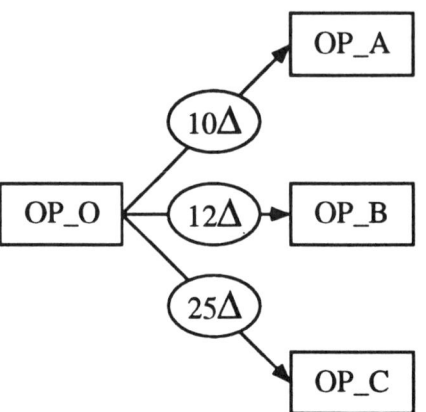

Fig. 6.7: Operator with Fan-out.

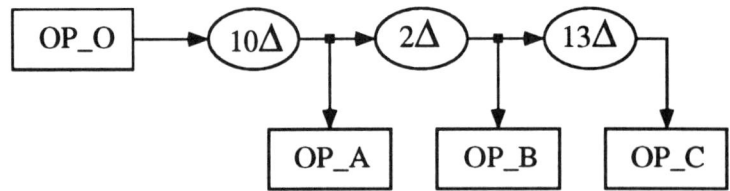

Fig. 6.8: Fan-out without sharing delays.

6.2. SOLVING THE PROGRAMMING PROBLEM

Now, denoting by D_x the maximum delay that needs to be applied to an operand $OPNDx$, of width w_x, we see that the total cost function to be minimized is

$$\text{Cost} = \sum_x D_x w_x \tag{6.2}$$

where the sum is over each operand node in the circuit.

For each node as shown in Fig. 6.6, we have the constraint 6.1, which we now repeat

$$S_j - S_i \geq t_i - t_j \ . \tag{6.3}$$

We have a further constraint that the delay from OP_i to OP_j is less than the maximum delay, D_x :

$$S_j - S_i - (t_i - t_j) \leq D_x \ . \tag{6.4}$$

The constraints (6.3) and (6.4) and the cost function (6.2) to be minimized together make up a linear programming problem, which when solved will provide the minimum cost scheduling for the circuit.

6.2 Solving The Programming Problem

Since it is possible to insert synchronous delays in a circuit that will delay a signal for whole numbers of clock cycles only, it is necessary that the scheduled times, S_i, for operator nodes be integer values. Thus, to find a least cost solution to the scheduling problem, it is necessary to find a solution in which S_i, S_j and D_x in (6.3) and (6.4) are integral values – an integer programming problem. Since in general, integer programming problems are harder than linear programming problems in which the solutions are allowed to be unconstrained real numbers we may anticipate difficulties in solving this problem. Fortunately, however, as will be shown next, a real valued solution to the problem (6.3), (6.4) and (6.2) can easily be transformed into an integral solution with the same cost. This integral solution must therefore be the minimum-cost integral solution.

First, we provide an example to show that there are minimum cost solutions to the problem (6.3), (6.4) and (6.2) in which not all the S_i, S_j are integral. Consider Fig. 6.9 which computes the expression $(a << 2) * ((a + b) + c)$. We assume that all operators have latency one clock cycle. Assuming that the first adder is scheduled at time 0, we see that the second adder must be scheduled at time 1 and the final multiplication at time 2. However, the shift operator may be scheduled at any time between 0 and 1. Fractional scheduled times are possible according to the constraint equations. Note, however, that by moving the shift operator back to time $t = 0$ or forward to time $t = 1$ provides an optimal integral scheduling. We will see that such an operation is possible in the general case. In fact, we prove a theorem identifying a class of linear programming problems that will always have integral solutions.

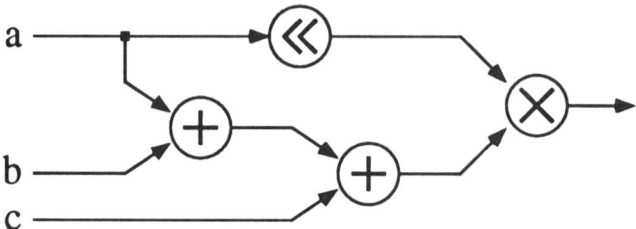

Fig. 6.9: Circuit with non-integral optimal schedule.

Theorem 6.2. *Consider a linear programming problem of the form : Minimize*

$$C = \sum_{i=1}^{N} w_i \, a_i \tag{6.5}$$

subject to constraints

$$a_{i_k} - a_{j_k} \geq b_k \tag{6.6}$$

for $k = 1, \ldots, M$. Suppose that $\sum_{i=1}^{N} w_i = 0$, and that all b_k are integers. Let $\{a_i : i = 1, \ldots, N\}$ be a minimum cost solution to this problem. Then for each real number r, the set of values $\{\lfloor a_i + r \rfloor : i = 1, \ldots, N\}$ is a minimum cost integral solution to the linear programming problem. (Here $\lfloor . \rfloor$ denotes the greatest integer, or floor function.)

Before proving this theorem a few remarks are appropriate. First of all, if we take $r = 0$, then $\lfloor a_i + r \rfloor = \lfloor a_i \rfloor$ is the result of truncating a_i. Thus given any real solution to the linear-programming problem, an integral solution may be obtained by truncation. On the other hand, choosing $r = 0.5$, we see that an integral solution may also be obtained by rounding any arbitrary solution.

Secondly, the $\geq$ inequalities in (6.6) may arbitrarily be changed individually to equalities or $\leq$ inequalities without affecting the conclusion of the theorem. This is because an inequality $a_{i_k} - a_{j_k} \leq b_k$ is equivalent to the inequality $a_{j_k} - a_{i_k} \geq -b_k$ and an equation $a_{i_k} - a_{j_k} = b_k$ is equivalent to a pair of inequalities $a_{i_k} - a_{j_k} \leq b_k$ and $a_{i_k} - a_{j_k} \geq b_k$.

The theorem will be proven in a number of steps, starting with some properties of the floor function.

Lemma 6.3. *If a_1 and a_2 are two real numbers, then $\lfloor a_1 - a_2 \rfloor \leq \lfloor a_1 \rfloor - \lfloor a_2 \rfloor$.*

Proof. If $a_2 \geq a_1$ then subtracting an integer from a_2 diminishes both sides of the required inequality by the same integer, and hence does not affect its truth or falsehood. Consequently, we may assume that $a_2 < a_1$. In this case, $\lfloor a_1 \rfloor - \lfloor a_2 \rfloor$ is equal to the number of integers that lie in the half-open interval $(a_2, a_1]$. Since an interval $(a_2, a_1]$ of length L must contain at least $\lfloor L \rfloor = \lfloor a_1 - a_2 \rfloor$ integers, we see that $\lfloor a_1 - a_2 \rfloor \leq \lfloor a_1 \rfloor - \lfloor a_2 \rfloor$. □

6.2. SOLVING THE PROGRAMMING PROBLEM

Lemma 6.4. *Let a_1 and a_2 be arbitrary real numbers satisfying $a_1 - a_2 \geq b$ where b is an integer. Then for any real number, r, it is true that $\lfloor a_1 + r \rfloor - \lfloor a_2 + r \rfloor \geq b$.*

Proof. By Lemma 6.3,
$$\lfloor a_1 + r \rfloor - \lfloor a_2 + r \rfloor \geq \lfloor (a_1 + r) - (a_2 + r) \rfloor .$$
However,
$$\lfloor (a_1 + r) - (a_2 + r) \rfloor = \lfloor a_1 - a_2 \rfloor \geq \lfloor b \rfloor$$
and since b is an integer, $\lfloor b \rfloor = b$. Putting all these relations together yields the desired result. □

Lemma 6.4 has shown that if $a_i : i = 1, \ldots, N$ is a solution to the equations (6.6), then $\lfloor a_i + r \rfloor : i = 1, \ldots, N$ is also a solution. It remains to show that if a_i is a minimum cost solution, then so is $\lfloor a_i + r \rfloor$.

Lemma 6.5. *Let a_i, b_i and w_i be real numbers, $1 \leq i \leq N$, satisfying $b_i < a_i$ for all i, and let*

$$C = \sum_{i=1}^{N} w_i (a_i - b_i) . \tag{6.7}$$

Suppose for every real number r,

$$C \leq \sum_{i=1}^{N} w_i (\lfloor a_i + r \rfloor - \lfloor b_i + r \rfloor) . \tag{6.8}$$

Then the above inequalities are in fact equalities. That is, for each r,

$$C = \sum_{i=1}^{N} w_i (\lfloor a_i + r \rfloor - \lfloor b_i + r \rfloor) . \tag{6.9}$$

Proof. It is easily verified that $\lfloor a_i \rfloor - \lfloor b_i \rfloor$ is equal to the number of integers that lie in the half-open interval $(b_i, a_i]$. Similarly, $\lfloor a_i + r \rfloor - \lfloor b_i + r \rfloor$ is equal to the number of integers in the interval $(b_i + r, a_i + r]$. This in turn is equal to the number of values of the form $n - r$, where n is an integer, in the interval $(b_i, a_i]$.

We now consider the mapping $f : x \mapsto \exp(2\pi j x)$ which maps the real line around the unit circle. The above remarks show that $\lfloor a_i + r \rfloor - \lfloor b_i + r \rfloor$ is equal to the number of times the image under f of the interval $(b_i, a_i]$ covers the point $f(-r)$. By counting each interval $(b_i, a_i]$ a total of w_i times and summing, the inequality (6.8) may be interpreted as stating that the weighted intervals $(b_i, a_i]$ together cover the point $f(-r)$ at least C times. Since this is true for each r, it follows that the weighted intervals $(b_i, a_i]$ together cover the whole circle at least C times. However, by (6.7), the total weighted length of the intervals is equal to C. It must follow, therefore, that the weighted intervals cover the circle exactly C times. This is expressed by relation (6.9) which was to be proven. □

The next step in the proof is a generalization of Lemma 6.5.

Lemma 6.6. *Let a_i and w_i be real numbers, $1 \le i \le N$, satisfying $\sum_{i=1}^{N} w_i = 0$. Suppose that*

$$C = \sum_{i=1}^{N} w_i\, a_i \qquad (6.10)$$

and that for every real number r,

$$C \le \sum_{i=1}^{N} w_i \lfloor a_i + r \rfloor \;. \qquad (6.11)$$

Then for each r,

$$C = \sum_{i=1}^{N} w_i \lfloor a_i + r \rfloor \;. \qquad (6.12)$$

Proof. Choose a value b less than each a_i, and set $b_i = b$ for $i = 1,\ldots,N$. Since $\sum_{i=1}^{N} w_i = 0$ it is easily seen that the hypotheses of Lemma 6.5 hold with this choice of a_i and b_i. Consequently,

$$C = \sum_{i=1}^{N} w_i \left(\lfloor a_i + r \rfloor - \lfloor b + r \rfloor \right) = \sum_{i=1}^{N} w_i \lfloor a_i + r \rfloor$$

which is the same as (6.12). □

We are now in a position to complete the proof of Theorem 6.2.

Proof. Let $\{a_i\}$ be a minimum cost solution to the inequalities (6.6) and let the cost be C. According to Lemma 6.4, for any real number, r, it is true that $\{\lfloor a_i + r \rfloor\}$ is also a solution to the inequalities (6.6), and therefore must have a greater or equal cost. The hypotheses (6.10) and (6.11) of Lemma 6.6 are therefore satisfied and we conclude by (6.12) that the solution $\{\lfloor a_i + r \rfloor\}$ also has cost C. This was the desired conclusion of the theorem, and the proof is complete. □

We now show that Theorem 6.2 applies to the scheduling problem formulated by constraints (6.3) and (6.4) and cost function (6.2). As formulated the constraints and cost function do not satisfy the hypotheses of the theorem, but a simple change of variable will put (6.2) – (6.4) into the desired form. For each node $OPND_x$ in the circuit, with notation as in Fig. 6.6, we introduce a new variable, $T_x = S_i + D_x$ and use it to eliminate D_x from the expression. The formulae then become

$$\text{Cost} = \sum_{x} w_x (T_x - S_i) \qquad (6.13)$$

$$S_j - S_i \ge t_i - t_j \qquad (6.14)$$

$$T_x - S_j \ge t_j - t_i \qquad (6.15)$$

6.3. PREVIOUS VALUES AND FEEDBACK LOOPS

which satisfy the hypotheses of the theorem. Therefore, we can state the following corollary.

Corollary 6.7. *If $\{Si\}$ are the scheduled times for operators found by finding the (least cost) solution to the linear programming problem (6.2) - (6.4), then a minimum cost integer scheduling can be obtained by replacing each S_i with $\lfloor S_i + r \rfloor$ for any fixed real number r.*

Exercise 6.8. Show that Theorem 6.2 and Corollary 6.7 also hold if, in their statement, the floor function $\lfloor . \rfloor$ is replaced by the ceiling function, $\lceil . \rceil$.

Exercise 6.9. The *DIODES* high-density-interconnect system ([42]) for electronic design makes use of operator building blocks that contain optional built-in delays. In such a system, any number of delays up to a given maximum may be attached to the output, or independently to any of the inputs of each operator at no cost. Thus, delays between operators may (up to a given limit) be pushed back onto the output of the previous operator, or pushed forward onto the inputs of the succeeding operators without cost. Any remaining delays require additional hardware, and must be accounted for in the cost function. Show how to formulate a linear programming problem for the optimal scheduling of such a circuit in such a way that Theorem 6.2 may be applied to obtain an optimal integer solution.

6.3 Previous Values and Feedback Loops

A common element in many DSP designs is the one-sample delay, or z^{-1} operator. Scheduling of the z^{-1} operator was considered in section 1.3.5 where it was indicated that the way to handle this operator is to treat it as a null-operator passing its inputs directly to its output with a nominal delay of $-W$, where W is the sample period. The reasoning behind this is as follows. If data x arrives at a z^{-1} operator at time $t = 0$, then the previous value of the data was output W clock cycles earlier, at time $-W$. Hence the z^{-1} operator has a negative latency of $-W$ cycles.

For circuits with feed-back loops it is not always possible to find a schedule. A condition for a schedule to exist is easily found.

Proposition 6.10. *A schedule exists for a circuit with feed-back loops if and only if the total latency delay around any loop in the circuit is non-positive.*

The method for scheduling a circuit with z^{-1} operators outlined in section 1.3.5 is as follows.

1. Schedule the circuit in the normal manner, assuming that the z^{-1} operator has latency $-W$. Be prepared for the case where no solution exists.

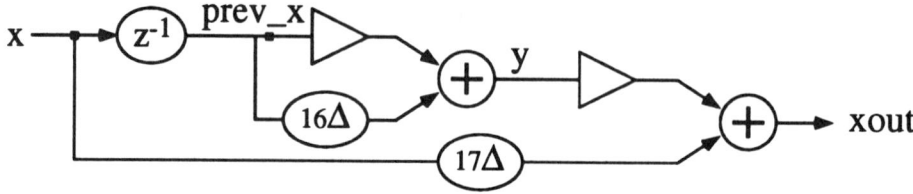

Fig. 6.10: Scheduling of circuit with z^{-1} operator.

2. Insert appropriate scheduling delays in the circuit.

3. Remove the z^{-1} operators.

6.3.1 Alternative Scheduling Method For z^{-1} Operators

The method described above for scheduling of z^{-1} operators may in some instances lead to non-optimal delay placement because of the fact that the z^{-1} operators are removed after the scheduling and delay placement is done, instead of before. If we consider the simple circuit described by the equations

$$\begin{aligned} prev_x &:= x[-1]; \\ y &:= prev_x * a + prev_x; \\ xout &:= b * y + x; \end{aligned}$$

Where a and b are constants, and it is assumed that $W = 16$ and constant multiplication has a latency of 16 cycles. The data flow diagram for this circuit is shown in Fig. 6.10 along with scheduling delays. The triangular symbols represent constant multiplication. After the z^{-1} operator is removed it is seen that the two delays of 16 and 17 cycles may be amalgamated, but this was not possible previously, since they were applied to apparently different signals, x and $x[-1]$. It is possible to find more complicated examples in which the scheduling of hardware operators will be affected in a non-optimal way by the presence of the z^{-1} operators during the scheduling.

A preferable method is to incorporate the word delays right into the linear programming equations. This is done by modifying equations (6.2) – (6.4) describing the linear programming problem slightly to take account of the presence of z^{-1} operators. Specifically, Fig. 6.11 shows a general situation in which the output of some operator OP_i undergoes a z^{-n} transformation before being used as input to the operator OP_j. The equations describing scheduling constraints, corresponding to (6.3) and (6.4) are

$$S_j - S_i \geq t_i - t_j - nW \qquad (6.16)$$
$$S_j - S_i - (t_i - t_j) + nW \leq D_x \qquad (6.17)$$

6.4. EARLIEST-POSSIBLE SCHEDULING

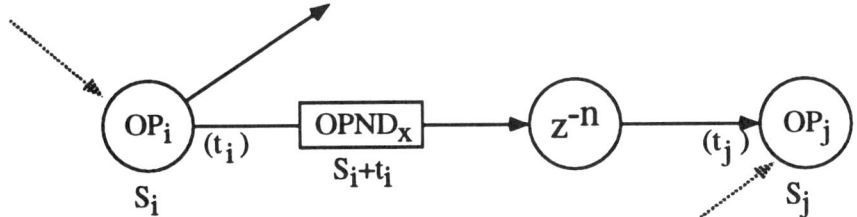

Fig. 6.11: General node to be scheduled.

where W is the number of clock cycles in a word. In this case, as before, $S_j - S_i - (t_i - t_j) + nW$ is the delay to applied to the connection shown in the diagram, and D_x is the maximum delay applied to this node. Constraints (6.16) and (6.17) may be put in a form to satisfy the hypotheses of Theorem 6.2, and we may solve the linear-programming problem to find an integer scheduling solution.

6.4 Earliest-Possible Scheduling

The linear programming scheduling method described in section 6.1 gives the optimal scheduling if a schedule exists for the circuit. Unfortunately, if the linear programming problem does not have a solution then no schedule for the circuit exists. In this case the linear-programming approach will give very little information as to the reason for the failure. In order to diagnose the reason for a scheduling failure, a different approach may be used based on *earliest-possible scheduling*. For circuits without feed-back loops, this method amounts essentially to doing a topological ordering of the nodes of the circuit. The earliest possible scheduling may be done treating the z^{-1} operators like and other operator, having latency equal to $-W$, where W is the sample period, as explained previously.

Given an operator node, OP_j as in Fig. 6.6, the "input operators" for OP_j are the operators such as OP_i which produce an output that feeds to an input of OP_j. It is convenient to introduce an extra node called "*INPUT*". The *INPUT* node will itself have no inputs but one output. All operator nodes that are connected to the input of the circuit will be connected to the output of the node *INPUT*, so that no operator in the circuit has hanging inputs.

If all the input operators for OP_j have already been scheduled, then the earliest possible scheduled time for OP_j may be deduced using (6.3) to be the minimum of $S_i + t_i - t_j$ (with notation as in Fig. 6.6) as S_i runs over all the input operators OP_i.

A technicality arises here in that there may exist operator nodes in the circuits which have no inputs. An example of this is a constant generator. Such an operator continually generates a digit-serial constant value. It has no inputs and a single output. Such a node has no earliest-possible scheduled time, since it may be scheduled arbitrarily far in the past. The adopted solution is to schedule it at a very large negative time, $t = -MAXINT/2$, where $MAXINT$ is the largest representable integer value. Such a scheduled time is sufficiently far in the past that it cannot force any of the other nodes forward beyond their earliest possible scheduled time. A time of $-MAXINT/2$ is chosen instead of $-MAXINT$, since there may be operators with negative latency in the circuit, and applying these to the output of an operator scheduled at time $t = -MAXINT$ could cause scheduled times of subsequent operators to exceed the range of representable integers and wrap around to a positive value.

These considerations lead to the following rule :

- In computing the earliest possible scheduled time for an operator, no operator is to be scheduled before time $-MAXINT/2$

The earliest possible scheduling algorithm for circuits without feed-back loops may now be described as follows:

Schedule all operators at time $t = -MAXINT/2$.
Schedule the $INPUT$ operator at time $t = 0$.
do

> {
> Select an operator OP_j for which all the input operators are already scheduled.
>
> Schedule OP_j to the earliest possible time consistent with its inputs.
> }

until (all operators have been scheduled)

For circuits without feed-back loops, this algorithm will terminate after scheduling each node exactly once, and it is easily seen that each node is scheduled at its earliest possible time subject to the restriction that the $INPUT$ node is scheduled at time $t = 0$.

Circuits with feed-back: For circuits with feed-back loops the above algorithm will not work, since the situation will arise that there exist nodes still to be scheduled, but none of the remaining operators may be scheduled, because at least one of its input operators has not been scheduled yet. In particular, it will never be possible to schedule any of the nodes in a loop, since each one must wait for its predecessor to be scheduled first. A way out of this deadlock is to select an arbitrary one of the remaining nodes and assign to it some scheduled time so that scheduling may continue. It may become necessary later to reschedule the chosen node, and possibly its output nodes as well. An algorithm based on this idea may be expressed as follows:

6.4. EARLIEST-POSSIBLE SCHEDULING

Schedule all operators at time $t = -MAXINT/2$.

Schedule the $INPUT$ node at time $t = 0$.

Place all operators in a stack of nodes to be scheduled.

while (the stack is non-empty)
{
Take an operator OP from the stack and recompute its earliest possible scheduled time consistent with the scheduled times of its input operands using inequality (6.3).

if (the new scheduled time of OP is greater than the old time)

{
Set the scheduled time of OP to the new scheduled time.

Place all the output operators of OP in the stack, if they are not already there.
}
}

It is possible that a node may be rescheduled several times in the course of this algorithm. If this algorithm terminates, then it finds the earliest possible schedule for the circuit, as expressed in the following proposition :

Proposition 6.11. *If the above algorithm terminates, then it results in each node being assigned its earliest possible scheduled time subject to conditions*

1. No node is scheduled at a time previous to $-MAXINT/2$

2. The INPUT node is scheduled at time $t = 0$.

Proof. The proposition is easily proven by induction over the number of moves. Suppose that T is a valid schedule for the data-flow graph satisfying the two subsidiary hypotheses. Let S_n be the schedule (not necessarily a valid schedule) found after n steps of the algorithm. What needs to be proven is that if O is any operator, then O is scheduled no later under S_n than under T. The proof is by induction on n. In the starting schedule, S_0, all operators are scheduled at the earliest possible time consistent with the hypotheses of the theorem, and so earlier than they are in T. Now consider step $n + 1$. Our inductive hypothesis is that S_n is an earlier schedule than T for all operators. Suppose that O is the operator to be rescheduled at this step to create a new schedule, S_{n+1}. Since all inputs of O are scheduled no later in S_n than in T, and O is rescheduled to the earliest time consistent with the ready-times of its inputs, it follows that O is scheduled no later in S_{n+1} than in T, and so S_{n+1} is an earlier schedule than T. Now at every step of the algorithm the scheduled time of some operator is increased. It follows that if a valid schedule T exists, then the algorithm must eventually terminate, and indeed with the earliest possible schedule.

□

It is not really our goal to find the earliest possible schedule, however, since it will not in general be optimal, and the linear-programming algorithm of section 6.1 gives the optimal scheduling. Rather it is our purpose to detect circuits that may not be scheduled. If there are feed-back loops with total latency greater than zero, according to Proposition 6.10 no schedule exists. In this case, the algorithm will go into an infinite loop circling round and round the loop in the circuit as the scheduled times of the nodes in the loop increase without bound. It is necessary to detect this condition and break out of the loop. This may be done as follows.

Each operator node keeps a pointer to the node that is forcing it to be rescheduled, as well as count of the length of the chain of forcing nodes. Originally, the pointer is null and the count is zero. When the scheduled time of a node OP_j is recomputed in the above algorithm, the new scheduled time is computed to be $\max(S_i + t_i - t_j)$ where the maximum runs over all the input nodes OP_i (see Fig. 6.6). Call the input node OP_i which realizes this maximum the "forcing node", since it is the one which is forcing the scheduled time of the operator OP_j to is new maximum value. If the new scheduled time is greater than the old one, then the pointer associated with OP_j is set to point back to the forcing node and the count associated with OP_j is set to one greater than the count of the forcing node. Thus, at any point it is possible, by following the pointers, to trace back along the path of nodes causing the node to be rescheduled. For instance:

> Node A is forced to be rescheduled by node B which is forced to be rescheduled by node C which was forced to be rescheduled ....

The count field will hold the length of this path until we get to a node with a null-pointer. If at any time in this path, we should come back to the original node, that is if we have

> Node A is forced to be rescheduled by node B which ... by node X which is forced to be rescheduled by node A

then there is an impossible feedback loop in the circuit, and the circuit cannot be scheduled. Instead of tracing back along the path to see if the node reoccurs, it is easier to detect this by using the *count* field. In particular, if the length of the path becomes longer than the total number of nodes in the circuit, then one of the nodes must be repeated in the path, and there is a loop. Once a loop has been detected, it is easy to find by tracing back along the path looking for a repeated node.

Example 6.12. Here is an example of a circuit description that contains a loop.

6.5. SWAPPING MULTIPLIER INPUTS

```
    specification
        x := a * b;
        c := x + d[-1];
        d := a * c + b;
    end;
```

The loop in question contains one multiplication and two additions as well as a z^{-1} delay. Because the combined latency of a multiplication and two additions exceeds one sample period, the loop is not feasible. The generated error message is as follows.

```
Error encountered by compiler
Infeasible feedback loop detected :

File test.bsl line 25 :
  c := x + d[-1];
       ^

File test.bsl line 26 :
  d := a * c + b;
       ^
```

6.5 Swapping Multiplier Inputs.

In Parsifal, the earliest-possible scheduling algorithm is run first in order to detect cases of infeasible feedback loops and give an informative error message. Since the earliest-possible scheduling algorithm does not in general give an optimal scheduling it is necessary to run the linear programming scheduling algorithm as well. Running the earliest-possible scheduling algorithm first provides an opportunity for certain optimizations of commutative and associative operations to be made.

It has been noted that a parallel-serial multiplier may be designed to receive one of its operators one sample period before the other. The timing diagram for a parallel-serial multiplier is shown in Fig. 6.4. It shows that one of the inputs is timed to arrive before the other. Since multiplication is commutative, it may be that swapping the inputs of the multiplier will allow the multiplication operation to be scheduled at an earlier time. Consequently, whenever a commutative multiplication operation is scheduled or rescheduled in the earliest-possible scheduling algorithm (Section 6.4), a check is made as to whether swapping its two inputs will result in an earlier scheduled time for the operator. If so the two inputs are swapped. This transformation has two benefits.

- By swapping the inputs to schedule the multiplication at the earliest possible time it is possible on occasions to make feedback loops feasible that would otherwise be infeasible. This is because the latency of the multiplication with respect to one of its inputs is $2W$ clock cycles, whereas with respect to the other input it is only W clock cycles. Since a feedback loop is feasible if and only if the total latency around the loop is less than or equal to zero, it is preferable to include the smaller latency in the loop.

- A plausible heuristic that has also been verified empirically is that the smaller the latency of the complete circuit, the smaller will be the total number of delays in the optimal schedule. Thus decreasing the total latency of the circuit will generally allow the linear programming algorithm to find a more economical schedule for the circuit.

6.6 Trees of Associative and Commutative Operators

A further instance in which the data-flow graph may be arranged in order to achieve a better schedule is the case of adder/subtractor trees, or in general trees of associative, commutative operators [56][44]. We consider addition and subtraction as typical associative commutative operators, though the same algorithms will apply to other operators, such as multiplication and division. Nevertheless, addition and subtraction are particularly suited to this sort of optimization, since the result of adding a number of variables represented in two's complement format is independent of the order in which the operations are done. This is true whether or not overflow occurs in some intermediate step of the calculation.

This observation may be justified as follows. Suppose that arithmetic is carried out on n-bit two's complement values, which we take for the moment to represent integers. The range of representable numbers is $[-2^{n-1}, 2^{n-1} - 1]$. Two's complement addition (possibly overflowing) may be considered as equivalent to addition in the ring $Z/2^n Z$ of integers modulo 2^n. Since addition in this ring is associative and commutative, so is two's complement addition of integers with possible overflow. For this reason, we may rearrange adder/subtractor trees at will without worrying about introducing overflow the result of the computation will be unchanged by the rearrangement.

As an example of rearrangement of adder trees, consider Fig. 6.12. It is clear that Fig. 6.12b is a preferable way of summing the operands than Fig. 6.12(a). We assume here that addition has latency of one clock cycle. The advantages of Fig. 6.12(b) are two-fold. First, there are fewer delays required, and secondly, the latency of the output is decreased. The example of Fig. 6.12 is somewhat simple – in the general case, the inputs will not all be available at the same time and best topology will not be a balanced tree.

6.6. TREES OF ASSOCIATIVE AND COMMUTATIVE OPERATORS

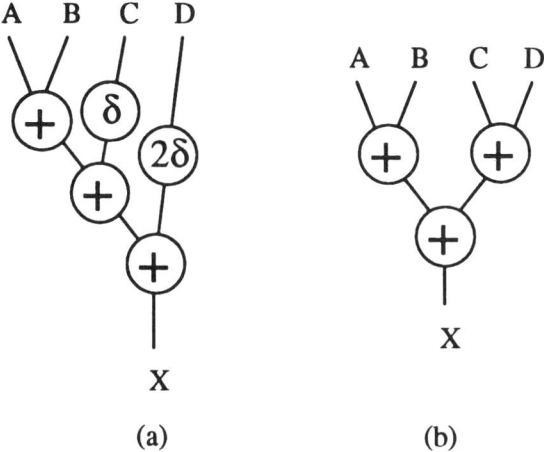

Fig. 6.12: Adder rearrangement.

Rearranging adder trees may make it possible to schedule cyclic data-flow graphs which may not otherwise be scheduled. Consider Fig. 6.13 which represents a recursive filter. Assume that this computation is carried out using bit-serial operators with word-length 16 in which addition has latency 1, constant multiplication latency 15.[1] Fig. 6.13(a) represents the filter described by the equation

$$y = y[-1] * a_1 + y[-2] * a_2 + y[-3] * a3 + x;$$

As seen in the diagram, the feedback loop through the multiplier a_1 has length greater than 0, and so scheduling is not possible. However, in Fig. 6.13(b), the adder tree has been rearranged and now the feedback loop has total latency 0 (including the z^{-1} operator) and so may be correctly scheduled.

To carry this example a little further, consider the case where $a_3 = 0.375$, $a_2 = -0.75$ and $a_1 = 0.625$, so that the filter is represented by the equation

$$y = 0.625 * y[-1] - 0.75 * y[-2] + 0.375 * y[-3] + x;$$

Consider a bit-serial implementation of this filter. With this choice of coefficients, the multiplications may be conveniently implemented using shift and addition (or subtraction) operators. For instance, $0.625 * y = y >> 1 + y >> 3$, and $-0.75 * y = y >> 2 - y$. Here, the operation $>>$ represents right shift, or division by the indicated power of 2. Bit- and digit-serial shift operations

---

[1] The actual latency of a bit-serial constant multiplier may depend on the constant multiplier value.

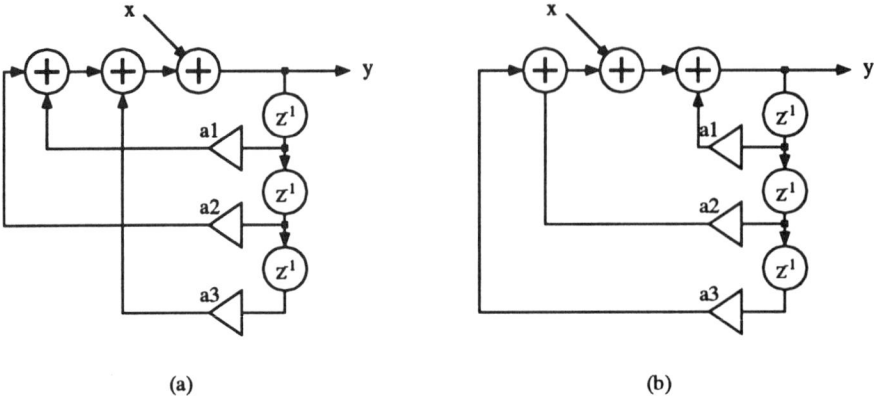

Fig. 6.13: Feasible scheduling obtained by use of associativity.

are discussed in detail in Section 2.3. This type of implementation of constant multiplication and particularly its use in filter designs is discussed in detail in Chapter 12. For the present, however, we consider this simple example. In shifting right n places, the first bit of output is equal to the $(n+1)$-st bit of the input. For that reason, the latency of an n-bit right shift is $n+1$ cycles if the output is latched as in Fig. 2.11. For digit-serial shift operators, the latency will be smaller, but still dependent on the number of bits shifted. Allowing for the latency of the adder, we may assume a latency of 5 cycles for multiplication by $a_1 = 0.625$, a latency of 4 cycles for multiplication by $a_2 = 0.75$ and a latency of 5 for multiplication by $a_3 = 0.375$. In this case, the loop though the multiplier a_1 in Fig. 6.13(a) has a total latency of 8 minus the latency of the z^{-1} operator. Since the latency of z^{-1} in a bit-serial implementation with word-size W is equal to $-W$, the total latency of the loop is $8-W$. In order for scheduling to be possible, the loop latency must be non-positive, which means that the word size W must be at least equal to 8. If on the other hand the configuration of Fig. 6.13(b) is used, then the loop latency is only $6-W$, so the minimum word size is 6. For less simple filter coefficients the multiplier latency may be higher, and larger word-sizes will be necessary.

6.7 Algorithm for Optimizing Single Adder Trees

We consider first the algorithm for scheduling a single adder tree such as that shown in Fig. 6.14. (Ready-times shown in parentheses.) We assume that an adder has two inputs and latency L.

6.7. ALGORITHM FOR OPTIMIZING SINGLE ADDER TREES

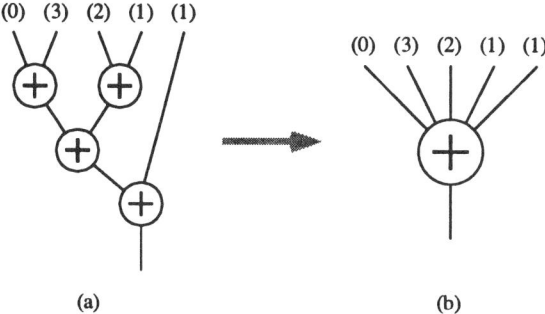

Fig. 6.14: Collapsing of adder trees.

First step: Collapsing adder trees. The first step is to collapse the tree to a single multiple-input adder as shown in Fig. 6.14(b). This is justified by the associativity of the addition operator. As shown in Fig. 6.14, each input to the composite adder has an associated ready-time, which is the earliest time that that input is available. The time between when an input is ready and when it is used must be filled by shimming delays. Necessary shimming delays will be ignored for the present. By commutativity, the order of the inputs does not matter and they may be ordered according to their ready time.

Second Step: Iterative splitting: The second step starts with a single composite adder with its inputs ordered. A new adder tree is formed by repetitively splitting off the two earliest scheduled inputs until only two inputs are left. The action of splitting off the two earliest inputs is illustrated in Fig. 6.15, where it is assumed that a and d are the two earliest inputs, ready at times t_1 and t_2 (shown as 0 and 1 in Fig. 6.15). A new adder is formed which has a and d as its inputs and the output of which is ready at time $t_2 + L$. The original composite adder now has one input fewer. Its inputs may be reordered and the next too earliest inputs split off, and so on.

It is a notable fact that the method given here simultaneously gives the best results in terms of latency of the output and total shimming delay, provided all the inputs represent different signals.

Theorem 6.13. *Given a composite adder with distinct inputs scheduled at times $t_1, t_2, \ldots, t_m$, the adder tree generated by repetitively splitting off the two earliest adder inputs (iterative splitting) as described above gives the adder tree with the smallest latency and the smallest total shimming delay.*

Proof. The proof is by induction on the number of inputs to the adder tree. Thus suppose that no tree with $m-1$ inputs has smaller latency or fewer shimming delays than the tree constructed by iterative splitting. Now consider an

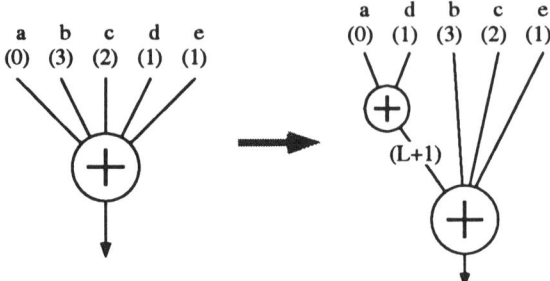

Fig. 6.15: Iterative splitting of collapsed adder.

adder tree T with inputs $a_1, \ldots, a_m$ ready at times $t_1, t_2, \ldots, t_m$ and internal nodes (temporary operands) $a_{m+1}, \ldots, a_{2m-1}$ ready at times $t_{m+1}, \ldots t_{2m-1}$. Suppose that T minimizes the latency or the number of shimming delays. Suppose without loss of generality that a_1 and a_2 are the two earliest-ready operands (but not necessarily in that order).

Case 1: If a_1 and a_2 are added together at time $\max(t_1, t_2)$ to produce a_j then we can split off a_1 and a_2 and apply the induction hypothesis to the adder tree with $m-1$ inputs $a_3, \ldots, a_m, a_j$ to deduce that the tree generated by iterative splitting has no greater latency or number of shimming delays than T.

Case 2: If a_1 and a_2 are directly added together in T at some time later than $\max(t_1, t_2)$, then moving the adder backwards to time $\max(t_1, t_2)$ will reduce the number of shimming delays (since an adder has two inputs, but only one output) without increasing the latency. Furthermore, this transformation reduces us to case 1 above.

Case 3: Suppose finally, that a_1 and a_2 are not directly added in T, but that a_1 is added to a_i at time s_1 and a_2 is added to a_j at time s_2. We assume (interchanging the roles of a_1 and a_2 if necessary) that $t_i \leq t_j$. (These are the ready times of a_i and a_j respectively.) The operands a_i and a_j may be either inputs to the adder tree or internal operands. Now consider the tree identical with T except that adder inputs a_2 and a_i are swapped. Thus a_1 and a_2 are added at time s_1 and a_i and a_j are added at time s_2. It may be verified that this is possible, since all adder inputs are ready before the scheduled time of their respective adders.

In particular, we argue as follows. Since a_1 and a_i are added at time s_1, we have $t_1 \leq s_1$ and $t_i \leq s_1$. Further, by assumption, a_1 and a_2 are the two earliest inputs, so $t_2 \leq t_i$. From this it follows that $t_1 \leq s_1$ and $t_2 \leq s_1$, so a_1 and a_2 may be added at time s_1.

Similarly, since a_2 and a_j are added at time s_2 we have $t_j \leq s_2$. By assumption we also have $t_i \leq t_j$ from which it follows that $t_i \leq s_2$. Thus, a_i and a_j may be added at time s_2 as required. The tree resulting from this switching operation has the same latency as the original tree.

We consider the number of shimming delays. One computes the number of shimming delays applied to a_1, a_2, a_i and a_j in the original tree to be equal to

$$(s_1 - t_1) + (s_1 - t_i) + (s_2 - t_2) + (s_2 - t_j) = 2(s_1 + s_2) - (t_1 + t_2 + t_i + t_j) \ .$$

After the swap, the number of shimming delays is

$$(s_1 - t_1) + (s_1 - t_2) + (s_2 - t_i) + (s_2 - t_j) = 2(s_1 + s_2) - (t_1 + t_2 + t_i + t_j) \ .$$

Thus, the number of shimming delays remains unchanged. The tree resulting from the swap has the same latency and number of shimming delays. This operations reduces the situation to that of case 2. This completes the proof.
□

6.8 Rearranging General Data-Flow Graphs

In a general situation, we have data-flow graphs which contain more circuitry than a single composite adder. The algorithm given in this chapter is a method whereby the output latency and number of shimming delays may be simultaneously minimized. This algorithm, however, does not allow for gains which may be made by rearranging the adder trees. There seems to be no way of directly modifying the linear-programming approach to account for rearrangements of the adder trees, however. The approach taken here is to modify the topology of the adder trees in a pre-processing stage, and then to apply linear programming to the modified graph.

Step 1: Collapsing adder trees: The first step in modifying the data-flow graph is to collapse trees of adders into multiple-input adders. This is done by iteratively collapsing adders until no more collapsing is possible. In particular, the algorithm may be described as follows.

> **while** not finished
> {
> Find an operator O_1 that is an adder.
> **if** O_1 has fanout equal to 1,
>   and if the single fanout operator is also an adder, O_2,
>   then delete O_1 and add its inputs to the inputs of O_2.
> }

Note that we do not collapse adders that have fanout greater than 1. The output of such an adder is a value that is used in more than one place in the data-flow graph, and may be thought of as a common sub-expression. As such, it is essential that it be computed explicitly, which it would not be if the adder tree in which it is involved were rearranged. We make no attempt here to optimize the choice of common sub-expressions (assuming that common sub-expressions may be chosen in more than one way).

Step 2: Finding the earliest schedule: It is our goal to find a valid schedule for the data-flow graph, and indeed the one in which all operators are scheduled at the earliest possible time (subject to all inputs being scheduled at time zero). The algorithm is the same as the earliest scheduling algorithm given in section 6.4, with a modification to take account of composite adders.

We explain the method of rescheduling O a little more closely. If O is not a multiple-input adder, then as in the earliest scheduling algorithm in section 6.4, the earliest time that O may be scheduled consistent with its inputs is given by (6.16). If O is an adder, then the method of iterative splitting finds an expansion of the composite adder that ensures (according to Proposition 6.11) the smallest possible latency, and hence the earliest possible ready-time for the output of the composite adder. Of course the determination of this scheduled time for the composite adder can (and must) be carried out without actually expanding the composite adder at this stage.

In the earliest possible scheduling algorithm, a given operator may be rescheduled many times, particularly if there are loops in the data-flow graph. Each time a composite adder operator is rescheduled the connectivity of the corresponding adder tree may be rearranged.

As with the earliest possible scheduling algorithm, it may be proven that this algorithm constructs the earliest possible schedule for every operator in the graph subject to the conditions of Proposition 6.11. As has been remarked, the schedule found by this algorithm does not however minimize the total number of shimming delays.

Step 3: Iterative splitting: The next step in the algorithm is to expand the composite adders to form adder trees again. The algorithm for doing this has already been described. Because of the schedule of the input operands of the adder trees, the topology of the data-flow graph will have been changed, and indeed in such a way as to minimize the latency of the adder trees.

Step 4: Minimization of shimming delays: The final step in the algorithm is to minimize shimming delays in the new graph topology. The method used is the linear-programming method described in section 6.3.

Let us summarize what this algorithm does. The first three steps of the algorithm rearrange the topology of the adder networks to find a preliminary schedule for the whole graph. This is done in such a way as to minimize the latency of the outputs of all adder networks. In addition, for graphs with loops, if there is a valid schedule for the graph which may be found by rearranging the adder networks, then this method will find it. Finally, step 4 reschedules the whole graph in its new topology to minimize the total number of shimming delays. This is done without changing the latencies of the outputs of the complete circuit.

What this algorithm does **not** do is make any attempt to optimize the choice of common sub-expressions to minimize a network of adders. For instance, suppose it is required to compute the sums $a + b + c$, $a + b + d$, $a + c + d$ and $b + c + d$. It is possible to do this with a network of six adders, but this may be done with several different choices of sub-expressions, some of which may be better than others for scheduling purposes. Common sub-expressions are represented in the graph by operands with multiple fanout. Nevertheless, within the constraint of preserving the common sub-expressions, this algorithm will minimize the latency of the outputs.

6.9 Handling Subtractors

The present discussion has so far been confined to adders only. It is very easy to include subtractors as well as adders by making slight changes to steps 1 and 3 of the algorithm. In step 1, the collapsing of adder trees is modified to allow collapsing of adders or subtractors. It is necessary only to keep track of the type (+ or −) of each of the inputs to the composite adder/subtractor. Then, in the step of iterative splitting, the correct operator, (add or subtract) must be chosen and the correct type propagated to the output according to the following table.

Types of inputs	Operator type	Type of output
+ +	Adder	+
− +	Subtractor	+
+ −	Subtractor	+
− −	Adder	−

It is guaranteed that the type of the final output is +.

6.10 Examples

The algorithm described above has been implemented in Parsifal. Examples of its performance are given below. They are mainly drawn from the realm of filter design. The first simple example is a bit-serial FIR filter described by the input

$$y := a0 * x + a_1 * x[-1] + a_2 * x[-2] + a3 * x[-3];$$

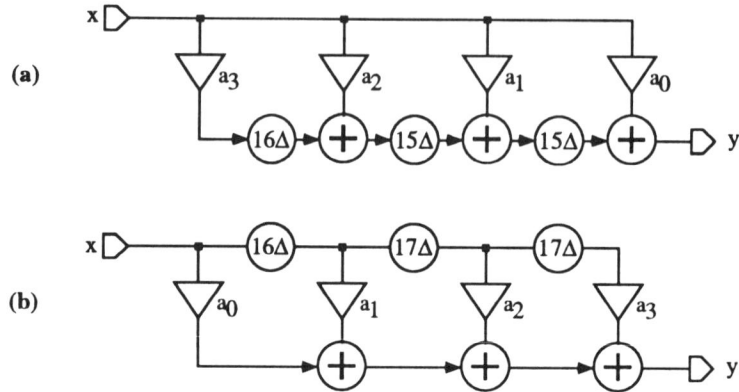

Fig. 6.16: Two bit-serial realizations of a 4-tap FIR filter.

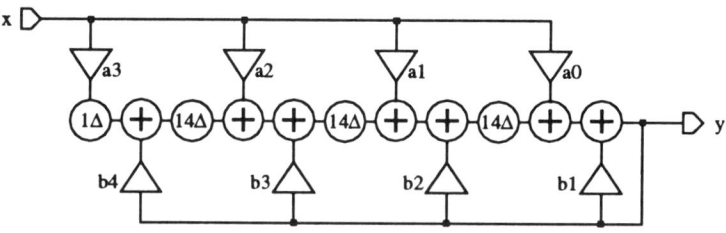

Fig. 6.17: Bit-serial fourth-order IIR filter.

In this particular example, addition has latency 1 and constant multiplication 15 in a word-length of 16. The compiler chooses the implementation in Fig. 6.16(a) rather than Fig. 6.16(b) as described. Fig. 6.16(a) has output latency 16 and 46 shimming delays rather than latency 18 and 50 delays in Fig. 6.16(b).

The second example is a recursive filter described by the equation

$$w := a0 * x + a1 * x[-1] + a2 * x[-2] + a3 * x[-3];$$

$$y := w + b1 * y[-1] + b2 * y[-2] + b3 * y[-3] + b4 * y[-4];$$

With the same operator latencies as in the previous example, the program found the implementation in Fig. 6.17. Note that it is critical that the addition of the $b_1 * y[-1]$ term appear last in the adder tree, or the feedback loop will be too long. Note that the total of 43 delays and latency of 17 are the least possible.

With different (somewhat unrealistic) latencies, adder = 2, multiplier = 1 with a word-length of 3, the previous topology is not possible, since the

6.10. EXAMPLES

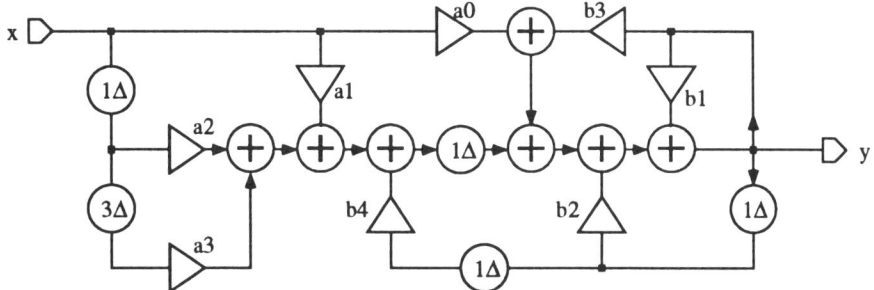

Fig. 6.18: Bit-serial fourth-order IIR filter for different latencies.

combined latency of two adders exceeds one word delay. The compiler found the circuit of Fig. 6.18 which has 7 delays in total and a latency of 9.

The final example is a digit-serial example – the implementation of a FIR filter using two simultaneous computations with 6-bit digits in a 12-bit word. Addition has latency one. This circuit is discussed also in section 7.4 of chapter 7. As described there, the parallel input stream is split into even and odd samples by the parallel/serial (PS) converter, which streams are recombined into one parallel stream by the serial/parallel (SP) converter. The chip description is

```
/* Serialize the inputs */
(x0, x1) := ps(X; 0, 1);

/* Compute the filter */
y0 := x1[-2] * a3 + x1[-1] * a1 + x0[-1] * a2 + x0 * a0;
y1 := x1[-1] * a2 + x1 * a0 + x0[-1] * a3 + x0 * a1;

/* Parallelize the outputs */
Y := sp(y0, y1; 0, 1);
```

The resulting circuit is shown in Fig. 6.19, and has only two six-bit wide delays.

The algorithm described in this section for minimizing the height of add/subtract trees, though very simple in concept, makes significant improvements to designs involving trees of adders, for instance digital filters. It significantly relieves the design burden by finding solutions to the scheduling and latency minimization problems. The discovery of optimal and correct schedules allowing topological changes to the circuit is quite involved and best performed by computer. The solutions found were often surprising and significantly better than those discovered without the aid of this tool.

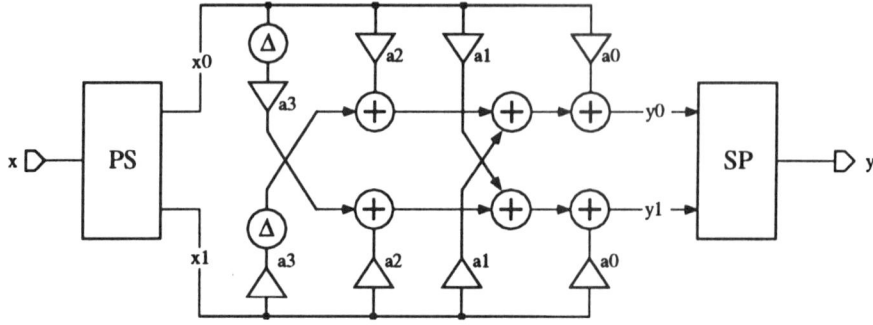

Fig. 6.19: Interleaved digit-serial FIR filter.

Although specifically directed to the scheduling of synchronous circuits where minimization of shimming delays is important, the algorithms described here are also applicable to rearranging adder trees in order to minimize the signal propagation delay through combinational adder networks so as to achieve optimum clock speed.

6.11 Why not other optimizations ?

Several authors have considered the scheduling of DSP circuits in which the goal is to perform appropriate sharing to minimize the number of operators used [57][16][44]. Sharing is possible between operators used in different mutually exclusive branches of an algorithm or when input data does not arrive in every clock cycle. These algorithms are normally applied to circuits using parallel data, and their applicability to digit-serial circuits is limited.

As a simple example, consider a circuit containing a number of adders, implemented using parallel arithmetic so that each addition operation occupies one clock cycle. For simplicity we assume that the circuit has no feedback loops. Suppose that this circuit accepts a new set of input data in each clock cycle and that each adder A_i is scheduled at time t_i. Now suppose that the data rate is changed so that valid input is received only every second clock cycle, at times $t = 0, 2, 4 \ldots$. Then each adder A_i scheduled at time t_i carries out a useful operation at times $t_i, t_i + 2, t_i + 4, \ldots$, whereas at times $t_i + 1, t_i + 3, \ldots$ it is operating on invalid or useless data. We say that the adder is *active* when it is carrying out a useful operation and is *idle* otherwise. It is clear that if two adders are scheduled at times t_i and t_j where $t_i - t_j$ is an odd number then one adder is active when the other is idle. The two adders may then potentially be amalgamated into one adder which is multiplexed between the two operations. On the average since half the adders will be scheduled at even time instances and half at odd time instances, every adder may potentially be multiplexed

6.11. WHY NOT OTHER OPTIMIZATIONS ?

with half the other adders in the circuit. An additional task that now falls to the scheduling algorithm is to ensure that the adders (and other operators in the circuit) are scheduled in such a way that a maximum amount of sharing takes place.

Compare this with what happens in a serial circuit. To emphasize the problems faced, we consider bit-serial computation with a word-length of 16. Suppose that A_i is scheduled at time t_i. Now, suppose that the throughput requirements are changed so that the circuit receives valid data only every second sample period, that is, at times $t = 0, 32, 64, \ldots$. In this case, an adder scheduled at time t_i will be active for blocks of 16 clock cycles starting at times $t_i, t_i + 32, t_i + 64, \ldots$ and it may be seen that the active periods of one adder correspond with the idle periods of another adder only if t_i and t_j differ by a multiple of 16. Without rescheduling, an adder may potentially be multiplexed with only one sixteenth of the other adders in the circuit. Unless the circuit contains a large number of adders, the pairing cannot be done very efficiently without rescheduling the circuit. Since rescheduling will lead to the insertion of extra shimming delays it is unlikely that the gains achieved by the elimination of the relatively small serial adders will outweigh the cost of the necessary extra delays and multiplexing circuitry. Greater potential for saving exists in sharing large operators such as serial multipliers; however, the fact that operations extend over several clock cycles (a complete sample period) introduces complications into the task of operator sharing that does not exist for parallel arithmetic.

Chapter 7

Digit-Serial Performance

A variety of measures can be used to evaluate the efficiency of processor architectures [58]. These should consider both the area occupied by the circuit and the time required to perform computations. Time measures in synchronous systems may consider either the number of clock cycles required to process data, or the rate at which the clock can be run. The performance of systolic networks can be measured by their latency, the time required for output data to appear after the corresponding input data is read, and throughput, the rate at which input samples are read and output samples are produced. Throughput is of primary importance in signal processing systems. Reducing latency can also be of benefit – in some applications latency is critical.

Typical efficiency measures are products of the form $A^k T^m$ where A is a measure of area, T is a measure of time, and k and m are real constants. Here, we use $k = m = 1$ to formulate the efficiency metric. This choice dictates an equal weighting of area and time, that is, if two processors require half as much time as one processor to perform a task, then they have an equivalent AT measure. The reciprocal of the AT measure represents the efficiency of a silicon implementation of an algorithm in terms of amount of throughput per unit area of silicon.

There are two basic possibilities for implementing operators in serial hardware. In the simplest case (as represented by the operations described in chapter 2), the operators are bit-sliced, and wider or narrower processors are constructed by adding or deleting bit-slices. In this case, the change in area with digit width is linear. Also, the change in clock period that results from changing the digit-size is linear since the time taken for partial results to ripple through the bit slices is proportional to the number of bit-slices. An alternative situation exists when operators are used that employ look-ahead techniques. These generally require $O(N \log N)$ area but operate in $O(\log N)$ time. Examples are carry look-ahead [26] and hierarchical carry-select [27] adders. Multipliers that operate in

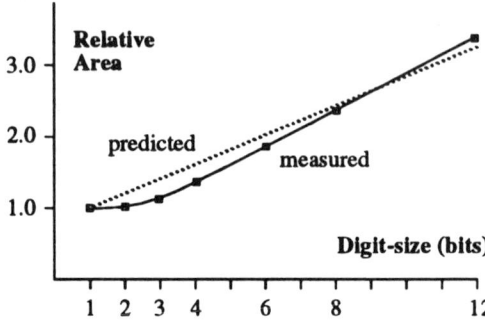

Fig. 7.1: Area versus digit-size.

log N time are less common. The most well known example, the Wallace-tree multiplier is difficult to lay out, and therefore usually not suitable for VLSI implementation. Operators employing both ripple and look-ahead techniques will nevertheless be examined here as the effects of digit-serial processing under each situation are different. The following analyses may be interpreted with reference to the digit-serial adder of Fig. 1.15, our basic digit-serial operator model.

7.1 Ripple-Carry Operators

Consider first the bit-sliced case. To determine the area, note that a constant amount of area, a_0, is required, independent of the digit-size, to perform the serial overhead functions, and to distribute power and clock lines to the processors. Also, each bit-slice requires area a_s. The total area is thus $a_0 + Na_s$. A reasonable estimate is that $a0 \approx 4a_s$, based on examination of the hardware involved in some typical operators. Fig. 7.1 shows a plot of predicted processor area versus digit-size using the estimated factor of 4. Also plotted on the same graph are the actual chip size versus digit-size measurements for a sample algorithm as compiled by Parsifal. The actual plot lies appreciably below the predicted value for low digit-sizes. This is because certain parts of the circuit (such as control and constant generation circuitry) actually decrease in size for higher digit sizes. Note that for a digit-size of two, the chip is insignificantly larger than for a digit-size of one (bit-serial), and yet it has twice the throughput. Even for a digit-size of four, the circuit is only a little larger. This is a persuasive argument for the use of digit-serial processing.

Next we consider the throughput of digit-serial circuits. The clock period is composed of an overhead component t_0 plus t_s per bit adding up to a total clock period of $t_0 + N t_s$. If W is the word-length, then one sample period which

7.1. RIPPLE-CARRY OPERATORS

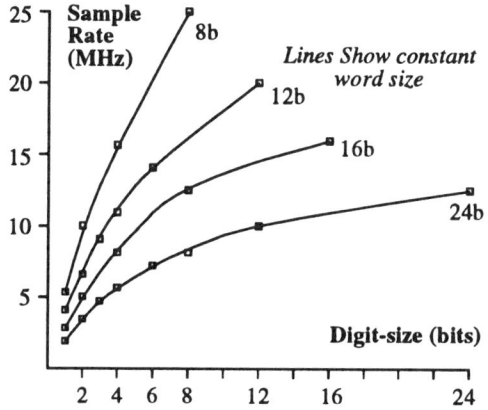

Fig. 7.2: Throughput versus digit-size.

consists of W/N clock cycles, is given by

$$T = (t_0 + Nt_s)W/N \ .$$

In our designs, we aimed for a ripple propagation time $t_s = 2.5$ns and an overhead allowance $t_0 = 20$ns for clock-skew, cycle dead-time, overhead computation time and propagation time of signals from one operator to the next, over possibly large interconnect distances. On this basis, a reasonable estimate is that $t_0 \approx 8t_s$. Fig. 7.2 is a plot of throughput rate versus digit-size for a number of word lengths. Throughput rate, shown in these graphs, is the reciprocal of sample period. Note that the throughput rate grows sub-linearly with the digit-size because of slowing of the clock-rate caused by increased carry-propagation path lengths.

As noted above, the reciprocal of AT is a measure of throughput efficiency of the design. Since $A = a_0 + Na_s$, and the total time per sample is $T = (t_0 + Nt_s)W/N$, the reciprocal efficiency value, AT, is:

$$AT = W(a_0 + Na_s)(t_0 + Nt_s)/N \ .$$

Note that AT is proportional to W, and hence the minimum value of AT is independent of the word-length. Fig. 7.3 is a plot of AT versus digit-size for a word-length of 24, scaled relative to bit-serial computation (digit-size = 1). For other values of the word-length, the graph has the same shape. It may be determined analytically that the minimum value of AT is achieved for a digit-size of

$$N = \sqrt{\frac{a_0 t_0}{a_s t_s}} \ .$$

Thus, the most efficient processing is done when the digit-size lies in the range of 4 to 8 bits, in other words, at an intermediate value between 1 (bit-serial)

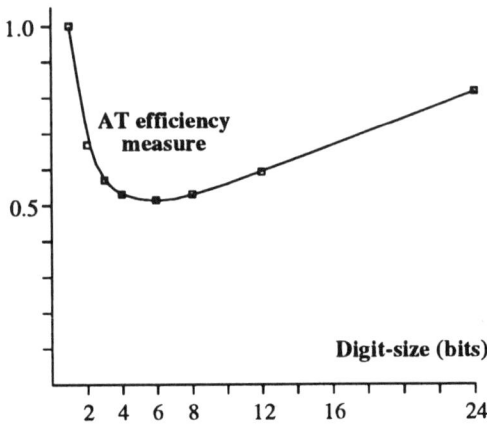

Fig. 7.3: Area-Time product for ripple carry operators.

and W bits (bit-parallel). For our assumed ratios of $a_0 = 4a_s$ and $t_0 = 8t_s$, the optimal efficiency, independent of word-length, is achieved at $N \approx 5.6$.

7.2 Look-Ahead Operators

The look-ahead operators can be analyzed in a similar fashion, but the change of area and time with digit-size cannot be estimated as accurately as in the linear case. Choosing
$$A = a_0 + a_s N \log 2N$$
and
$$T = (t_0 + t_s \log 2N) W/N$$
gives values equivalent to the linear case for digit-size of 1, and also gives the desired dependence of area and time on digit-size. (Here, logarithms are base 2.) Figures 7.4, 7.5 and 7.6 are the analogs of Figures 7.1, 7.2 and 7.3 for the look-ahead case. Fig. 7.6 is a plot of the reciprocal efficiency curve for a 24-bit word-length. Once more the shape of the curve and the position of the minimum is independent of the word-length as seen from the formula
$$AT = W.(a_0 + a_s N \log 2N).(t_0 + t_s \log 2N)/N \ .$$

For the estimated ratios given above, the minimum of this expression, independent of W, can be evaluated graphically at $N \approx 2$, though this optimum digit-size depends strongly on the actual parameter values. Therefore, in both the linear and logarithmic case, there is strong justification for using digit-serial processors to achieve efficient processing. While the values here are based on particular assumed ratios of a_0 to a_s and t_0 to t_s, the results with other ratios

7.3. IMPROVED PERFORMANCE THROUGH UNFOLDING

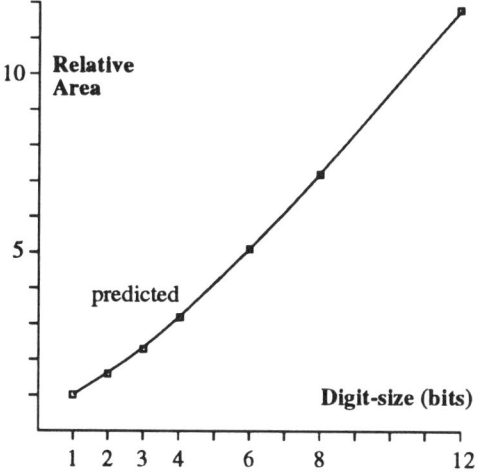

Fig. 7.4: Area versus digit-size for look-ahead operators.

will be similar. This can readily be seen in the linear case, where the optimal digit-size of 5.6 depends only on the square root of the area and time ratios, and therefore is relatively insensitive to changes in these ratios. In its simplest implementations, a digit-serial system will have a lower throughput than a bit-parallel system. However, this loss is mitigated by an increase in the clock rate. Overall efficiency is increased as there is a reduction in both operator and interconnect area.

7.3 Improved Performance Through Unfolding

As was seen in the previous section, the improvement in clock rate achieved by using a digit-size smaller than the full word-length is one of the primary reasons for the use of digit-serial as opposed to bit-parallel processing. On the other hand, throughput suffers, because the rate of the pipeline is divided by the number of digits in a word $P = W/N$. Therefore, overall performance will be slower in any digit-serial design than it will be in the bit parallel version, even though the clock rate will be higher. This loss can be overcome by replicating the calculations between P concurrent digit-serial streams, a method referred to here as *unfolding*. This allows the higher clock rate achieved by using digits of size $N = W/P$ to be used while still computing one result every clock cycle as in the bit-parallel version. In essence, the method requires the use of word-serial bit-parallel to word-parallel digit-serial and word-parallel digit-serial to word-serial bit-parallel converters to create and then merge a number of simultaneous computation streams, as shown in Fig. 7.7. Using these converters, it

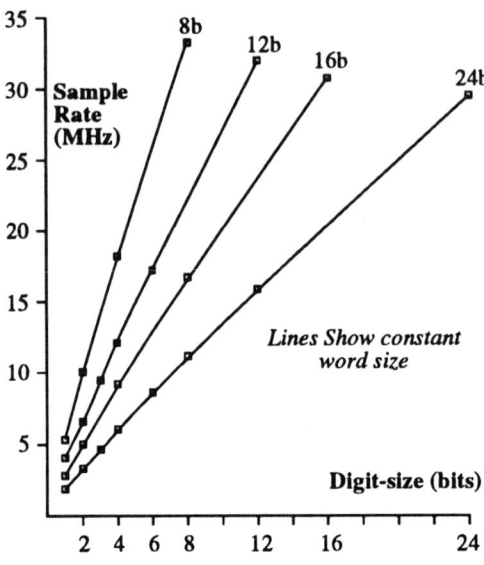

Fig. 7.5: Throughput versus digit-size for look-ahead operators.

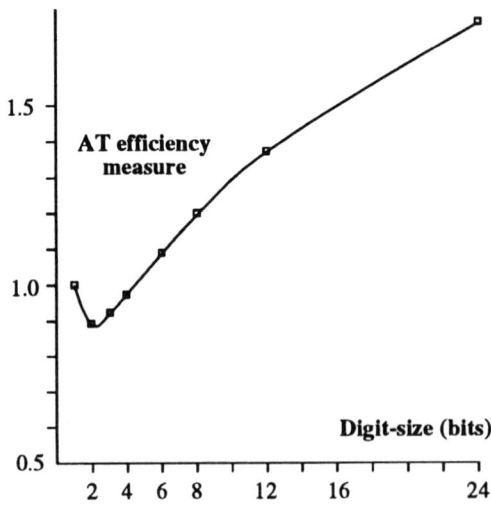

Fig. 7.6: AT versus digit-size for look-ahead operators.

7.3. IMPROVED PERFORMANCE THROUGH UNFOLDING

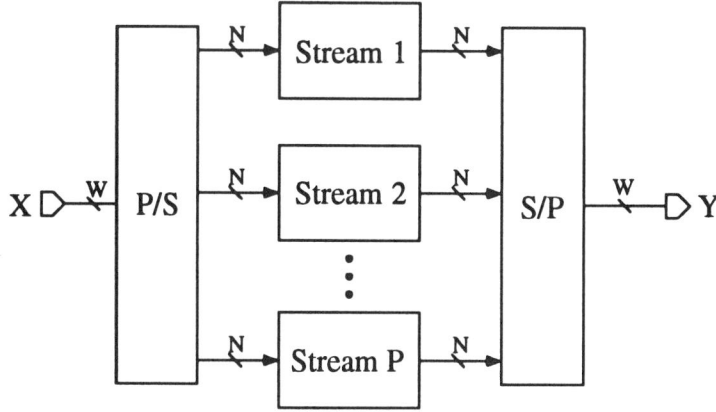

Fig. 7.7: Parallel computation streams.

is possible to divide a word into P digits and carry out P (or fewer) simultaneous computations, merging the results prior to output. The design of these converters was discussed in section 4.4.2. Note that this operation of splitting and merging is invisible from the outside world.

The parallel/serial converter block marked P/S in Fig. 7.7 actually consists of P separate parallel-in/digit-serial-out registers connected to the same input bus. Each such converter operates by latching the value on the bus and then sending it out one digit at a time during the next P clock cycles, after which the new input value is latched. The P separate converters are enabled in cyclic order so that input signals are distributed among the P streams in cyclic order, similar to dealing a pack of cards. Consecutive digit-serial output signals are staggered in time by one clock cycle. The S/P converter block (see section 4.4.3) carries out the dual function of merging and format conversion. P separate digit-serial input streams are fed into serial-in/parallel-out registers and passed in cyclic-order onto a common tri-state output bus. Consecutive digit-serial input streams lag each other by one clock cycle so as to maintain proper synchronization between the serial input signals and the bus-driver enable signals for the individual serial/parallel converters. The proper timing for synchronization of the data streams is handled automatically by Parsifal (see Chapter 6).

In the above description it was assumed that the number of computation streams is equal to the number of digits in each word, P. This is not necessary. The number of computation streams may be any number less than P. In this case, data will not be latched from the input bus in every cycle, but rather intermittently. Similarly, output data will not be placed on the output tri-state bus in every clock cycle but will be held over from the previous cycle. A common case is that in which the number of streams is a divisor of P and

Table 7.1: Layout Comparison

digit-size	Area (sq. mils)	Time per sample (ns)	Sample rate (MHz)	$1/AT$
1	16286	22.5 × 16	2.7	170.6
2	16788	25.0 × 8	5.0	297.8
4	19720	30.0 × 4	8.3	422.6
8	26972	40.0 × 2	12.5	463.5
16	42430	60.0 × 1	16.7	392.8
2 × 8	47391	40.0 × 1	25.0	527.5
4 × 4	59843	30.0 × 1	33.3	557.0

the internal clock rate of the chip is greater than the inter-chip data rate. For instance, with a word-length of 24 and digit-size of 3, P is equal to 8. With 2 concurrent computations, data is read from the input bus every 4 clock cycles. Thus, the on-chip sample period is 8 clock cycles (per data stream) and the chip-to-chip parallel data sample period is 4 clock cycles. Alternatively, four concurrent computations would allow a new input every second clock cycle.

General case: In the general case of arbitrary data-flow graphs, the situation is not quite so simple as that shown in Fig. 7.7 and does not usually result in independent computation streams as shown there. In general the concurrent computation streams will interact with each other. This is illustrated by the example design discussed in the next section.

7.4 An Example

By considering a particular example of an FIR filter, we illustrate the trade-off between speed and area afforded by variable digit-size and the unfolding technique described above. For the purposes of illustration, we choose a 4-tap filter with a word-length of 16 bits. In order to allow for programmability of the filter, the tap coefficients are loaded over a parallel data bus into internal registers. Consequently, full variable by variable multiplications are necessary for computing the filter expressions. Table 7.1 gives the result of compiling the chip using different digit sizes. The areas are actual areas of compiled chips (including pads). The speed rates are estimates. Note for the single stream implementations the way that efficiency ($1/AT$) peaks for an 8 bit digit-size and then decreases for the 16-bit digit size.[1] This is despite the fact that for the 16 bit or fully parallel version of the chip, the parallel-serial converters were not included, being unnecessary. In fact, for long word-length, we have met a situation of diminishing returns where chip area increases substantially, but

---
[1] Units of efficiency are samples per sq.mil per second.

7.4. AN EXAMPLE

throughput does not. Note also the minor increase in size between digit-sizes one and two.

The second last line of Table 7.1 refers to an unfolded design where two 8-bit digit-serial streams are used to implement the filter. It is also possible to use four parallel streams 4 bits wide, achieving a throughput of 33.3 MHz. The last line of Table 7.1 refers to such a design. We consider next in some detail the unfolded design using two concurrent 8-bit data streams. This same design was also considered in Section 6.10 from the point of view of scheduling.

In FIR filtering, it is not possible for the even and odd numbered samples to be processed in completely independent streams, since each output value in general depends on all recent inputs, and not alternate inputs. The two digit-serial streams must interact with each other. How this is done is now described. Denote by x the input and by y the output of the filter. We may write equations for the even and odd numbered outputs as follows.

$$y(2n) = a_0 x(2n) + a_1 x(2n-1) + a_2 x(2n-2) + a_3 x(2n-3)$$
$$y(2n+1) = a_0 x(2n+1) + a_1 x(2n) + a_2 x(2n-1) + a_3 x(2n-2).$$

Now, the input data stream is split up into two separate data streams consisting of the even and odd numbered samples of x. Let the even and odd data streams be denoted by x_0 and x_1. The j-th sample in data stream x_i or x will be denoted by $x_i(j)$ or $x(j)$. It can be seen that $x_0(i) = x(2i)$ and $x_1(i) = x(2i+1)$. A similar notation can be used for the output, y. Using this notation we rewrite the above pair of filter equations as follows.

$$y_0(n) = a_0 x_0(n) + a_1 x_1(n-1) + a_2 x_0(n-1) + a_3 x_1(n-2)$$
$$y_1(n) = a_0 x_1(n) + a_1 x_0(n) + a_2 x_1(n-1) + a_3 x_0(n-1).$$

The flow diagram represented by these equations is shown in Fig. 7.8. In this figure, the z^{-1} operator denotes a full sample delay of two clock cycles, not a one clock-cycle delay. In such systems, a delay operator is also referred to as a *block delay operator* [59]. Note that there are 8 multipliers and 6 adders compared with only 4 multipliers and 3 adders for a standard representation of a 4-tap FIR filter. However, the multipliers are only about half the size of bit-parallel 16 by 16 multipliers and function at a higher clock rate.

The various combinations of digit-size and number of parallel streams which are possible for a 16 bit word-length, assuming one set of parallel input pins, are summarized in Table 7.2. Since designs with 8 or 16 streams are probably not feasible for reasonably sized chips, the designs represented by the last two lines in Table 7.1 indicate the best achievable throughput using this method. In summary, greater speed and efficiency can be achieved in many designs by using digit-serial rather than bit-parallel computation. The highest sample rates can be achieved by splitting the computation into parallel computational streams in order to use a digit-size in the optimal range of 4 to 8 bits.

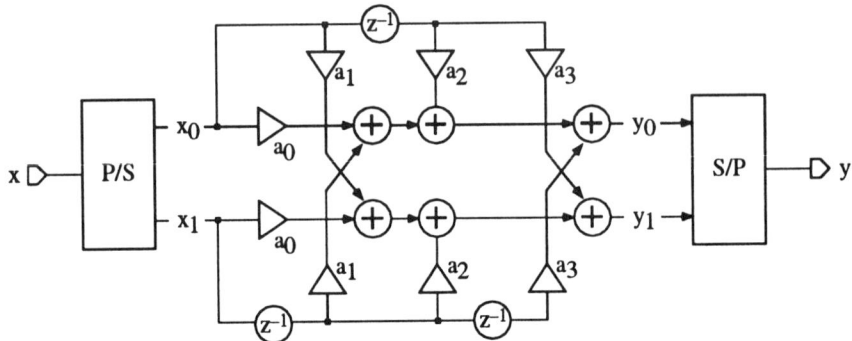

Fig. 7.8: FIR filter with two parallel streams.

Table 7.2: Achievable Throughput Rates

digit size	Achievable throughput (MHz sample rate)				
	number of streams				
	1	2	4	8	16
16	16.6				
8	12.5	25.0			
4	8.3	16.6	33.3		
2	5.0	10.0	20.0	40.0	
1	2.8	5.6	11.1	22.2	44.4

Chapter 8

Bit-Level Unfolding

8.1 Introduction

In chapter 7 a technique of word-level unfolding was described in which a single computational stream is unfolded into several computational streams carrying out the same computation independently or in an interleaved manner. In this chapter we describe a technique investigated by Parhi ([10][9]) called "bit-level unfolding". This method gives a systematic way of generating digit-serial designs from bit-serial designs. The method does not make fundamental use of the fact that the original design is a "bit-serial design", but can be applied to any synchronous design made up of combinational circuitry and delay latches. We make the assumption only that the circuit should be synchronous, that is, there are no unclocked feedback loops.

The unfolding technique described here may be used on individual bit-serial operators to derive digit-serial operators. It may also be applied to a complete bit-serial circuit to transform it to a digit-serial circuit for the purposes of throughput enhancement.

8.2 Description of the Technique

Consider a general synchronous circuit as shown in Such a circuit may be modeled as a block of combinational circuitry, M, with inputs $A, B, C, \ldots$ and outputs $X, Y, Z, \ldots$. There are also latched feedback loops, as shown, in which signals are delayed one clock cycle. It is clear that this model also accounts for circuits in which signals are subject to several cycles of delay. In such a case, the signal loops around several times through the combinational block, being passed straight through the block each time. We denote by $A(i), B(i), C(i), \ldots, X(i), Y(i), Z(i), \ldots$ the values on each of the input and output wires at time i relative to some common reference time $t = 0$.

148   CHAPTER 8. BIT-LEVEL UNFOLDING

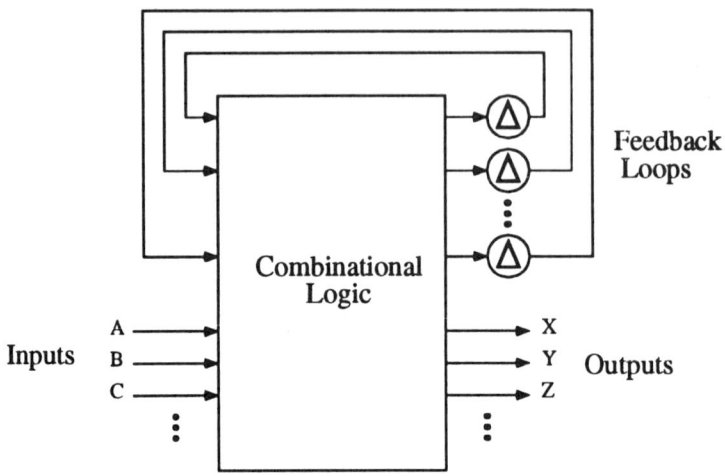

Fig. 8.1: General synchronous circuit.

Thus, the sequence of values carried on input A in successive clock cycles is $A(0), A(1), A(2), \ldots$. Fig. 8.1.

The principle of unfolding is shown in Fig. 8.2 where combinational circuitry block M is duplicated N times as shown. The delayed feedback loops are replaced by undelayed connections from block M_i to block M_{i+1} for $i = 0, 1, \ldots, N-2$. To complete the loop, M_{N-1} is connected back to M_0, this time through a latch as shown. The input and output ports to the combinational block M_i are labeled $A_i, B_i, \ldots$ as shown. The input signals $A(0), A(1), A(2), \ldots$ that are provided at input port A in the original circuit must be distributed among the ports $A_i, i = 0, \ldots, N-1$ in such a way that at time t, port A_i receives the signal $A(Nt+i)$. Thus, port A_0 receives inputs $A(0), A(N), A(2N), \ldots$, port A_1 receives inputs $A(1), A(N+1), A(2N+1), \ldots$, and so on for all the other ports.

In this configuration, assuming that the values stored in the latches at the start of the cycle $t = 0$ are the same in both the original and unfolded versions of the circuit, it may be verified that at any time $t \geq 0$, block M_i receives the same inputs as the block M in the original circuit receives at time $Nt + i$. This may easily be proven by induction on $Nt+i$, but is easily verified by inspection. Since block M_i is combinational, it follows that its outputs at time t must be the same as those of block M at time $Nt+i$. Accordingly, labeling the outputs of block M_i by $X_i, Y_i, Z_i, \ldots$, the values at output port X_i are related to those at port X in the same way that the values at input ports A_i and A are related, namely, port X_i carries value $X(Nt+i)$ at time t. Note how the serial inputs in the original "bit-serial" version are broken down into digits of size N in the unfolded "digit-serial" version. In particular, at time $t = 0$, the N inputs

8.2. DESCRIPTION OF THE TECHNIQUE

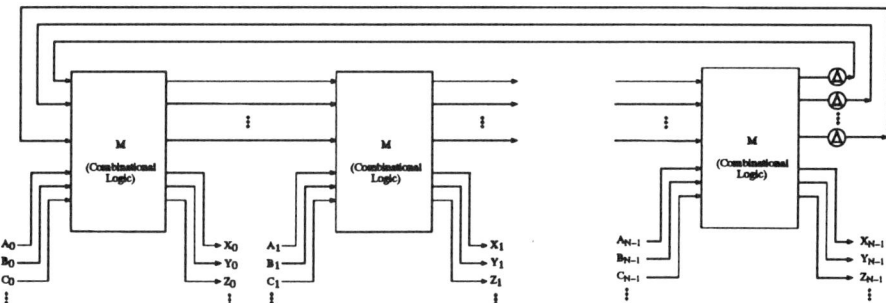

Fig. 8.2: Unfolded circuit.

$A(0), A(1), \ldots, A(N-1)$ are present at input ports $A_0, A_1, \ldots, A_{N-1}$. These N ports may be thought of as constituting a single N-bit wide port at which the successive N-bit digits of the input are presented in consecutive time intervals. The same holds for the output ports. In the most general case, however, the N bits making up a digit need not be consecutive bits of a serial word, or indeed bear any relationship to each other. For instance, the original circuit may itself be a digit-serial circuit in which the ports $A, B, C, \ldots$ may be divided into digit-serial input ports. Alternatively, the original circuit may be a circuit using parallel arithmetic in which ports $A, B, C, \ldots$ may be divided into words of parallel input data. However, our main interest will be in unfolding bit-serial data in which the data values at each port may be divided up into serial words of data.

To unfold an edge with multiple delays, we can formulate a simple algorithm. Consider an edge from node A to B with i delays. Then the result of the j-th iteration of A is needed for the execution of $(j+i)$-th iteration of B. First consider the case when the number of delays i is less than the unfolding factor N. In this case, we connect A_j to B_{j+i} for j ranging from 0 to $N-i-1$ (since $i+j \leq= N-1$). For j from $N-i$ to $N-1$, connect A_j to B_{j+i-N} with 1 delay. Next consider the case when $i \geq N$. For this case, connect A_j to $B_{(j+i) \bmod N}$ with $\lfloor (j+i)/N \rfloor$ delays where $\lfloor x \rfloor$ represents the largest integer less than or equal to x and $k \bmod N$ represents the remainder of k on division by N. Since

$$\lfloor i/N \rfloor + \lfloor (i+1)/N \rfloor + \ldots + \lfloor (i+N-1)/N \rfloor = i$$

unfolding of an edge preserves the number of delays.

Example 8.1. Consider the bit-serial system shown in Fig. 8.3(a). The corresponding unfolded architectures with unfolding factors 2 and 3 are shown in Fig. 8.3(b) and Fig. 8.3(c), respectively. Note that the number of delay elements are 5 in all three architectures as expected.

150 CHAPTER 8. BIT-LEVEL UNFOLDING

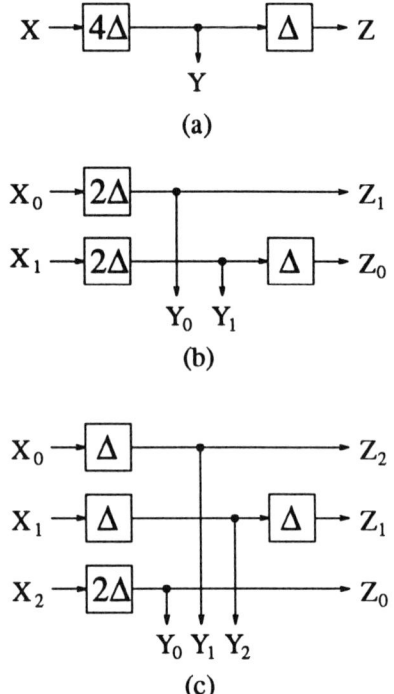

Fig. 8.3: Unfolding of edges with delays.

8.2. DESCRIPTION OF THE TECHNIQUE

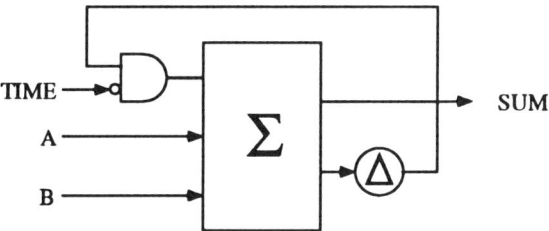

Fig. 8.4: Bit-serial adder.

Example 8.2. A digit-serial adder: We apply this method to the bit-serial adder shown in Fig. 8.4. The corresponding digit-serial unfolded circuit is shown in Fig. 8.5. The degree of unfolding or the unfolding factor, N, is equal to 3 in this example.

We will now consider this example in two particular cases. We consider first the case in which the digit-size (here 3) divides the word-length of the serial data. We assume for the sake of illustration that the word-length is equal to 12. In this case, the digit-serial circuit may be simplified.

Suppose the reference time $t = 0$ is taken to coincide with the first bit of a word of input data. Consider, now the timing signal $TIME$. For the bit-serial circuit to function properly, it is necessary that the $TIME$ signal have the appropriate form. In this case it must be a periodic signal with period W containing a high (1) pulse at time $t = 0, W, 2W, \ldots$. (Note that this notation of $TIME$ is different from that used in other chapters where $TIME$ was high during the most-significant-bit position.) Denote this signal for convenience by $\delta(t \bmod W)$, where δ is the function defined by

$$\begin{aligned}\delta(i) &= 1 & \text{if } i = 0 \\ &= 0 & \text{otherwise.}\end{aligned}$$

Now, the signal at port $TIME_i$ in the digit-serial circuit is equal to $\delta((Nt + i) \bmod W)$. Assuming that n divides W, we see that

$$\begin{aligned}\delta((Nt + i) \bmod W) &= 0 \text{ for all } t & \text{if } i \neq 0 \\ &= \delta(t \bmod W/N) & \text{if } i = 0 \:.\end{aligned}$$

This means that for $i \neq 0$, the port $TIME_i$ receives a constant 0 signal, whereas port $TIME_0$ receives a periodic signal with period equal to W/N, the number of digits in a word, otherwise known as the sample period. The fact that the input signals $TIME_i$ in Fig. 8.5 receive a constant 0 signal means that the circuit may be simplified by removing the inverted-input AND gates, which act as feed-throughs for the second input signal. The resulting circuit, shown in Fig. 8.6, is the standard digit-serial adder cell.

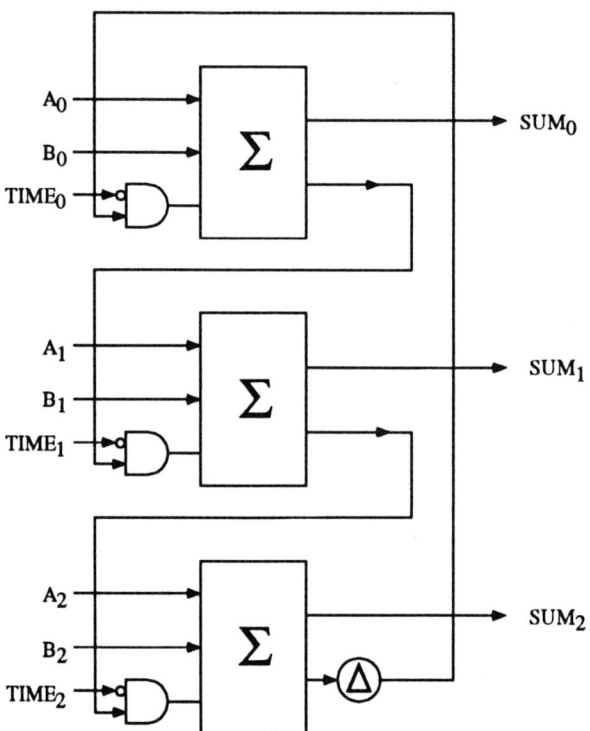

Fig. 8.5: Unfolded digit-serial adder.

8.2. DESCRIPTION OF THE TECHNIQUE

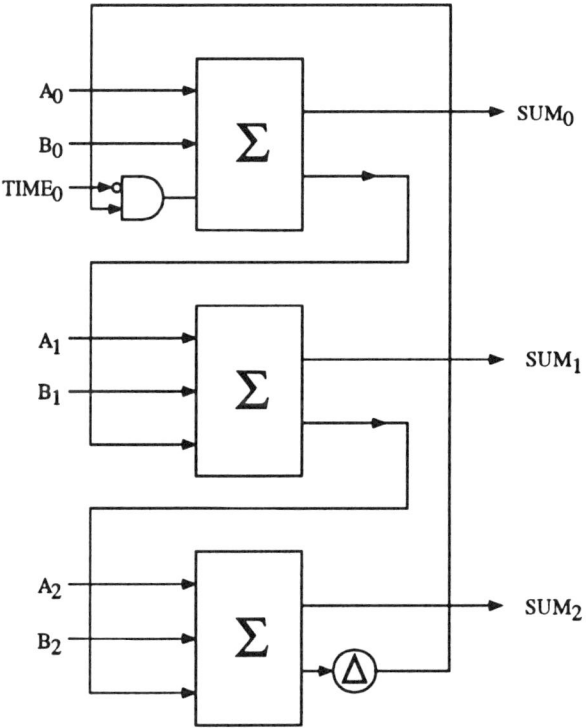

Fig. 8.6: Digit-serial adder cell.

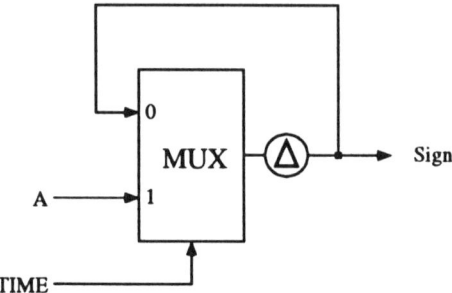

Fig. 8.7: Sign detector circuit.

Most circuit elements that receive a constant 1 or 0 signal may be eliminated from the circuit, or replaced by simpler elements. For instance, two input gates may be replaced by a feed-through or an inverter, multiplexors may be replaced by a feed-through of one of their inputs, and delay elements may be eliminated. This simplification may often lead to elimination of large parts of the unfolded circuit. For instance, consider the sign-detector circuit shown in Fig. 8.7. This circuit latches the sign bit of a bit-serial word and repeats it during the whole of the next sample period. The unfolded circuit is shown in Fig. 8.8. Assuming that the digit-size is a divisor of the word-length, most of the multiplexors in the figure will act as simple feed-throughs, allowing the circuit to be reduced to that shown in Fig. 8.9. Now, all but one of the inputs may be eliminated, since they are unused, and since all the outputs are connected, they may be amalgamated into one. The simple sample-and-hold circuit shown in this example is often used as a part of more complex circuits, for example an arithmetic comparator.

8.2.1 Operators With Latency

The example of the adder given in the last section illustrates the general method of unfolding. In the case of operators with non-zero latency, slight adjustments to the method are necessary. Consider the case of a bit-serial adder with a one-cycle output delay, as shown in Fig. 8.10. Assuming an unfolding factor of 3, and a word-length W, a multiple of 3, a straightforward implementation of the above unfolding method leads to a "digit-serial" adder as shown in Fig. 8.11. Comparing this with the digit-serial adder of Fig. 8.5 reveals that the output bits are rotated one place towards increasing significance with the high order bit of each digit being delayed until the next digit. Further, at time $t = 0$, the output bits $X(0), X(1)$ and $X(2)$ do not all belong to the same word. In fact, because of the delay in the bit-serial circuit, the output belonging to the input signals $A(0)$ and $B(0)$ does not appear until time $t = 1$, and is therefore signal $X(1)$. The signal $X(0)$ belongs to the previous word.

8.2. DESCRIPTION OF THE TECHNIQUE

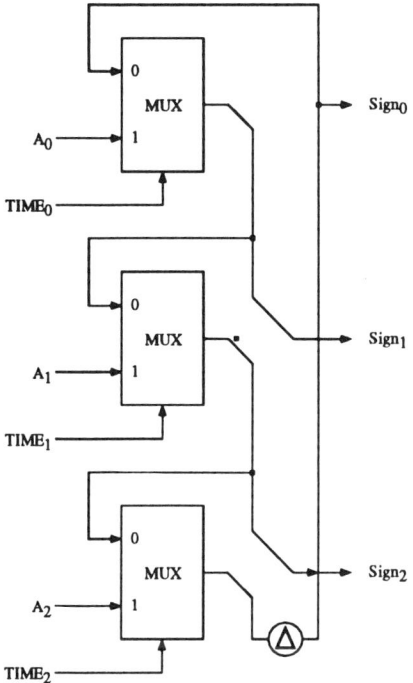

Fig. 8.8: Unfolded sign detector circuit.

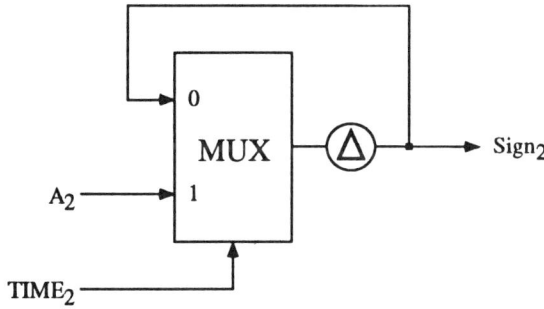

Fig. 8.9: Simplified digit-serial sign detector circuit.

156   CHAPTER 8. BIT-LEVEL UNFOLDING

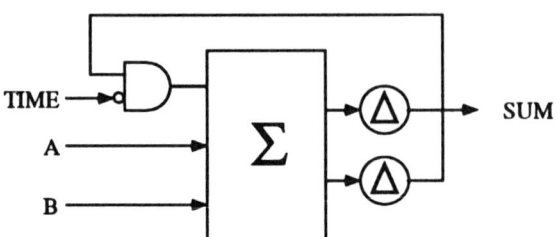

Fig. 8.10: Adder with one cycle delay.

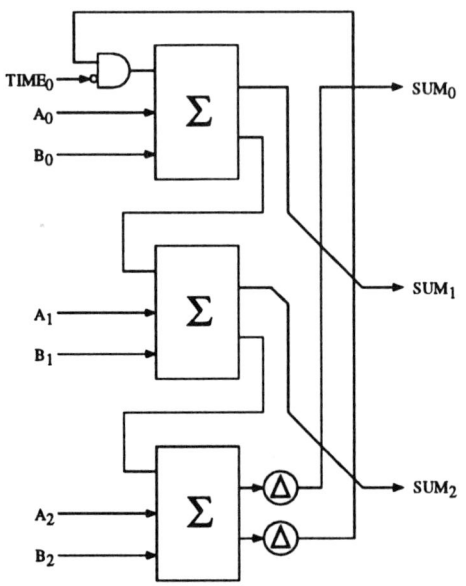

Fig. 8.11: Unfolded adder with one cycle delay.

8.3. COLLAPSING OUTPUTS

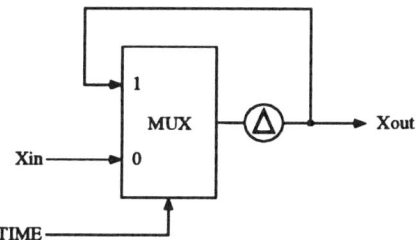

Fig. 8.12: Bit-serial shift-right circuit.

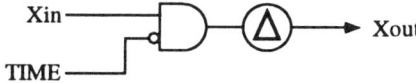

Fig. 8.13: Bit-serial shift-left circuit.

It is normally desirable in digit-serial operators that the division into digits should correspond with the division into words, so that words do not span digit boundaries. This can be ensured if a simple condition is respected.

> **Principle:** In order that unfolding of a bit-serial operator should produce a digit-serial operator in which the digits are synchronized with the word boundaries, it is necessary and sufficient that the digit-size (degree of unfolding) divide the word-length and that the timing values of the input and output signals should be multiples of the digit-size.

Normally, the condition described by this principle can easily be ensured by adding extra delays to outputs and possibly inputs.

Exercise 8.3. Consider the circuit shown in Fig. 8.12, which represents a bit-serial shift-right-one operator (with sign extension). What is the latency of the operator ? (Hint : It is not 1.) modify this circuit appropriately and apply unfolding to derive a digit-serial shift-right operator for digit-size 3. Verify the operation of the circuit, and ensure that the digits line up with the word boundaries.

Exercise 8.4. Same as exercise 8.3 for the shift-left circuit of Fig. 8.13.

8.3 Collapsing Outputs

In some cases, when a bit-serial operator is unfolded to a digit-serial operator, it is possible to collapse outputs that carry the same signal into one output.

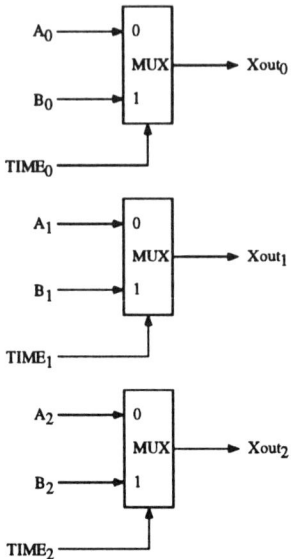

Fig. 8.14: Digit-serial multiplexor obtained by unfolding.

An example of this is the sign-detector circuit of Figs 8.7 and 8.8. As is shown, the three output signals $Sign_0$, $Sign_1$ and $Sign_2$ are connected together and are amalgamated into one signal in Fig. 8.9. Typically, the output of such a cell is used as the selector input of a multiplexor. Consider a digit-serial multiplexor obtained by unfolding (Fig. 8.14). Normally, the select input to a bit-serial multiplexor is intended to receive a signal that carries a constant 0 or 1 value during each sample period, such as the output of bit-serial sign-detector circuit, Fig. 8.9. If this is the case, then the three input signals $select_i$ of the digit-serial multiplexor may be combined. It is, however, impossible to deduce this fact without knowledge of the nature of the select input signal of the bit-serial multiplexor. If it is known, however, that multiplexor input comes from the output of a sign-detector such as Fig. 8.9, then as we have seen, the fact that the signals may be amalgamated falls naturally out of the unfolding process. This illustrates the fact that unfolding is best carried out on large circuit blocks rather than individual operators, so that each operator may be optimally unfolded according to its context.

8.4 Automatic Unfolding

The method described here for unfolding a bit-serial circuit may be applied either to individual bit-serial cells or to a complete bit-serial circuit described in

8.5. UNFOLDING TO ARBITRARY DIGIT SIZES

terms of elementary circuit components, that is, gates, multiplexors and delay elements. Additional input to the algorithm is a specification of the known periodic input signals, such as the *TIME* signals in the previous examples. The algorithm will consist of steps

- If required, modify the latencies of IO ports of the chip to be multiples of the desired digit-size, so as to ensure correct digit-word alignment.
- Formally unfold the circuit by duplicating all circuit elements except delays (details above).
- Compute the form of the unfolded periodic signals.
- Eliminate redundant circuit elements and amalgamate identical signals.

Parhi ([9]) reports having implemented such an algorithm. Such a design style would be particularly suited for physical implementation using gate-arrays. A description of bit-serial implementations of basic DSP operators would be kept in a library. A required design could be implemented in bit-serial, and then the complete design can be unfolded to generate a digit-serial design to satisfy the required performance parameters.

8.5 Unfolding to Arbitrary Digit Sizes

For the most part, it has been assumed in the above discussion of unfolding that the chosen digit-size is a divisor of the word-length. Although desirable, this is not really necessary. If one is prepared to give up the requirement of digit-word alignment, then it is possible to choose arbitrary digit-sizes. To see how this may be done, let us see how a sequence of 8-bit words may be transmitted in a 3-bit wide digit stream. Consider the following diagram.

d6	d3	d0	c5	c2	b7	b4	b1	a6	a3	a0
d7	d4	d1	c6	c3	c0	b5	b2	a7	a4	a1
e0	d5	d2	c7	c4	c1	b6	b3	b0	a5	a2

Each row of the table represents one of the bits of a 3-bit digit-serial stream. The heavy lines represent the boundaries between individual words. As may be seen, the break between the words rotates bit-positions.

Now, as an example, consider unfolding a bit-serial adder to make a 3-bit wide digit. Assume that the word-length is 8. The original bit-serial adder is shown in Fig. 8.4. When unfolded, as before, the digit-serial adder of Fig. 8.5 is obtained. The three timing inputs, $TIME_0$, $TIME_1$ and $TIME_2$ will be equal to the periodic signals $\delta(3t \bmod 8)$, $\delta(3t+1 \bmod 8)$, and $\delta(3t+2 \bmod 8)$. By contrast with the 12-bit word-length discussed in Section 8.2, none of these input signals is a permanently low signal. Rather, they are different phases of the periodic signal $\delta(t \bmod 8)$. In particular[1], it may be seen that

$$\delta(3t \bmod 8) = \delta(t \bmod 8)$$
$$\delta(3t+1 \bmod 8) = \delta(t+3 \bmod 8)$$
$$\delta(3t+2 \bmod 8) = \delta(t+6 \bmod 8).$$

Now, since none of the $TIME$ signals is permanently low, none of the AND gates used for resetting the carry is redundant. Consequently, the circuit may not be reduced as much as it could in the case of a 12-bit word-length.

The advantages of using arbitrary digit-sizes are that it may be possible to match the throughput requirements of the application and the architecture more precisely. This may be particularly useful in the case where the word-length is a prime, and so there exists no exact divisor suitable as a digit-size. Compared with bit-serial computation, the use of digit-serial computation with a non-divisible digit-size may still represent an efficiency gain, since the proportion of delay elements is reduced.

On the other hand, in the case of digit-sizes which are not divisors of the word-length there will in general not be much scope for reduction of the hardware through elimination of redundant operators. This is illustrated in the case of the adder just discussed. It becomes particularly apparent in considering the circuit of Fig. 8.8 (sign detector). As was shown, in the case where digit-size divides word-length, large amount of hardware may be eliminated. This will not be possible in the case of a non-dividing digit-size. There will be a cost in extra cell circuitry, as well as extra cell-to-cell routing area since all three outputs of the cell in Fig. 8.8 are now necessary.

8.6 Parallel to Digit-Serial Converter

In order to take advantage of the possibility of using digit-sizes which are not divisors of the word length, it is necessary to have converters which reorder the input words into the appropriate format. This section shows how this may be done. As usual, a parallel-to-digit-serial converter may be derived from a parallel-to-bit-serial converter by unfolding. Consider first a parallel-to-bit-serial converter as shown in Fig. 8.15. In this example, we assume a word-length of 4, and a desired digit-size of 3. In this circuit, a parallel signal is input on the parallel bus $P<0:3>$. It is assumed that the input signal is held on the bus for four consecutive clock cycles before changing. Once in every four cycles, the parallel bus is sampled and fed into the serial shift-register. The sampled value is fed out serially during 4 clock cycles on the serial output pin, S.

---

[1] In general if $nn' \equiv 1 \pmod{W}$, then $\delta((Nt+i) \bmod W) = \delta((t+N'i) \bmod W)$. The value N' is known as the inverse of N modulo W, and it will always exist if N and W are coprime.

8.6. PARALLEL TO DIGIT-SERIAL CONVERTER

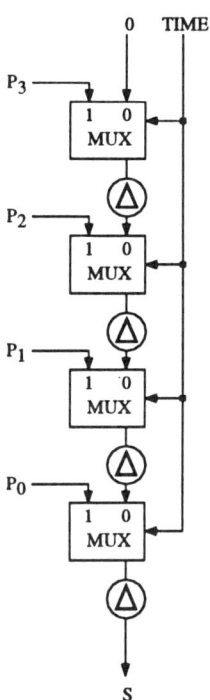

Fig. 8.15: Parallel-to-bit-serial converter.

The unfolded digit-serial circuit is shown in Fig. 8.16. We have deviated slightly from the prescribed method of unfolding by not duplicating the parallel input bus. This is possible, because during three of the four clock cycles in every sample period, the input bus is unused in the bit-serial version. In this digit-serial circuit, the output rate is three words in four clock cycles. The signal on the parallel input bus is ignored during the fourth clock cycle of every sample period. It is interesting to see exactly how this circuit indeed transforms the parallel input signal to digit-serial format.

The following table shows where each output $S0, \ldots, S2$ and each latch $L0, \ldots, L3$ derives its value from in each clock cycle.

	$S0$	$S1$	$S2$	$L0$	$L1$	$L2$	$L3$
$t=0$	$L0$	$P0$	$P1$	$P2$	$P3$	0	0
$t=1$	$L0$	$L1$	$P0$	$P1$	$P2$	$P3$	0
$t=2$	$L0$	$L1$	$L2$	$P0$	$P1$	$P2$	$P3$
$t=3$	$L0$	$L1$	$L2$	$L3$	0	0	0

Using this table, we may compute the value at each output, and in each latch at each clock cycle. In the next table, the notation Pi_j denotes the value present at the input Pi at time $t = j$.

	$S0$	$S1$	$S2$	$L0$	$L1$	$L2$	$L3$
$t=0$	X	$P0_0$	$P1_0$	$P2_0$	$P3_0$	0	0
$t=1$	$P2_0$	$P3_0$	$P0_1$	$P1_1$	$P2_1$	$P3_1$	0
$t=2$	$P1_1$	$P2_1$	$P3_1$	$P0_2$	$P1_2$	$P2_2$	$P3_2$
$t=3$	$P0_2$	$P1_2$	$P2_2$	$P3_2$	0	0	0
$t=4$	$P3_2$	$P0_4$	$P1_4$	$P2_4$	$P3_4$	0	0
$t=5$	$P2_4$	$P3_4$	$P0_5$	$P1_5$	$P2_5$	$P3_5$	0

Notice that the sequence of output digits are in the correct digit-serial format. The input values at time $t = 3$ are ignored, never appearing at the output.

8.6. PARALLEL TO DIGIT-SERIAL CONVERTER

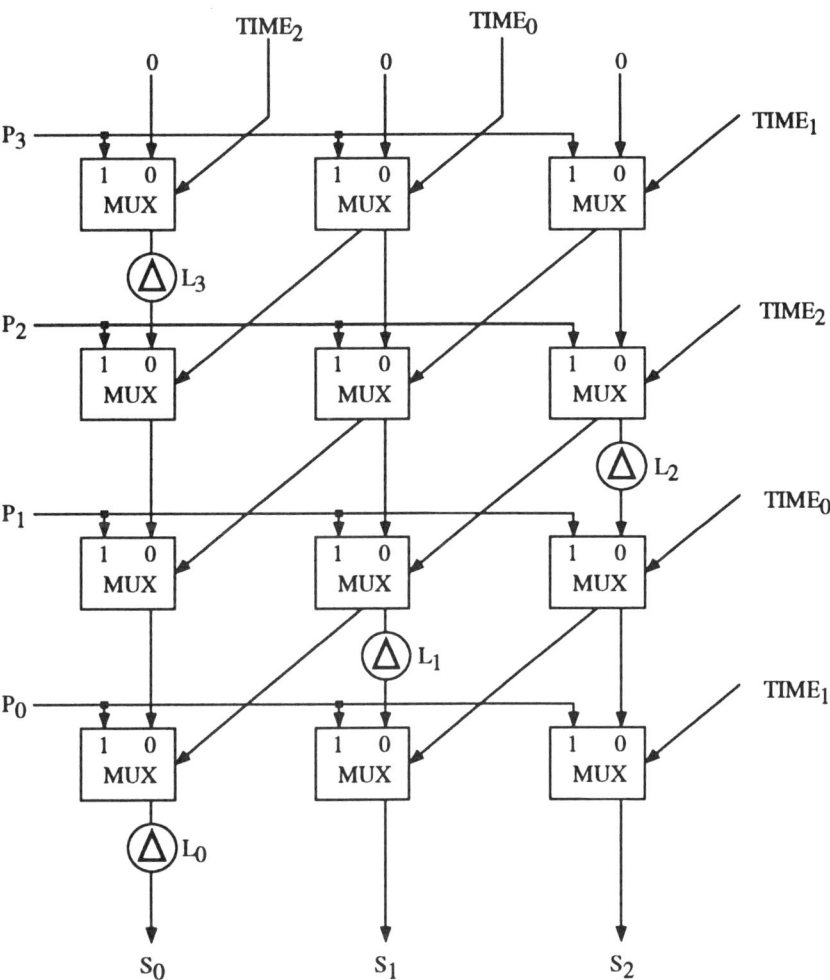

Fig. 8.16: Parallel-to-digit-serial converter.

Chapter 9

The Folding Transformation

9.1 Introduction

In Chapter 8, the unfolding technique was used to derive digit-serial architectures from bit-serial architectures. This chapter presents the *folding* technique which is the reverse of unfolding [60]. The folding technique is briefly described and is used to derive digit-serial architectures directly from bit-parallel architectures.

The folding technique is applicable to any arbitrary irregular data-flow graph for any arbitrary folding set. However, for the special case when the data-flow graph is regular (for definition of regularity, see Section 9.3) and the folding set can be described by a linear space-time mapping, the folding process reduces to the *systolic array design* methodology. While folding of bit-parallel to digit-serial architectures cannot be described by linear space-time mapping, folding of bit-parallel to bit-serial architectures can be. Therefore, another less-direct technique to derive digit-serial architectures from bit-parallel involves folding of bit-parallel architectures to bit-serial using systolic array design methodology and unfolding of the bit-serial architecture to digit-serial.

This chapter illustrates use of both these approaches of deriving digit-serial architectures from bit-parallel.

9.2 Digit-Serial Design Using Folding

The folding technique transforms an algorithm data-flow graph to an architecture data-flow graph where multiple algorithm operations are *folded* to a single hardware processor. The folding set specifies at what time partition and in which hardware processor an algorithm operation is executed. Let N algorithm operations be folded to a single hardware processor. Then all time instances can be described by N time partitions numbered from 0 to $(N-1)$ where the time partition corresponds to the time instance modulo N.

Before we consider folding, we illustrate the representation of folding sets. Consider a DFG with two types of operations, add and multiply. Let 4 operations be mapped to a single processor and $S_A = \{A2, A1, A4, A3\}$ and $S_M = \{M1, M3, \phi, M2\}$ represent the (ordered) folding sets for the add and multiply, respectively. In this representation, the hardware adder implements operation $A2$ in time partition 0 or in time instances $4l$ where l is a nonnegative integer. Similarly $A4$ is executed by the same processor in time partition 2 or time instances $(4l+2)$. Note that the multiply operator implements one null operation every $(4l+2)$-nd cycle. Equivalently, we can also write $A1 = \{S_A|1\}, M3 = \{H_A|1\}$, etc.

Consider an arc $U \to V$ with i delays as shown in the DFG in Fig. 9.1(a). Let U and V respectively be executed by operators H_u and H_v with folding order u and v. Folding of the arc $U \to V$ results in a communication link between H_u and H_v. The input to each operator is a N-way switch and the hardware operator H_u connects to one switch input of H_v with the switching instance $(Nl+v)$ with appropriate number of communication delays denoted as $D_F(U \to V)$ as shown in Fig. 9.1(b). In Fig. 9.1(b), the P_U delays are used for internal pipelining of H_U. These delays are not available to be shared by multiple communicating arcs.

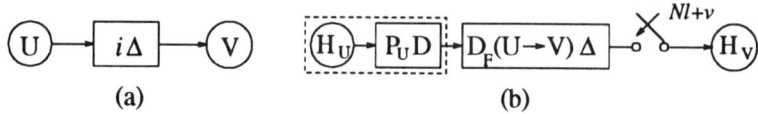

Fig. 9.1: Folding of an arc $U \to V$ with i delays.

Let the l-th iteration of the task U (i.e., U_l) start in cycle $[N_u l + u]$, where u is the execution, or folding order number of U within the set S_u. The operation of U_l is then completed in cycle $[N_u l + P_u + u]$ (where P_u is the level of pipelining of H_u). The result of U_l is used for the execution of the $(l+i)$-th iteration of the task of V, V_{l+i}, which starts in cycle $[N_v(l+i)+v]$, where v is the execution order number of V within the set S_v. When this arc is subjected to the folding algorithm, the output of H_u needs to be stored for $[N_v(l+i) - N_u l - P_u + v - u]$ cycles. In order for the number of delays to be iteration independent (i.e., independent of 'l'), N_u must be equal to N_v. If we let $N = N_u = N_v$, then the number of required storage units, or *folded arc delays* must be:

$$D_F(U \to V) = Ni - P_u + v - u .$$

Example 9.1. Consider the algorithm flow graph and the folding set specified

9.2. DIGIT-SERIAL DESIGN USING FOLDING

in Fig. 9.2(a). All the tasks in this flow graph are of one type (specified as type A), the task A_i is the i-th task of type A. We assume the use of a two-stage pipelined hardware "A" operator in the folded architecture. To fold this, we

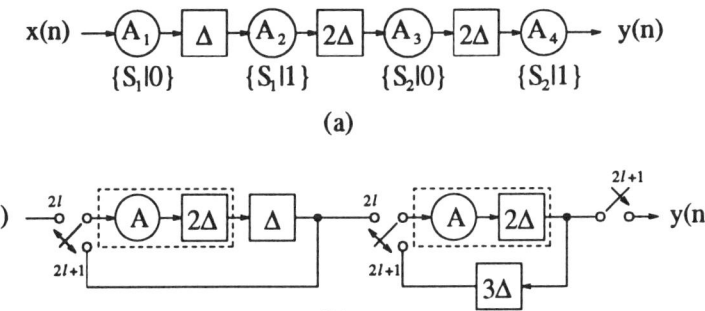

Fig. 9.2: Folding of an arbitrary data-flow graph.

need to calculate the folded arc delays:

$$D_F(A_1 \rightarrow A_2) = 2(1) - 2 + 1 - 0 = 1$$

$$D_F(A_2 \rightarrow A_3) = 2(2) - 2 + 0 - 1 = 1$$

$$D_F(A_3 \rightarrow A_4) = 2(2) - 2 + 1 - 0 = 3 .$$

The folded architecture is described by the data-flow graph in Fig. 9.2(b). It is useful to examine the operation of the folded architecture. In time step 0, $x(0)$ is sampled and the operation A_1 is carried out in the left hardware processor. Since this processor is pipelined by 2 stages, the output of this operation is available in time step 2. This output cannot be used immediately since the operation A_2 can only be scheduled in an odd cycle or time instance $2l + 1$. The output of A_1 is stored for one cycle and in time step 3 the operation A_2 is executed in the left hardware adder. The switches in the hardware architecture data-flow graph can be implemented as multiplexors.

If all the folded delays $D_F(.)$'s are non-negative, then these delays can be used to synthesize the folded architecture. However, if one or more of the folded delays are negative, then the algorithm DFG must first be preprocessed for folding using pipelining and retiming as described next.

9.2.1 Pipelining and Retiming For Folding

The objective of the preprocessing step is to insert the pipelining latches [61] and to retime [62] existing latches as appropriate such that the pipelined version of the DFG leads to non-negative folded arc delays for all arcs.

To automatically pipeline and retime for folding, we follow an approach similar to retiming. The retiming technique is a general approach to change the number of arc delays in a DFG. In [62], retiming was used to reduce the clock period. Pipelining can also be achieved in retiming by considering the input and output nodes as separate host nodes. In contrast, our objective is to pipeline and retime for folding such that the folded arc delays become positive.

We now derive the constraints for pipelining and retiming for folding. These constraints can be solved using any shortest path algorithm. Consider an edge $U \rightarrow V$ with i delays. The retiming and pipelining change the number of delays in this arc to i_r, given by,

$$i_r(U \rightarrow V) = i(U \rightarrow V) + r(V) - r(U)$$

where $r(V)$ and $r(U)$ represent the retiming values of nodes V and U. Our objective is reduced to the calculation of the retiming values for all nodes in the DFG, such that the folded path delays of all arcs $U \rightarrow V$ for the pipelined and retimed DFG are non-negative. Let $D_F(U \rightarrow V)$ denote the folded arc delays, for the arc $U \rightarrow V$, in the original DFG and $D'_F(U \rightarrow V)$ represent the folded arc delays in the retimed and pipelined DFG. Therefore, we need to satisfy $D'_F(U \rightarrow V) \geq 0$. This equation can be exploited to obtain constraints for pipelining and retiming for folding.

$$\begin{aligned} D'_F(U \rightarrow V) &= Ni_r(U \rightarrow V) - P_u + v - u \\ &= Ni(U \rightarrow V) - P_u + v - u + Nr(V) - Nr(U) \\ &= D_F(U \rightarrow V) + Nr(V) - Nr(U) \geq 0 \ . \end{aligned}$$
(9.1)

The last inequality can be satisfied by the constraints

$$r(U) - r(V) \leq D_F(U \rightarrow V)/N \ .$$

However, since retiming only considers insertion or movement of integer number of delays to or from any arc, the retiming variables $r(.)$'s are constrained to be integers. This is achieved by constraining the right hand side of the inequality to be an integer as:

$$r(U) - r(V) \leq \lfloor D_F(U \rightarrow V)/N \rfloor \ .$$

In the above notation, $\lfloor x \rfloor$ is the floor function which represents the largest integer less than or equal to x.

For any arbitrary DFG, the folded path delays can be calculated. These values can be used to solve the above set of constraints to obtain the retiming variables by using the Bellman-Ford or any other shortest path algorithm. The pipelined and retimed DFG is obtained by using the retiming variables and by calculating the delays in the retimed DFG. The retimed DFG can be easily

9.2. DIGIT-SERIAL DESIGN USING FOLDING

folded. For simplicity, we add a constant number to all the retiming variables such that the lowest retiming variable is zero.

It is important to note that a specified folding set may be infeasible. When this occurs no solution exists to the shortest path problem for preprocessing by pipelining and retiming for folding. Furthermore, many folded architectures can be obtained from a given algorithm by using several folding sets. All of these folded architectures can be unfolded to obtain many data-flow architectures which are pipelined and/or retimed versions of the original algorithm data-flow graph and of each other.

9.2.2 Folding of Bit-Parallel to Bit-Serial and Digit-Serial Architectures

We consider examples of folding to fold a two's complement bit-parallel multiplier architecture to bit-serial and digit-serial.

Example 9.2. Consider the multiplication operation described as a tabular operation in Fig. 3.7. In this table, each addition operation generates sum and carry outputs where the sum is used in an add operation in the next row and the carry is used in the add operation to the left in the same row. The sum outputs are represented by solid lines and the carry outputs are represented by dashed lines. Since carry output of an adder is rippled to the left adder in the same row, this addition operation is referred to as a carry-ripple multiplication. The dependence graph of the two's complement ripple-carry multiplication operation is shown in Fig. 9.3 for a word-length of 4. Consider the folding set description shown in Fig. 9.4 for the folding of the bit-parallel ripple-carry multiplier dependence graph shown in Fig. 9.3 to a bit-serial architecture. In Fig. 9.4, the folding sets are described for the nodes whose computations are omitted from this figure for clarity. The notation $\{A|1\}$ at a node implies that this operation is executed by processor A in time partition 1 or at time steps $4l + 1$ since the word-length is 4 and the computation is periodic with period 4. The diagonal cut set shown in Fig. 9.4 will need to be placed exactly at same location in Fig. 9.3 during the pre-processing step for folding to ensure that all folded arc delays are non-negative. The folded bit-serial architecture is shown in Fig. 3.8 which is redrawn in Fig. 9.5. We illustrate folding of one arc for the folding of the result signal which is accumulated in a diagonally down manner.

$$D_F((A|1) \rightarrow (B|1)) = 4(0) - 0 + 1 - 1 = 0 \ .$$

The reader can verify the folding of other arcs in the bit-parallel architecture. Note that the pipelining parameter of the adder has been assumed to be 0 in this example. The control signals T_0, T_1, and T_2 placed on top of the multiplexors in Fig. 9.5 are high at time instances $4l$, $4l+1$, and $4l+2$, respectively, and the top input signal is selected when the control signal is low and the bottom input

170   CHAPTER 9. THE FOLDING TRANSFORMATION

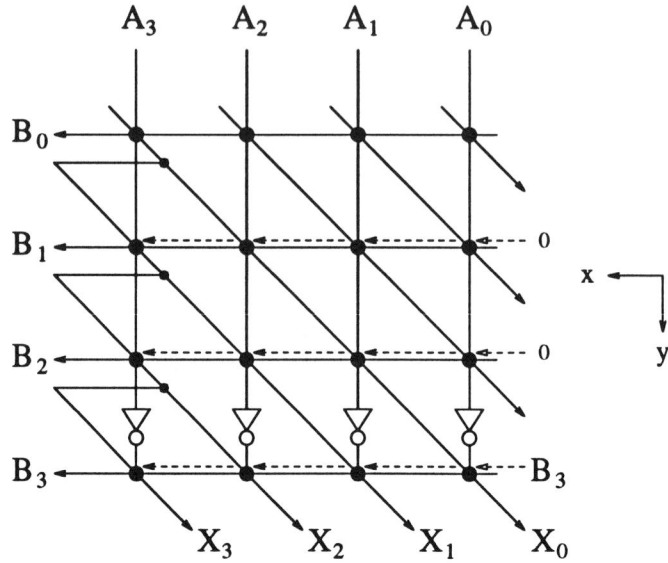

Fig. 9.3: Multiplication dependence graph with carry ripple operation.

signal is selected when the control signal is high. Note that the carry clearing circuit is not shown in the figure for clarity.

Example 9.3. Consider the folding set description shown in Fig. 9.6 for the folding of the bit-parallel ripple-carry multiplier dependence graph shown in Fig. 9.3 to a digit-serial architecture with digit-size 2. In Fig. 9.6, the folding sets are described for the nodes whose computations are omitted from this figure for clarity. Since the word-length is 4 and the digit-size is 2, each word can be processed in two clock cycles. The notation $\{A1|1\}$ at a node implies that this operation is executed by processor $A1$ in time partition 1 or at odd time steps. The diagonal cut set shown in Fig. 9.6 will need to be placed exactly at same locations in Fig. 9.3 during the pre-processing step for folding to ensure that all folded arc delays are non-negative. The folded digit-serial architecture is shown in Fig. 9.7. Let us calculate one folded arc delay using the pipelined version of the data-flow graph in Fig. 9.3 by placing latches at diagonal cutsets at various locations as shown in Fig. 9.6.

$$D_F((B_1|1) \to (B_0|0)) = 2(1) - 0 + 0 - 1 = 1 .$$

The above folding corresponds to placement of 1 latch from the carry-out of the B_1 processor to carry-in of B_0 processor The adder pipelining level is assumed to be 0 in this example. The details of control operation of the multiplexors and

9.3. FOLDING OF REGULAR DATA-FLOW GRAPHS

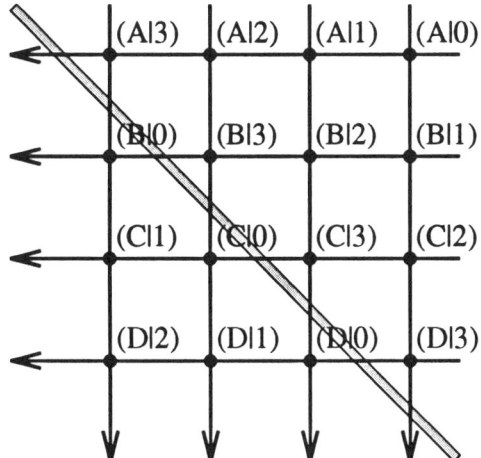

Fig. 9.4: The folding set specification for design of a bit-serial architecture.

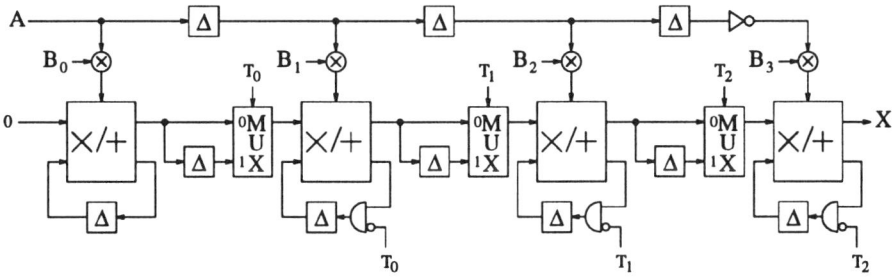

Fig. 9.5: Folded bit-serial two's complement multiplier.

carry clearing multiplexors have been omitted. The reader can verify systematic folding of other arcs to obtain the digit-serial architecture of Fig. 9.7.

9.3 Folding of Regular Data-Flow Graphs

In this section, we discuss folding of regular data-flow graphs to systolic architectures using algebraic techniques [63][64][65][66]. The approach illustrated in this section is general in the sense that some of the edges in the algorithm DFG can contain delays. The corresponding schedule inequalities are derived and made use of.

A data-flow graph is regular, if the existence of an edge $I_1 \to I_1 + e$ with i delays implies the existence of the edge $I_1 + I' \to I_1 + I' + e$ with the same number of i delays for all indices I' permissible within the index set of the

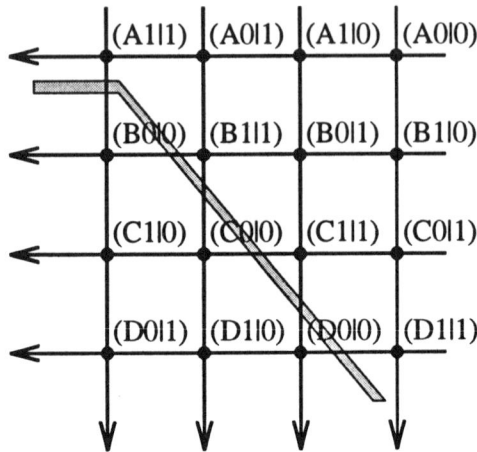

Fig. 9.6: The folding set specification for design of a digit-serial architecture.

data-flow graph. The notation I_1 represents an index location. Fig. 9.8 shows an example of a two-dimensional regular data-flow graph. The coordinates are represented by an index vector $I = [xy]^T$; the x and y axes are shown in the figure. In this data-flow graph, there are three *fundamental* edges: $e_1 = [01]^T$ with 0 delays, $e_2 = [10]^T$ with 0 delays, and $e_3 = [-1\ -1]^T$ with 4 delays. The edges e_1 and e_2 respectively represent the edges in the y and x directions, and the edge e_3 represents the diagonal edge.

The folding operation is described by an *iteration vector* d. In other words, all operations at index locations with index difference d are executed by the same processor. For example, an *iteration vector* $[01]^T$ implies the folding of all operations located in parallel to the y-axis (i.e., all operations with identical x index value but different y index values) into the same processor. The synthesized hardware architecture is also described by a regular data-flow graph with processors located along one dimension. A *processor space*, denoted by P^T, describes the mapping of the task to a processor. A third notation needs to be introduced: the *scheduling vector*, denoted as S^T. The task at index I is executed by the processor located at $P^T I$ at time unit $S^T I$.

Simple algebraic equations can be derived for consistency in choosing the parameters P^T and S^T for given d. First, consider the folding of two tasks, displaced by the iteration vector, to the same processor. The tasks at index locations I and $I + d$ are executed at time units $S^T I$ and $S^T I + S^T d$. Since these two tasks cannot be executed by the same processor simultaneously, the constraint $S^T d \neq 0$ must be satisfied. Furthermore, since a processor executes these two tasks with time interval of $|S^T d|$, the hardware utilization efficiency (HUE) of the architecture is given by $1/|S^T d|$. Since the tasks located at I and

9.3. FOLDING OF REGULAR DATA-FLOW GRAPHS

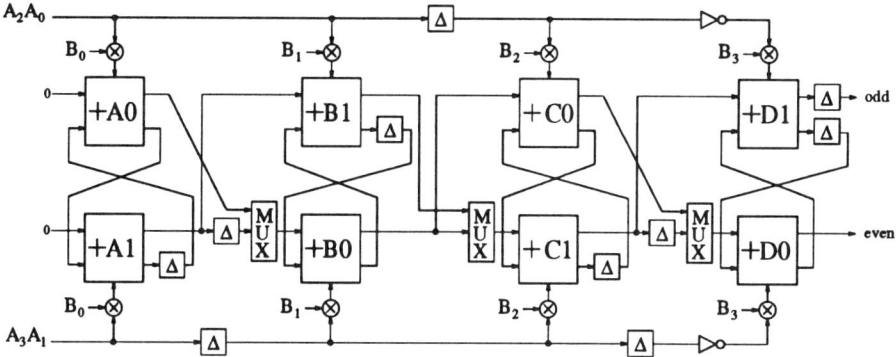

Fig. 9.7: A 2-digit-serial ripple-carry two's complement multiplier.

$\mathbf{I} + \mathbf{d}$ are executed by the same processor, we must satisfy $\mathbf{P}^T\mathbf{I} = \mathbf{P}^T(\mathbf{I}+\mathbf{d})$, or equivalently, $\mathbf{P}^T\mathbf{d} = 0$. In words, the processor space vector is orthogonal to the iteration vector.

Now we can describe the mapping of the edges in the algorithm regular data-flow graph to the hardware regular data-flow graph. An edge $\mathbf{e} = \mathbf{I}_2 - \mathbf{I}_1$ in the algorithm data-flow graph implies an edge from $\mathbf{P}^T\mathbf{I}_2$ to $\mathbf{P}^T\mathbf{I}_1$, i.e., in an edge in the direction $\mathbf{P}^T\mathbf{e}$ in the hardware data-flow graph.

To calculate the number of storage units associated with the edge in the hardware data-flow graph, we calculate the difference in the execution time units of two tasks connected by the edge in the algorithm data-flow graph. For simplicity, assume each task can be executed with one time unit. Consider the edge from $\mathbf{I}$ to $\mathbf{I}+\mathbf{e}$ with i delays. The 1-st task at $\mathbf{I}$ is executed at time instance $\mathbf{S}^T\mathbf{I}$ by the processor $\mathbf{P}^T\mathbf{I}$. The 1-st task at $(\mathbf{I}+\mathbf{e})$ is executed by the processor at $\mathbf{P}^T(\mathbf{I}+\mathbf{e})$ at time unit $\mathbf{S}^T(\mathbf{I}+\mathbf{e})$. Since two executions in any processor are separated by $|\mathbf{S}^T\mathbf{d}|$, the two iterations of any task are separated by $N'|\mathbf{S}^T\mathbf{d}|$, where N' is the number of nodes or tasks in the DFG which are mapped to the same processor. Therefore, the $(i+1)$-st iteration of the task at $(\mathbf{I}+\mathbf{e})$ is executed at time unit $\mathbf{S}^T(\mathbf{I}+\mathbf{e})+iN'\mathbf{S}^T\mathbf{d}$. The result of the 1-st task of $\mathbf{P}^T\mathbf{I}$ needs to be held until the $(i+1)$-st task of $\mathbf{P}^T(\mathbf{I}+\mathbf{e})$. Thus the number of storage units for the edge in the hardware data-flow graph is given by

$$\mathbf{S}^T(\mathbf{I}+\mathbf{e}) + iN'\mathbf{S}^T\mathbf{d} - \mathbf{S}^T\mathbf{I} = \mathbf{S}^T\mathbf{e} + iN'\mathbf{S}^T\mathbf{d} \ .$$

To ensure proper pipelining, the number of storage units must be greater than or equal to 1 (if non-pipelined designs are allowed then this should be greater than or equal to 0). In inequality form, we can write

$$\mathbf{S}^T\mathbf{e} + iN'\mathbf{S}^T\mathbf{d} \geq 1 \ .$$

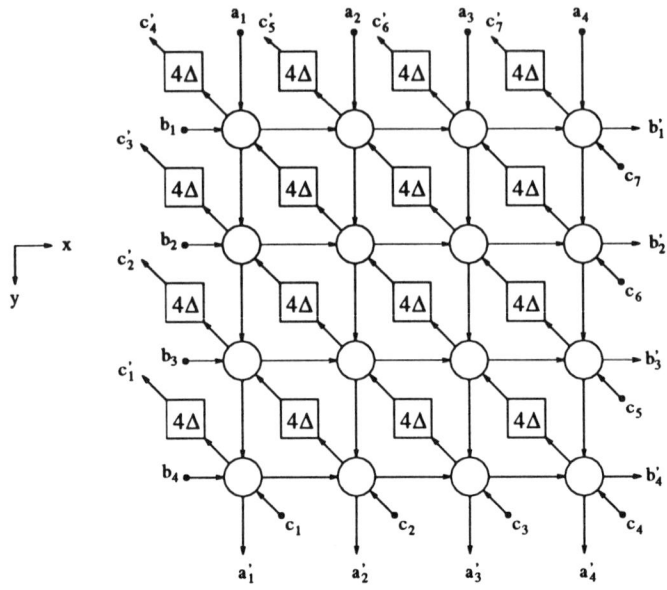

Fig. 9.8: A regular two-dimensional data-flow graph

These inequalities are referred to as *schedule inequalities*. In solving these inequalities, the components of the schedule vector $\mathbf{S}^T$ are chosen to be as small as possible to minimize $\mathbf{S}^T\mathbf{d}$ (which in turn maximize the HUE). This completes the theory of folding of regular data-flow graphs.

Example 9.4. We now illustrate the use of the folding algorithm described above using the DFG in Fig. 9.8. Let $\mathbf{S}^T = [s_1 s_2]$, and let $\mathbf{d}^T = [d_1 d_2]$. The schedule inequalities for the three edges can be expressed by:

$$\text{edge } (10)^T \text{ with 0 delays: } s_1 \geq 1$$

$$\text{edge } (01)^T \text{ with 0 delays: } s_2 \geq 1$$

$$\text{edge } (-1-1)^T \text{ with 4 delays:}$$

$$-s_1 - s_2 + 16(s_1 d_1 + s_2 d_2) \geq 1 \ .$$

We illustrate folding of the DFG in Fig. 9.8 with two different folding sets (or equivalently, for two different iteration vectors). Let $\mathbf{d}^T = [01]$, i.e., all tasks located on a line parallel to the y-axis are mapped to the same processor. The processor space should be orthogonal to the iteration vector space. Choose $\mathbf{P}^T = [10]$. Solving the above inequalities for $d_1 = 0$ and $d_2 = 1$ yields $s_1 = 1$ and $s_2 = 1$. The HUE, $\mathbf{S}^T\mathbf{d}$, is 1. This architecture, therefore, is fully hardware efficient. The edges in the algorithm DFG are then mapped to the

9.4. DIGIT-SERIAL ARCHITECTURES BY UNFOLDING

hardware DFG. For the edge $e_1 = [10]^T$ with $i_1 = 0$ delays, $P^T e_1 = 1$, and $S^T e_1 + i_1 S^T d = 1$. Therefore, in the hardware data-flow graph, we introduce an edge between two consecutive processors in the positive direction with one delay. Similarly for the edge $e_2 = [01]^T$ with $i_2 = 0$ delays, $P^T e_2 = 0$, and $S^T e_2 + i_2 S^T d = 1$. This implies a self loop in each processor with 1 delay. For the edge $[-1 - 1]$, the hardware DFG has an edge in the direction -1 with 14 storage units. The complete synthesized hardware architecture is shown in Fig. 9.9.

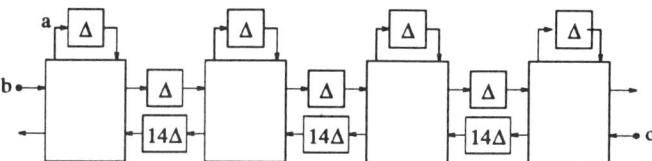

Fig. 9.9: Folded data-flow graph where all operations in a column are mapped to same processor.

Consider another folding set for the algorithm DFG in Fig. 9.8. Let the iteration vector be given by $d = [11]^T$. Choose the processor space as $P^T = [1-1]^T$ to satisfy $P^T d = 0$. The scheduling vector components can be obtained by solving the inequalities for $d_1 = d_2 = 1$, and are given by $s_1 = s_2 = 1$. The quantity $S^T d$ is 2, and this hardware architecture is utilized only half the time (since the HUE = 1/2). The edges are similarly mapped to obtain the final architecture in Fig. 9.10. A null operation is carried out between any two useful operations by this architecture. The reader can verify the calculation of the number of storage units.

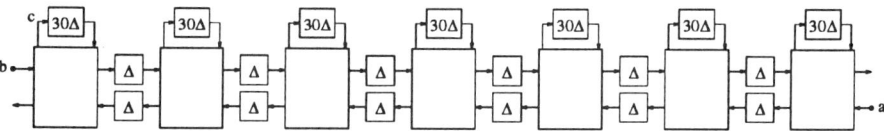

Fig. 9.10: Folded data-flow graph where all operations on a diagonal are mapped to the same processor.

9.4 Digit-Serial Architectures by Unfolding

In this section we illustrate the design of digit-serial architectures by first folding the bit-parallel architecture to bit-serial using the systolic array design methodology used for folding of regular data-flow graphs [67][68] and then by unfolding

[10] the bit-serial architecture with unfolding factor equal to the digit-size. In two's complement arithmetic operations, the sign bit is used in a slightly different way than the other bits and this introduces irregularity at the boundary cells which are different from other internal cells. However, we can perform the folding operation by the regular data-flow graph folding methodology by first treating all the cells to be identical and then by modifying the architecture manually to include the effect of the boundary cells.

Example 9.5. We can fold the two's complement ripple-carry multiplier shown in Fig. 9.3 by projecting the dependence graph such that all operations in a horizontal line are folded to the same processor. This leads to the bit-serial architecture shown in Fig. 9.5. For this folding, we select $\mathbf{d} = [10]$, $\mathbf{P}^T = [01]$ and $\mathbf{S}^T = [11]$. For the edges for B and $carry$, the processor displacement, $\mathbf{P}^T\mathbf{e}$, is 0 and the number of folded delays, $\mathbf{S}^T\mathbf{e}$, is 1. For the A edge, the processor displacement is 1 and the number of folded delays is 1. These numbers are 1 and 0, respectively, for the $result$ edge. The carry clearing operation is not included in this figure. Note that the carry in the last adder should be reset to B_3 whereas the carry is reset to 0 for all other adders.

The digit-serial architecture shown in Fig. 9.7 can be obtained by unfolding the bit-serial architecture shown in Fig. 9.5 by an unfolding factor of 2.

Example 9.6. We can also obtain another bit-serial architecture by folding the ripple-carry multiplier shown in Fig. 9.3 with a different projection direction where all operations on a column (or on lines parallel to y axis) are folded to the same processor. For this folding, select $\mathbf{d} = [01]$, $\mathbf{P}^T = [10]$ and $\mathbf{S}^T = [01]$. The processor displacements can be calculated to be 1, 1, 0 and -1 for $carry$, B, A and $result$ signals, respectively. The number of folded delays can be calculated to be 0, 0, 1 and 1 for the same edges, respectively. The folded architecture is shown in Fig. 9.11. In this figure, several control details have been omitted.

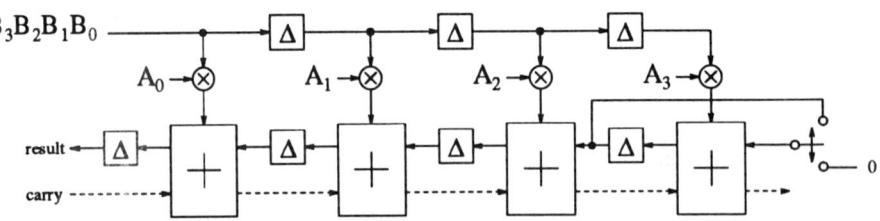

Fig. 9.11: Another folded bit-serial two's complement multiplier.

Note that the bits $\{A_i\}$ need to be inverted once every word-length number of cycles. In addition, the $carry$ input signal needs to be reset to 0 every 3 out of 4 cycles and to B_3 every 1 out of 4 cycles.

9.4. DIGIT-SERIAL ARCHITECTURES BY UNFOLDING

Example 9.7. We illustrate folding of a bit-parallel carry-save multiplier to a bit-serial one. Unlike carry-ripple multiplier where the carry output of an adder is rippled to the adder to the left in the same row, the carry output in a carry-save adder is saved and used in an adder in the next row. The tabular representation of a carry-save multiplication operation is shown in Fig. 9.12. The carry output of an adder is shown in dashed lines whereas the sum output is shown in solid lines. Unlike the carry-ripple multiplier, the carry-save multiplier

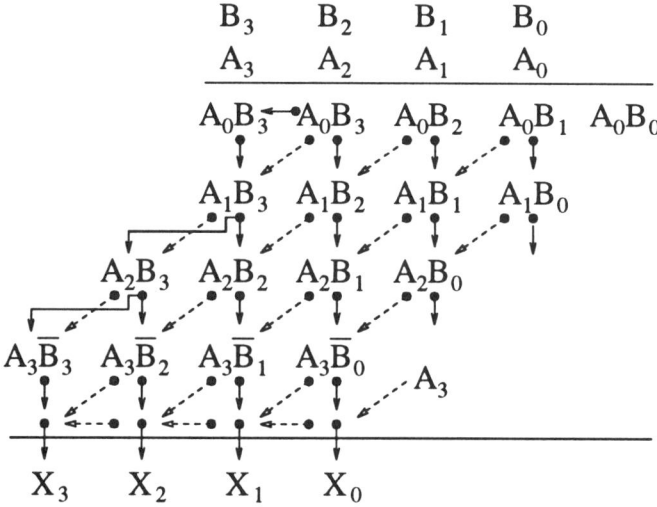

Fig. 9.12: Tabular representation of two's complement carry-save multiplication operation.

requires an extra row of add operation. Fig. 9.13 shows the dependence graph of the carry-save multiplication operation. The systematic regular data-flow graph methodology can be used to fold this dependence graph to bit-serial. Consider the folding operation where all operations in a column (i.e., in a line parallel to y axis) are mapped to the same processor. For this folding, $\mathbf{d} = [10]$. Select $\mathbf{P}^T = [01]$ and $\mathbf{S}^T = [11]$. The processor displacements for the edges A, B, $carry$, and sum are 0, 1, 1, and 1, respectively. The number of folded delays for the same edges can be calculated to be 1, 1, 1, and 0, respectively. The folded bit-serial architecture is shown in Fig. 9.14. Note the use of an extra adder in this architecture. The multiplexor control signals have been omitted for clarity.

In all folded bit-serial multipliers designed so far, the bit-serial signal is latched when moving from one adder cell to the next. These multipliers, referred to as *data-systolic*, can require many number of latches in a pipelined design. This is because the sum output is usually not latched and latches must be

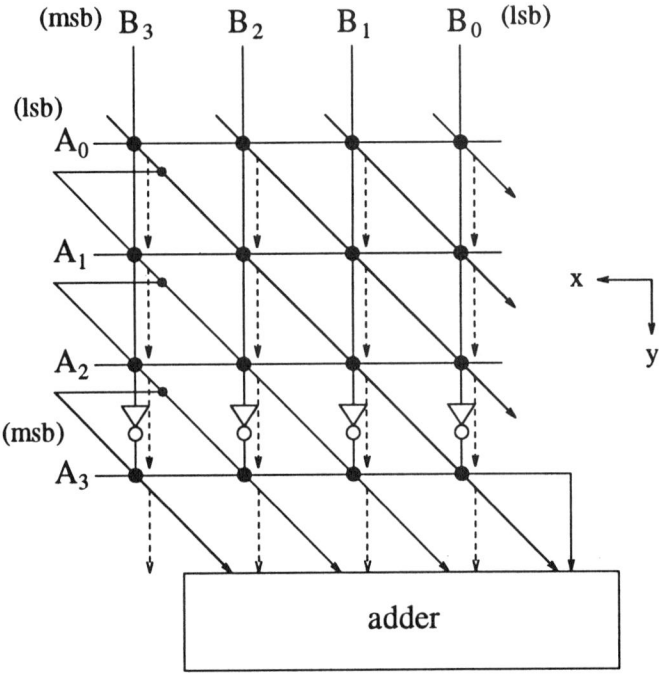

Fig. 9.13: Multiplication dependence graph with carry-save operation.

9.4. DIGIT-SERIAL ARCHITECTURES BY UNFOLDING

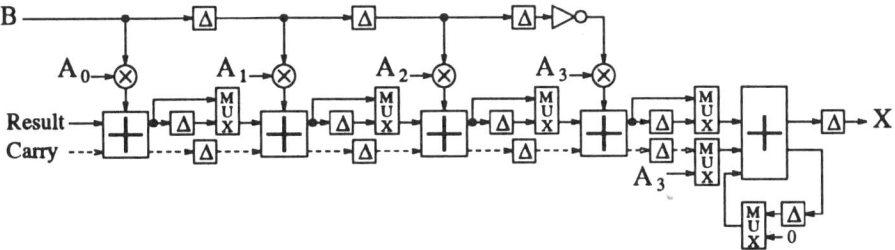

Fig. 9.14: Bit-serial carry-save multiplier.

placed at appropriate cutsets to pipeline the bit-serial adder at one-bit level. At a higher level, the bit-serial multiplier resembles the finite impulse response (FIR) filter shown in Fig. 9.15(a) where the input signal is latched from one tap to the next and the add operations are not pipelined. However, the same FIR filter can be implemented in a data-broadcast form using the structure shown in Fig. 9.15(b) where the data is broadcast to all tap multipliers simultaneously. In addition, the add operations are now automatically pipelined and no additional pipelining latches are needed to pipeline the circuit at multiply-add level. If the structure in Fig. 9.15(a) is to be pipelined at multiply-add level, then four additional latches must be placed at two feed-forward cutsets at locations shown in dashed lines. The data-broadcast FIR filter can be obtained by transposing the data-systolic multiplier [29]. In the *transpose* operation, the circuit is first represented as a signal flow graph and the direction of all edges are reversed and the input and output signals are interchanged. The reader can verify how the circuit in Fig. 9.15(b) can be obtained by transposing the circuit in Fig. 9.15(a). Note that the order of the coefficient multiplication operations are reversed in these two circuits during the transpose operation. This results in a low-latency multiplier. In the example below, we illustrate how a data-broadcast bit-serial multiplier can be designed.

Example 9.8. Consider the two's complement multiplication operation shown in a tabular manner in Fig. 9.16 using the Baugh-Wooley multiplication algorithm [69]. The dependence graph for this operation is shown in Fig. 9.17. To design a bit-serial multiplier by folding using this dependence graph, consider the folding parameters $\mathbf{d} = [10]$, $\mathbf{S}^T = [10]$ and $\mathbf{P}^T = [01]$. In this folding, all operations in a horizontal line are folded to the same processor. The processor displacements for the edges A, B, *carry*, and *result* are 0, 1, 0, and 1, respectively. The number of folded delays for these edges can be calculated to be 1, 0, 1, and 1, respectively. The folded bit-serial multiplier is shown in Fig. 9.18. The blocks marked "C" in this figure perform inversion operations at some clock cycles to generate the appropriate partial product and these control details have been omitted for simplicity. The switches can be implemented

180 CHAPTER 9. THE FOLDING TRANSFORMATION

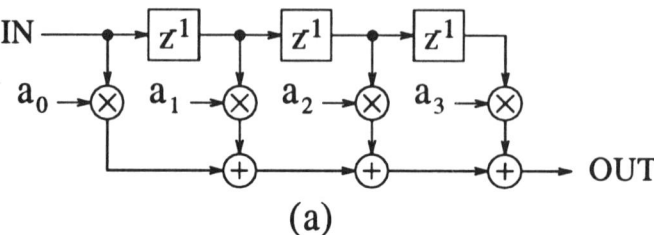

(a)

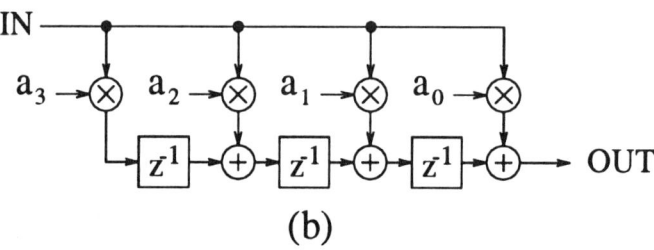

(b)

Fig. 9.15: Data-systolic and data-broadcast FIR digital filter structures.

as multiplexors. This multiplier requires 4 latches for each bit-serial stage as opposed to 5 required in Fig. 9.5.

9.4. DIGIT-SERIAL ARCHITECTURES BY UNFOLDING

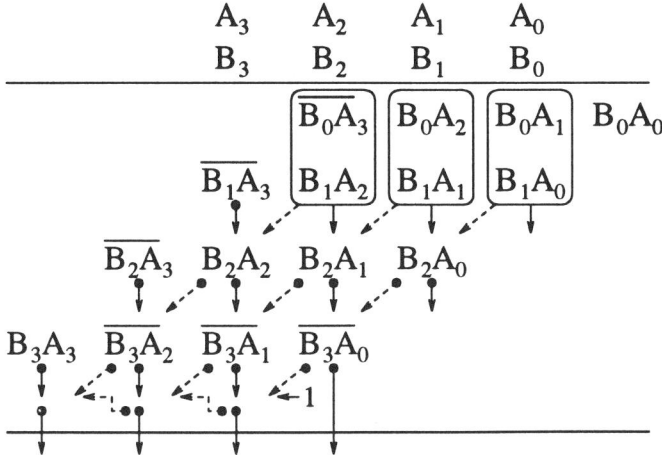

Fig. 9.16: Tabular representation of Baugh-Wooley multiplication.

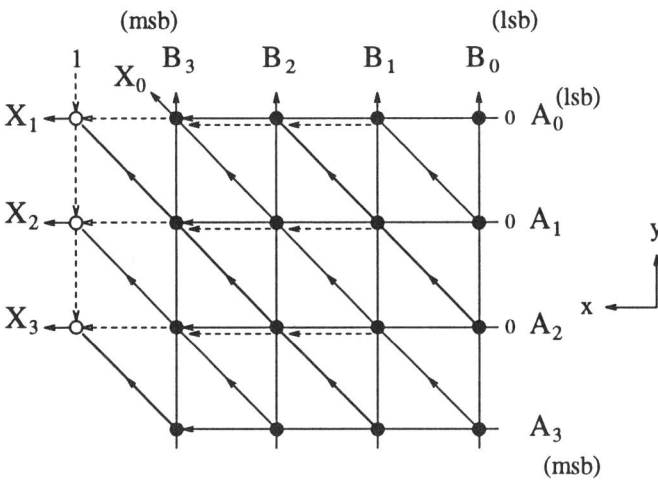

Fig. 9.17: Dependence graph of the Baugh-Wooley multiplication operation.

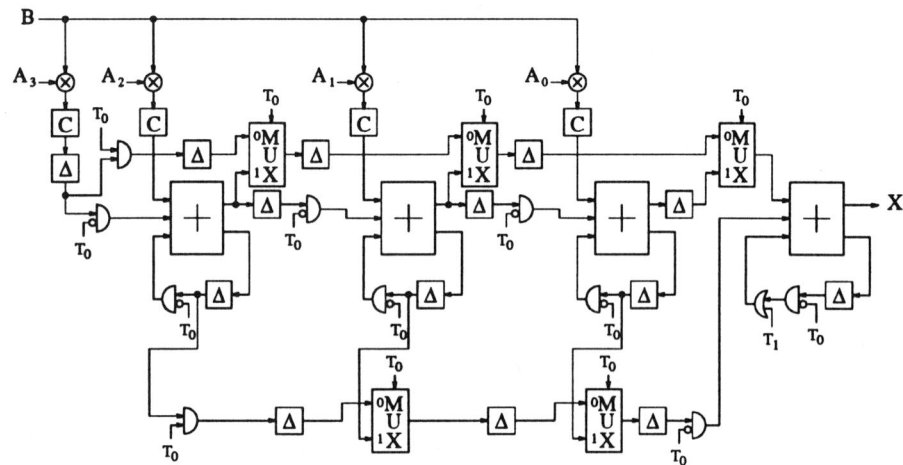

Fig. 9.18: Bit-serial data-broadcast multiplier.

Chapter 10

Wavelet Transform Architectures

10.1 Introduction

This chapter presents the design of discrete wavelet transform (DWT) architectures as an application of the digit-serial methodology. Wavelets, based on time-scale representations, provide an alternative to time-frequency representation based signal processing [70]. The shifting (or translation) and scaling (or dilation) are unique to wavelets. Wavelets are then represented by dilation equations, as opposed to difference or differential equations. Wavelets maintain orthogonality with respect to their dilations and translations, but loose the eigenfunction property of the differentiation operator. Orthogonality of wavelets with respect to dilations leads to multigrid representation. Wavelets decompose the signal at one level of approximation into approximation and detail signals at the next level. Thus, subsequent levels can add more detail to the information content. The perfect reconstruction property of the analysis and synthesis wavelets and the absence of perceptual degradation at the block boundaries favor use of wavelets in video coding applications [71].

The nature of wavelet computation makes it an ideal application for the digit-serial methodology [72]. With increase in the number of levels in a wavelet algorithm, the amount of detail (or resolution) increases. The computations in a m-level wavelet repeat with a period 2^m. In other words, the computations can be assumed to be frame-periodic where each frame consists of 2^m samples. The word-size to be used in a m-level wavelet is assumed to be a multiple of 2^m. Thus, for 3 or 4 wavelet resolution levels, the word-length should be multiples of 8 or 16 bits, respectively. In the digit-serial wavelet architecture, each level of the wavelet is implemented using a different digit-size processor. The digit-serial architecture requires a single clock, achieves complete hardware utilization and requires simpler routing. The use of different digit-sizes at different levels of the digit-serial architecture is the key to achieving simpler routing and complete hardware utilization.

10.2 The Wavelet Computation

Typically the transmitter end performs a wavelet analysis operation and the receiver end performs a wavelet synthesis operation. Consider the three-level wavelet analysis computation shown in Fig. 10.1. The synthesis wavelet is discussed in section 10.4.

The number of resolution levels in Fig. 10.1 is denoted as m and is three in this example. The computation of the 3-level wavelet is periodic with period 8, i.e., identical sets of computations are separated by a time index of 8. In general, the computations are periodic with period $M = 2^m$ for a m-level wavelet. In the wavelet of Figure 10.1, G is a high-pass filter and H is a low-pass filter. For simplicity, we assume these filters to be non-recursive FIR digital filters. For a set of M input samples, the G and H wavelet filters at level i (where i is numbered 1 through m) output $M/2^i$ samples. For these M input samples, the system outputs the samples generated by the G filter of levels 1 through m and the sample generated by the H filter of level m leading to a total of M output samples.

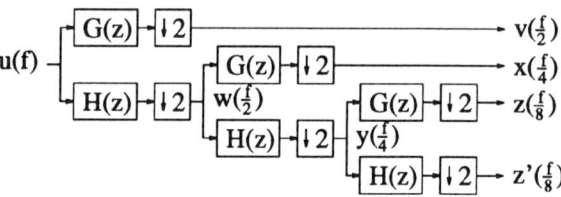

Fig. 10.1: A 3-level analysis wavelet structure.

Since each filtered output is decimated by a factor of 2, we need to only compute the signal samples which are not thrown away. Thus as we move from one resolution level to the next, the frequency of the computed samples decreases. Let us calculate the total number of filters needed in a m-level wavelet. For the first stage, each filter computes $M/2$ samples. So the total number of samples generated at the G and H filters of level-1 wavelet is M. Similarly, each filter in the second-level wavelet computes $M/4$ samples and the total number of samples computed at level-2 is $M/2$. In an m-level wavelet, the total number of samples computed at all levels is

$$M + M/2 + M/4 + \ldots + 2 = 2(M - 1) \ .$$

10.3. THE ANALYSIS WAVELET ARCHITECTURE

Since the wavelet computation is periodic with M samples, the number of samples computed every sample period is $2(M-1)/M$ or $2(1 - 1/M)$, which is upper bounded by 2. This implies that the maximum number of filters needed in a one-dimensional wavelet is 2. In other words, one low-pass and one high-pass filter will always be adequate for computation of one-dimensional wavelet transform. These two filters will be under-utilized $1/M$-th of the time.

The digit-serial architecture can overcome the under-utilization problem associated with bit-parallel implementation style. For the three-level wavelet, the four output signal wires, v, x, z, and z', in the analysis wavelet algorithm dataflow graph generate 4, 2, 1, and 1 output samples. With a period of 8 cycles, it seems natural to use 1/2-word, 1/4-word, 1/8-th word, and 1/8-th word digit-sizes in the processing of these four signal wires. This processing would always require eight clock cycles for transmission of the respective number of outputs. This type of processing can reduce the routing and interconnection cost (since this architecture only involves local interconnection) and can achieve complete hardware utilization. The drawback of this architecture is that the word-length must be a multiple of 8 or 16 for 3-level or 4-level wavelet cases, respectively. Another drawback is the increase in the system latency of the architecture introduced due to the various converters required to change the output format of a processor from one digit-size to the input format required by the next processor for another digit-size. Since the different wavelet levels are realized with different digit-sizes and a single clock is used in the entire architecture, this architecture can be considered a single-clock, non-uniform style architecture.

10.3 The Analysis Wavelet Architecture

The i-th level wavelet is implemented using digit-size $W/2^i$ where W is the word-length and $i = 1, 2, 3$ for a 3-level wavelet. With a single clock, conversion of a word-parallel data format to a half-word data format at first seems impossible. This is because the data format converter must maintain identical data rates at the input and output [73], where the *data rate* corresponds to the number of units of data per input/output clock cycle period. With a constant clock cycle period at input and output, the number of units of data at both input and output of the converter must remain the same. But the input processes one word every cycle and the output generates a half-word every cycle; therefore, the data-rates at the input and the output of the converter are not the same. This poses a difficulty in designing the digit-serial architecture.

This problem can be solved by recognizing that half the computations are not needed (due to the decimation operations) and that the computation can be reformulated as follows:

$$g_0 u(2k) + g_1 u(2k-1) + g_2 u(2k-2) + g_3 u(2k-3)$$
$$= [g_0 u(2k) + g_2 u(2k-2)] + [g_1 u(2k-1) + g_3 u(2k-3)] \ .$$

The converter can now convert one-word-serial bit-parallel format to two-word-parallel half-word-serial format, i.e., we generate two words in one cycle and only half-word of each word is generated in each cycle. This approach makes the digit-serial computation feasible by maintaining constant data rate at the input and output of the converter. The converter can process the serial signal $u(n)$ and generate two half-words of $u(2k)$ and $u(2k-1)$ in parallel. In this computation, $u(2k-2)$ can be obtained from $u(2k)$ with one word delay (or two delays at half-word level) and $u(2k-3)$ can be obtained from $u(2k-1)$ in a similar manner. The output computation can now be broken into even and odd components with each component containing half of the filter coefficients. The result of the even and odd components can be added to obtain the desired output. The high-level block diagram of the digit-serial three-level wavelet architecture is shown in Fig. 10.2(a). A possible floor-plan is shown in Fig. 10.2(b). Note that the local interconnections associated with this floor-plan lead to a compact layout. The digit-serial wavelet architecture closely resembles the algorithmic structure; this is somehow quite appealing!

The basic modules in the digit-serial architecture of Fig. 10.2 are the converter modules and the digit-serial filtering blocks. The converters in Fig. 10.2 can be designed using systematic *life-time analysis* and *register allocation* schemes [73] [74]. Life-time analysis is used to count the number of variables which are *live* at any clock cycle. A variable which is no longer needed can be considered *dead* and need not be stored any further in registers. The maximum number of live variables in any clock cycle represents the minimum number of registers needed to design the converter circuit. The live variables need to be allocated to the registers using a *register allocation* scheme such as "forward-backward" [73] or "two-dimensional" allocation [75]. In formulating a register allocation scheme, it is important to guarantee completion of allocation of all live variables to available registers such that the inter-frame pipelining rate can be sustained. Since the computation is periodic, the register allocation scheme must also be periodic.

Consider the design of a converter $(1, W/2) \rightarrow (2, W/4)$ in which the input bits to the converter belong to 1 word and contain $W/2$ bits of that word and the output bits belong to 2 different words and contain $W/4$ bits of each word. Thus the input format corresponds to word-serial digit-serial (with digit-size $W/2$) and the output format corresponds to word-parallel digit-serial (with digit-size $W/4$). The design of this converter is illustrated using life-time analysis and a forward-backward register allocation scheme [73].

10.3.1 Life-Time Analysis

Life-time analysis can be carried out either by analytical or by graphical approach. The design of the converter $(1, W/2) \rightarrow (2, W/4)$ is now presented using both approaches. Note that this converter reformats the signal w (see Fig 10.1)

10.3. THE ANALYSIS WAVELET ARCHITECTURE

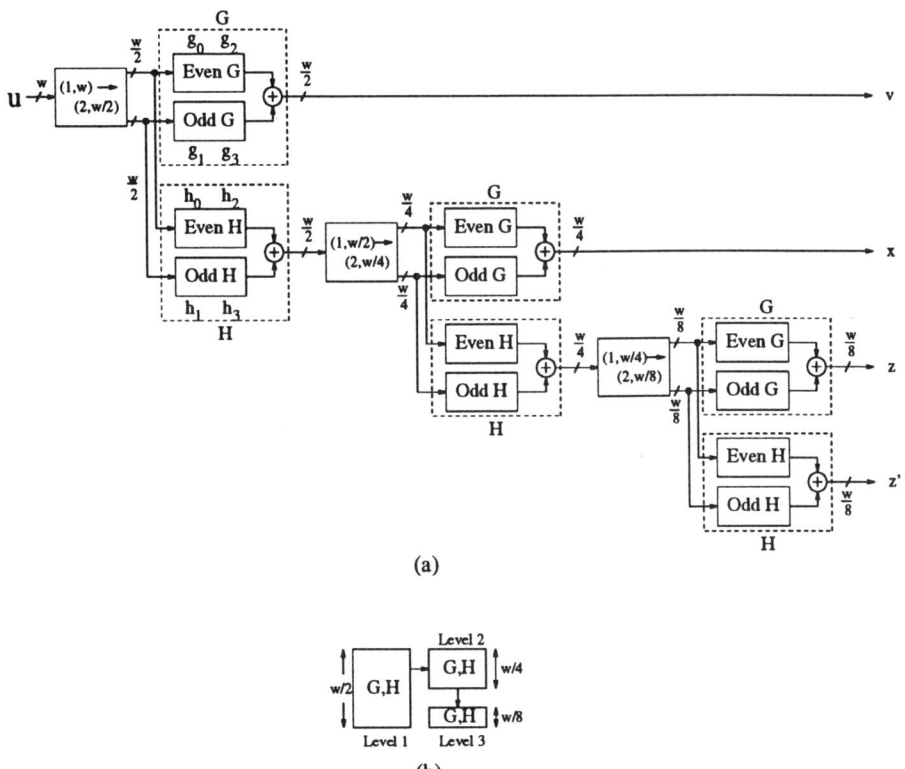

(a)

(b)

Fig. 10.2: Digit-serial architecture for a 3-level analysis wavelet.

188 CHAPTER 10. WAVELET TRANSFORM ARCHITECTURES

for even samples only. The computation of odd w samples is unnecessary due to the decimator.

Analytical method: This converter processes $W/2$ bits of one word in one clock cycle. Thus, one complete word is processed in two consecutive cycles. Consider the bit q of word p. This bit is input to the converter in time unit $2p + \lfloor q/(W/2) \rfloor$. Assuming the data format conversion to be carried out with no latency, this bit would be output in time unit $\lfloor (p/2) \rfloor 4 + \lfloor q/(W/4) \rfloor$. Let the difference be denoted as T_{diff} and is given by:

$$T_{diff} = \lfloor (p/2) \rfloor 4 + \lfloor q/(W/4) \rfloor - 2p - \lfloor q/(W/2) \rfloor .$$

The negative of the maximum negative T_{diff} is referred to as the *latency* of the converter. Note that while p can take any non-negative value, q must lie between 0 and $(W-1)$. It is easy to verify that the maximum negative value of T_{diff} is -2 and its negative is 2. This implies that the converter latency is 2 time units. Ideally, in the $(2, W/4)$ output format, $w_0(0)$, $w_1(0)$, $w_0(2)$ and $w_1(2)$ would need to be output in clock cycle 0 where the notation $w_i(j)$ represents the bit i of word $w(j)$ and bit 0 is the least significant bit. But the bits $w_0(2)$ and $w_1(2)$ are input in clock cycle 2. Since the converter can output any bit only after it has been input (to satisfy causality), the converter must introduce a latency of 2 clock cycles (or time units). A life-time chart of a single frame for the converter is shown in Fig 10.3(a). The life-time chart is periodic with period 4 time units and two words can be input and output in 4 clock cycles. Note that the life-time of consecutive frames can overlap and this makes the register allocation process challenging.

After the latency is determined, we consider any time unit greater than the latency. Consider the time unit $(2 + l)$ where l is a non-negative integer. The number of bits input to the converter by this time unit is $4(2 + l)$ and the number of bits output by this time unit is $4l$. The number of live bits at time unit $(2 + l)$ is 8 which is independent of l. Thus, the maximum number of live bits over all possible values of l is 8. Therefore, this converter requires a minimum 8 registers.

Graphical approach: The minimum number of registers needed for the converter can also be obtained by graphical approach using the life-time chart. Fig. 10.3(a) shows the *linear* life-time chart of the converter. In this case, the period of computation of the converter is 4 and the converter life-time chart spans 6 time units. Thus, a maximum of $\lceil 6/4 \rceil$ or 2 frames of computations overlap at any time. The maximum number of live variables for time units 0 to 3 are seen to be 0, 4, 8 and 8, respectively. The number of live variables for time unit 4 due to current frame must be added to the number of live variables at time unit 4 due to the next frame to obtain the total number of live variables at time unit 4. Due to periodic computation, the number of live variables at

10.3. THE ANALYSIS WAVELET ARCHITECTURE

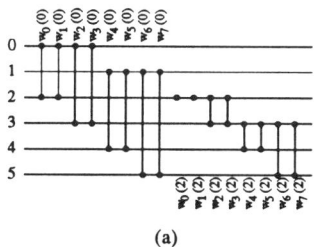

(a)

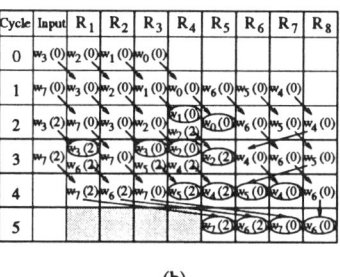

(b)

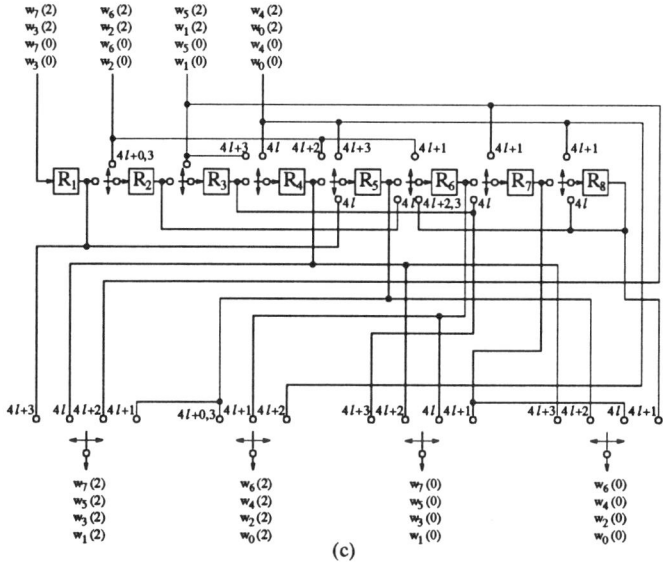

(c)

Fig. 10.3: Design of a $[1,W/2] \rightarrow [2,W/4]$ converter.

time unit 4 due to next frame is same as the number of live variables at time unit 0 due to the current frame. Thus the total number of live variables at time unit 4 is the sum of the live variables at time units 4 and 0 and is $8 + 0 = 8$. Similarly, the number of live variables at time unit 5 is the sum of the number of live variables at time units 5 and 1 and is $4+4 = 8$. The maximum number of live variables at any time unit after taking frame overlap into account is 8. We can also reach the same conclusion by considering a *circular life time chart* as shown in Fig. 10.4. In the circular life-time chart, the circle is divided into period number of time partitions. In this case, the circular life-time chart contains 4 time partitions. Note that time partition i (where i lies between 0 and $P-1$ for period P) represents all time units $(i+lP)$ where l is a non-negative integer. The linear life-time chart is easily mapped to the circular life-time chart. If the life duration of a variable is greater than the period, its representation in the circular life-time chart exceeds one full circle. The number of live variables at each time partition is counted and the maximum number of live variables at any time partition is the minimum number of registers needed to design the converter. In Fig 10.4, the number of live variables at all time partitions is 8 and the number of registers needed to design this converter is 8.

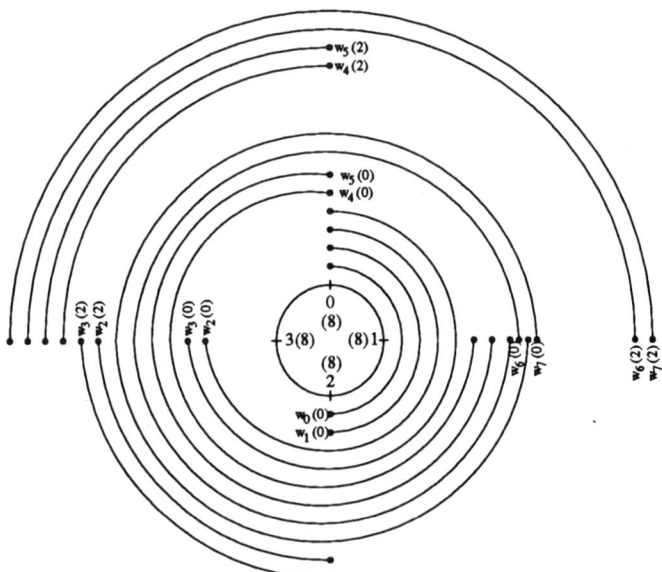

Fig. 10.4: Circular life-time chart of the variables used in the digit-serial converter. Note that the variable names are indicated at the tail end of the life duration.

It is now easy to conclude that the minimum number of registers needed to design a $(1, W/2) \rightarrow (2, W/4)$ converter is W. For a formal proof of this, the reader is referred to [73].

10.3.2 Register Allocation

The live variables can be allocated to the minimum number of registers using many possible register allocation schemes. The register allocation should be periodic in nature, i.e., a variable of the current frame and the corresponding variable of the next frame must be allocated to the same register and be displaced in time by the period of the computation. The allocation scheme must be designed such that inter-frame pipelining rate can be sustained. Furthermore, the allocation scheme must run to completion, i.e., all live variables should be allocated to the minimum number of registers.

While several feasible allocation schemes can be designed, we use the *forward-backward* allocation scheme [73]. to design the $(1, W/2) \rightarrow (2, W/4)$ converter. In this scheme, we allocate the variables all the way forward, and then allocate it to an appropriate backward register (thus the name "forward-backward"). The forward-backward register allocation scheme proceeds as follows:

1. First we perform a life time analysis, and calculate the minimum number of registers.

2. All variables with life time less than or equal to the number of registers are allocated in a forward manner until these are either dead or these reach the last register. In forward allocation, if the register i holds the variable in the current cycle, then the register $(i+1)$ holds the same variable in the next cycle. If multiple variables are input in a given time unit, these are allocated to multiple registers such that the variable with the longest life time is allocated to the initial or first available register and other variables are allocated to consecutive or next available registers in decreasing order of life time.

3. Since the allocation is periodic, the allocation of the current frame also repeats itself in the subsequent frame. Thus if R_j is occupied with a variable in time unit l, then the R_j would occupy the corresponding variable of next frame in time unit $(l + P)$ where P denotes the periodicity of the allocation. Thus R_j will not be available for a backward allocation at time unit $(l + P)$ for any variable in the current frame. To prevent any illegal backward allocation, we hash the variable entry position for R_j at time unit $(l + P)$ for all j and l.

4. The *remaining life period* is calculated, and these variables are allocated to a register in a backward manner in a first-come first-served manner. To perform the backward allocation of a variable, we first obtain all backward registers which are available in the next cycle. If one register is available

for backward allocation, then the variable is allocated to this register in the next cycle. If multiple registers are available for backward allocation, then the allocation is performed in a manner which minimizes (i) the number of feedback lines (from the last register to the backward registers), and (ii) which has the least *sufficient* number of available forward registers from the backward register to the last register. The allocation is done such that the condition (i) is given higher priority. If multiple backward registers satisfy (i), then ties are broken using the condition (ii). If no backward register satisfies (i), then allocation is carried out using (ii).

5. If the allocation is not yet complete, then the hashing to prevent illegal allocation and backward-forward allocation is continued by repeating steps 3 and 4 until the allocation is complete.

In each step of the forward-backward allocation, we maintain a list of interconnects from last register to backward registers. Condition (i) attempts to minimize the number of interconnects (which represent feedback paths). This minimizes the control circuits associated with the switch implementation at the input of the backward register. The least sufficient number of available forward registers as used in condition (ii) requires more explanation. For example, if a variable needs to be stored for 5 more units of time, then we consider all the registers which are at least five units away from the last register and from these we choose the register closest to the last register. The hashing requirement in step 3 prevents any illegal allocation. While the allocation in most converters would run to completion at the end of step 4, some converters need to execute step 5 for completion. The forward-backward register allocation and the synthesized converter architecture for the $(1, W/2) \rightarrow (2, W/4)$ converter are shown in Fig. 10.3(b) and Fig. 10.3(c), respectively. Although the synthesized converter looks complex, one should note that each wire in the converter is a one-bit wire, as opposed to a bus of width same as word-length as required in a uniform style bit-parallel architecture. The samples $w(0)$, $w(2)$, ..., etc. are input to the converter. The converter was synthesized using only two samples, and the schedule repeats periodically with period 4 cycles.

10.3.3 Digit-Serial Building Block

The implementation of one digit-serial building block is now illustrated. Consider the computation of $x(0)$ and $x(4)$ in the G filter using w's. For simplicity, assume the coefficients of the G filter to be $[1/2, 1/2, 1/2, -1/2]$. The even coefficients are $[1/2, 1/2]$ and the odd coefficients are $[1/2, -1/2]$. Fig. 10.5 shows the $W/4 = 2$-digit-serial implementation for computation of x for these sets of coefficients. Note that, for $W/4$-digit-serial structures, one word-level delay is equivalent to 4 digit-level delays. In this stage of the multi-rate wavelet, each word-level delay represents two sample delays (because of parallel processing

10.4. THE SYNTHESIS WAVELET ARCHITECTURE

data format) of w. But w is already decimated by a factor of 2, i.e., $w(n)$'s for odd n have not been computed. Thus 4 latches in this realization represent 4 delays at the sample delay of input rate. This digit-serial structure can also be first realized as a bit-serial structure, and then can be unfolded by 2 to obtain the digit-serial structure.

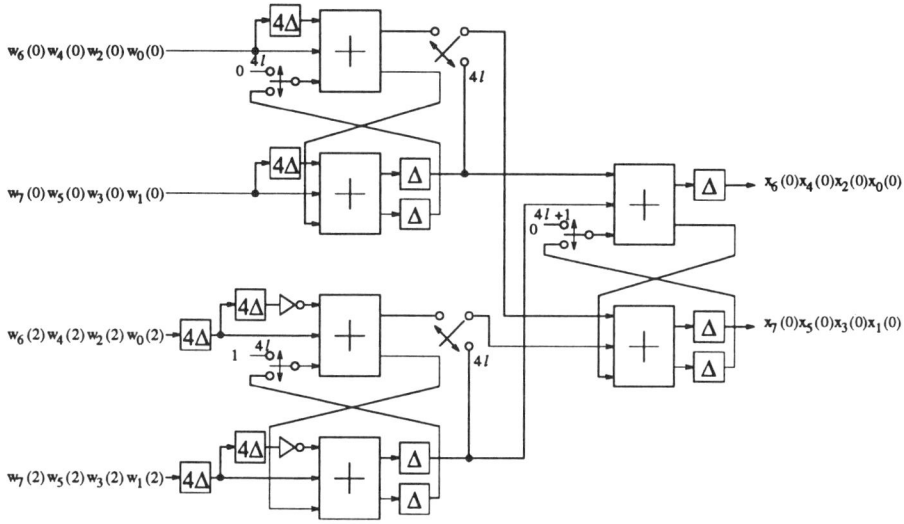

Fig. 10.5: A digit-serial module realization.

The digit-serial wavelet architecture achieves complete hardware utilization. While bit-parallel architectures requires 2 word-level filters, the digit-serial architecture exactly requires $(2 - 2/M)$ word-level filters. The local interconnection and simpler multiplexed control circuit make this architecture very attractive for VLSI implementation. One drawback of this architecture is the constraint on word-length (since, for m-level wavelet, the word-length should be multiple of $2^m = M$). However, for practical values of $m = 3$ or 4, the word-length of 8 or 16 is quite practical. It is concluded that matching different digit-sizes to different levels of wavelet in an implementation results in a naturally efficient architecture.

10.4 The Synthesis Wavelet Architecture

The synthesis wavelet reconstructs the original signal as shown in Fig. 10.6. The digit-serial synthesis wavelet architecture can be realized in the same way using different digit-sizes at different levels. The converters in this case perform the reverse operation as compared with the analysis wavelet architecture. The block diagram of the digit-serial synthesis wavelet architecture is shown in Fig 10.7.

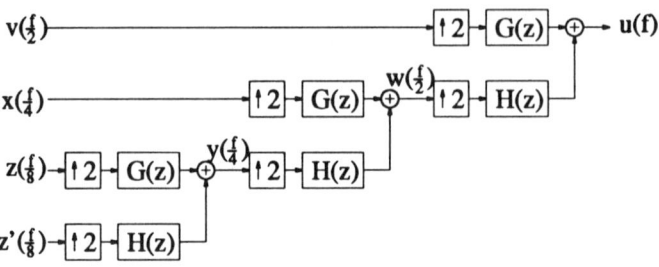

Fig. 10.6: A 3-level synthesis wavelet structure.

Note that the even parts of G and H filters are added together to realize an even filter. Similarly the odd parts of G and H filters are added together to realize the odd filter. The modules in this architecture are similar to the analysis filter and are not discussed further.

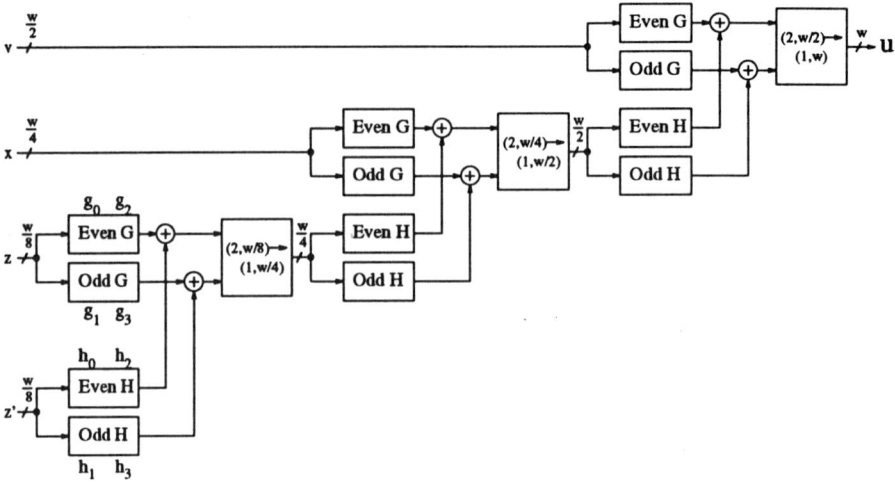

Fig. 10.7: Digit-serial architecture for the 3-level synthesis wavelet.

Chapter 11

Digit-Serial Systolic Arrays

In this chapter we describe a method of transforming systolic architectures using bit-parallel arithmetic into arrays using digit-serial arithmetic. This chapter is derived largely from a paper written by Peter Corbett in collaboration with one of the authors of this book ([76]). As shown in chapter 7, digit-serial computation is an area-time efficient method of doing high-speed arithmetic calculations, having the advantage through appropriate choice of digit-size and word-length of allowing throughput capacity to be matched to design needs. For a certain class of systolic arrays, however, digit-serial arithmetic allows a further very significant benefit, by transforming arrays in which processors are under-utilized into arrays with 100% processor utilization. As an example, converting a well-known band-matrix multiplication array to digit-serial is calculated to give an improvement of more than 3 times in area-time efficiency.

11.1 Introduction

Systolic arrays have achieved widespread use in a number of signal and image processing applications. An important reason for this success is that signal processing algorithms which perform repeated similar operations on large amounts of data can often be partitioned so that many operations are performed in parallel, and these parallel operations are repeated a number of times. Keys to systolic array design are that all communications are local (i.e., the length of communication paths is constant, independent of the array size), and that the system is fully pipelined, or systolized. What this means is that all combinational paths are localized within processors, and all data transfer between processors is done through delay registers. Digit-serial arithmetic would seem to be particularly suited to use in systolic arrays because of the natural use of pipelining that is an intrinsic part of digit-serial computation.

In translating a systolic array using parallel arithmetic to one implemented using digit-serial arithmetic, one must be careful that the functionality of the array is unchanged. The different timing characteristics of digit-serial operations as compared with bit-parallel operations means that the timing of the array may be changed, affecting functionality. This chapter gives a simple method for translating an array to digit-serial with a chosen digit-size. As long as a simple criterion is satisfied identical functionality between the original bit-parallel array and the digit-serial version is guaranteed.

Despite their high efficiency of computation, systolic arrays have some potential disadvantages. One of these is low processor utilization. A frequent occurrence in systolic designs is that processors cannot be provided data on every clock cycle. Data arrives on only one of every a cycles where a is some integer greater than one. Therefore, the utilization of the processor is reduced to $1/a$. However, the processor itself must still meet all the performance requirements of a fully utilized processor since, whenever it is required to process data, it must do so in the same time as a fully utilized processor would. In such arrays, it is impossible to maintain a data rate of one sample per systolic cycle in the data pipelines and it is necessary to separate valid data by dummy or zero-valued data in order to meet the systolic requirements.

The low utilization of processors arises because of cycles in the data flow graph of the array. Typically, the projection of the data dependency graph onto a reasonable number of processors leads to an array in which there are loops with duration greater than one systolic cycle. In order to implement the systolic algorithm correctly, it is necessary to separate the data by dummy or zero values.

Examples of systolic arrays which make less than optimal use of the computational elements are the arrays $W1$ and $R1$ in Kung's paper, [63]. In these arrays, valid data must be interleaved with zero data values in order for the data to be properly synchronized. This means that every second cycle, the cells are unproductively employed multiplying zero values together. The R1 example will be discussed in more detail below. A further example, that of band-matrix multiplication, requires a data rate of one sample every three clock cycles only. This example is contained in [64], page 200. The standard approach is to separate the data with two dummy values, thereby achieving only a one-third utilization of the hardware.

Several techniques have been proposed to utilize the unused cycles in these arrays. In one technique, the array hardware remains much the same, but is utilized on every cycle. This utilization is achieved by interleaving problem instances on a single array. These may be multiple copies of identical problem instances to achieve some measure of tolerance for transient faults in the network. This is accomplished by comparing two or more outputs which should be identical in the fault free case. The $a = 2$ case allows fault detection. Cases where $a > 2$ allow fault correction through voting.

Another alternative is to increase the overall throughput of the systolic ar-

11.2. DIGIT SERIAL SYSTOLIC ARRAYS

ray by interleaving different problem instances onto the array. Both of these alternatives can require additions to the processor hardware to separate data belonging to different problem instances. They also require more complex external interfaces to control the interleaving of problem instances. Another technique is to multiplex data streams through the processors. This requires many registers and multiplexors, as well as control, and therefore adds considerable complexity.

This chapter proposes an alternative approach which increases processor utilization to unity. Rather than increasing the utilization of the standard processor hardware, the technique proposed here allows throughput to be maintained while reducing the hardware by a factor approaching a. The technique is to divide the data words into a digits, and to process these digits serially. Digit-serial data inherently has a data rate that is an integer submultiple of the parallel data rate and the digit-size may be chosen so as to meet the data rates required by the systolic algorithm. The advantages of this approach are two-fold. First, the number of wires required for the transmission of data is decreased, and second, the computational elements may be smaller, digit-serial computational elements replacing the standard parallel operators. Related to the second point above is the fact that the computational elements are now fully utilized.

11.2 Digit Serial Systolic Arrays

In the diagrams of this chapter digit-serial operators will be shown as if they had zero latency, and a delay symbol $k\Delta$ will follow the operator indicating the number of cycles, k, of latency. Thus the operator will be abstracted as if it consisted of a zero-latency operator followed by a number of cycles of delay.

The application of digit-serial computation to underutilized systolic arrays is as follows. Suppose we have a systolic network in which computational units are utilized only one cycle out of a ($a > 1$). Our method is to apply digit-serial computation to this array in such a way that data words are divided into a digits, each of size $N = W/a$ where W is the word-length. The fully parallel processors are replaced by digit-serial processors and full width communication links between processors are replaced by digit wide communication links. When and why such an array will have the same function as the original array will be discussed in detail in this chapter. The digit-serial array will be shown to have throughput equal to or even greater than in the fully parallel case. At the same time, the size of the hardware forming the array is reduced greatly.

If the systolic array is fully utilized, the application of digit-serial computation to the array no longer has the advantage of increasing processor utilization. However, it is still (usually) possible to modify the array so that digit-serial computation may be used to implement it. In this case, the high efficiency of digit-serial operators (see Chapter 7) may make such a transformation cost-

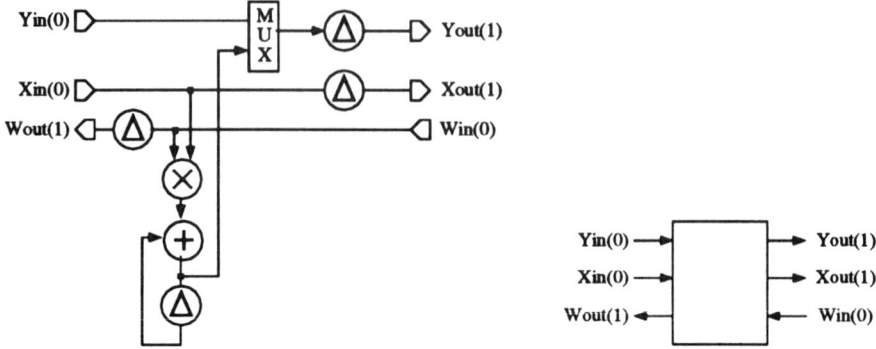

Fig. 11.1: Convolution array cell and its timing diagram.

effective. Finally, the possibility of transforming a parallel systolic array to a digit-serial array with an appropriately chosen digit-size allows the designer to trade-off throughput for circuit area to match the constraints of the present design.

11.3 Design Examples

Systolic arrays are naturally represented in graphical form, with vertices representing cells or processors and edges representing communication links. Each cell in the array may be represented schematically by a box with input and output ports. Each port is labelled with a number giving the relative timing of the input and output signals. This timing diagram will be described more fully later. The Δ operator represents a one-cycle delay.

A number of examples of systolic arrays are given in [63] and [64]. We consider in this section arrays that make less than full utilization of their hardware.

11.3.1 Convolution

Design R1 in Kung's paper [63] is an example of a design where the approach proposed here will work well. It is shown in Fig. 6 of that paper and Fig. 11.1 of the present chapter. Coefficients (Win) and inputs (Xin) move from right and left and the results remain fixed. The coefficients must be continually repeated from the right. There will be as many cells in the array as the number of coefficients. When the complete set of coefficients have passed a given cell, the computation at that point is complete. The output data is then latched into a shift register and shifted out systolically (Yout). The latching is triggered by the last coefficient passing that cell. An extra latch signal, not shown, must be shifted with the coefficients to control the multiplexor. For the sake of

11.3. DESIGN EXAMPLES

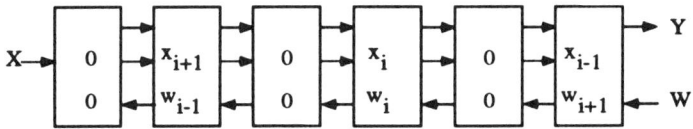

Fig. 11.2: Convolution array, parallel arithmetic

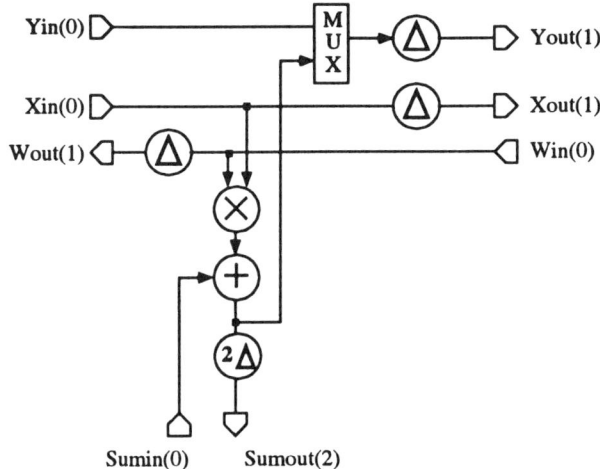

Fig. 11.3: Modified convolution cell.

simplicity, Fig. 11.1 and similar figures to appear later do not show the circuitry necessary for selecting the multiplexor or for resetting the accumulator. Since the coefficients move left and the outputs move right, it may be seen that the output values, similar to the inputs, will be separated by unused values.

Fig. 11.2 shows the convolution array with positions of data indicated. Note how it is necessary for the X and W data to have interleaved values in order that succeeding coefficients and input data meet correctly.

Careful consideration of the array in Fig. 11.1 reveals that the dummy interleaved samples actually must be zeros. We prefer a slight alteration to this array shown in Fig. 11.3 obtained from Fig. 11.1 by the addition of an extra delay in the feed-back loop. In order to avoid feed-back loops inside the processor cell, we show the feed-back loop broken and assume a feed-back loop outside the cell, as shown in Fig. 11.4. The reasons for preferring this configuration will become more apparent later. The modified array Fig. 11.3 has the advantage over that of Fig. 11.1 in that the interleaved dummy values may be arbitrary, and in fact two independent interleaved computations may take place on the array.

The fact that this design requires a data rate of one sample every two systolic

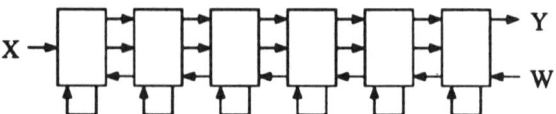

Fig. 11.4: Modified convolution array

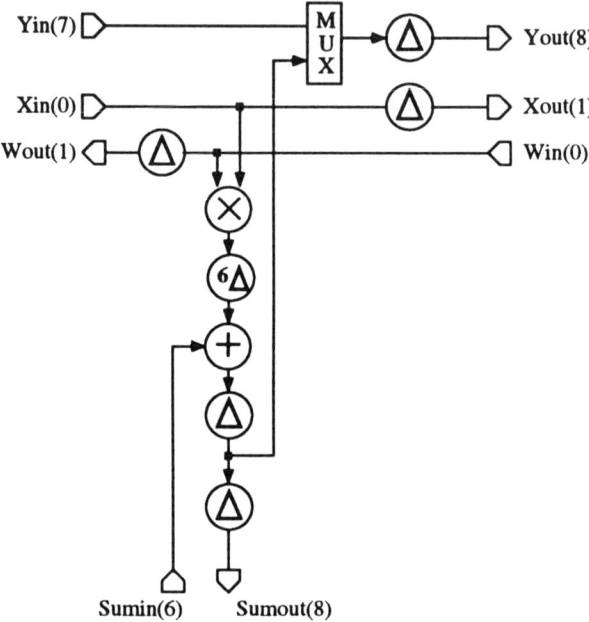

Fig. 11.5: Convolution cell implemented in digit-serial arithmetic.

cycles is an indication that it may be implemented in digit-serial arithmetic using a digit-size equal to one half the word size, thereby achieving the required data rate of two clock cycles per data value. The difference is that instead of the data being propagated in parallel with an unused cycle in which dummy data is propagated, the data transmission is spread out over the two available clock cycles, one digit in each cycle.

Fig. 11.5 shows the convolution cell implemented in digit-serial arithmetic with two digits per word, and Fig. 11.6 shows the complete array indicating positions of digits within each cell. We assume a latency of 6 clock cycles for multiplication and 1 cycle for addition and multiplexing. Superscripts 0 and 1 indicate the two digits of each data word. Note that as the weights move right and the coefficients move left, the corresponding digits of the consecutive inputs and coefficients meet allowing the computation to be carried out on digit-serial data. The output latches are enabled when either of the two digits

11.3. DESIGN EXAMPLES

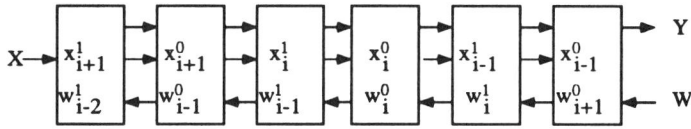

Fig. 11.6: Digit-serial convolution array showing positions of digit-serial data.

of the completed result are present in a cell. This will be 7 and 8 cycles after the arrival of the final coefficient, since the output will be delayed by the latency of the digit-serial computation. In two consecutive clock cycles the two digits of the result are latched on to the output shift register. One cell is outputting the second digit of its result while the cell on its left is already outputting the first digit. In the next clock cycle, the "latch window" has moved one position left and the output shift register values one position right. It may be verified that the output words will be shifted out sequentially in a continuous stream in digit-serial format.

The difference in timing between this array and the bit-parallel one is due to the latency of the multiplication operator (assumed to be 6 cycles). This latency delays the output of the adder, and ultimately of the Yout output.

Let us consider the advantages of the digit-serial approach. First, the computational elements are smaller, since the digit-serial computations work on only half the bits at a time. Size of digit-serial operators was discussed in Chapter 7. For a large word-length, the area saving will approach a factor of 2, because we have avoided the waste associated with unutilized cells. Secondly, the clock-rate that may be achieved will be faster, since the individual computations are smaller. For instance, the time taken to do a $W \times W$ bit multiplication may be as much as twice the time required for the $W \times W/2$ bit multiplication required in each clock cycle of a digit-serial multiplication with digit-size $W/2$. This effect is also considered in Chapter 7. Thirdly, the amount of interconnect will be halved, since data lines are digit-wide instead of word-wide. Taken all together, this adds up to an area-time saving which will exceed 2 and may even approach 4 for long word-lengths.

11.3.2 Band Matrix Multiplication

Band matrix multiplication is a fundamental systolic application. It is appropriate for the demonstration of digit-serial techniques since in one systolic form $a = 3$. Fig. 11.7 shows the band matrix multiplier cell and Fig. 11.8 shows the band matrix multiplier array for a pair of matrices with a band of width 3. Note that it is necessary to separate the valid data with two dummy or zero values so that the computation is correctly timed.

To convert this to a digit-serial array, we proceed by replacing each multiplier-adder by a digit-serial multiplier-adder of width $N = W/3$. We

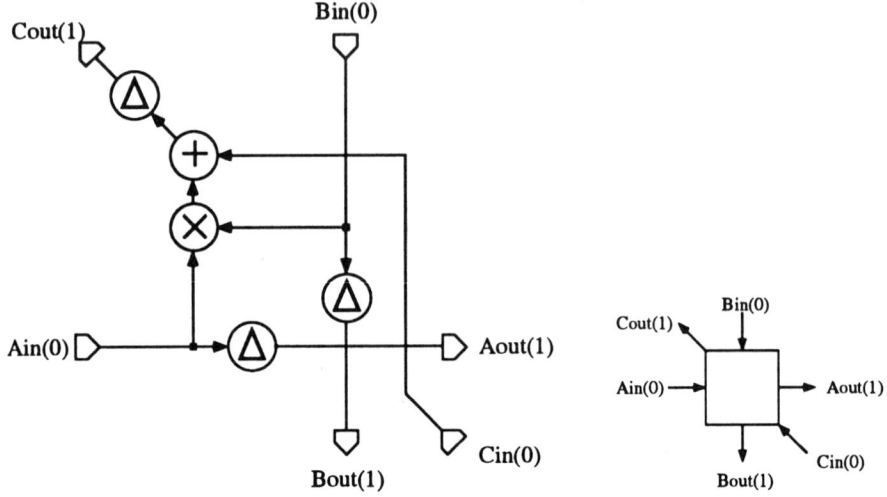

Fig. 11.7: Band-matrix multiplier cell and its timing diagram.

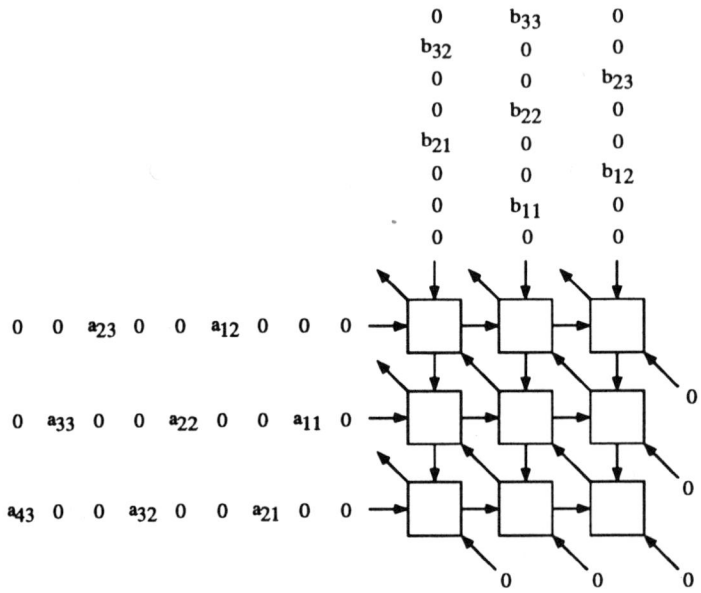

Fig. 11.8: Band-matrix multiplier array.

11.4. THE GENERAL CASE - THEORY

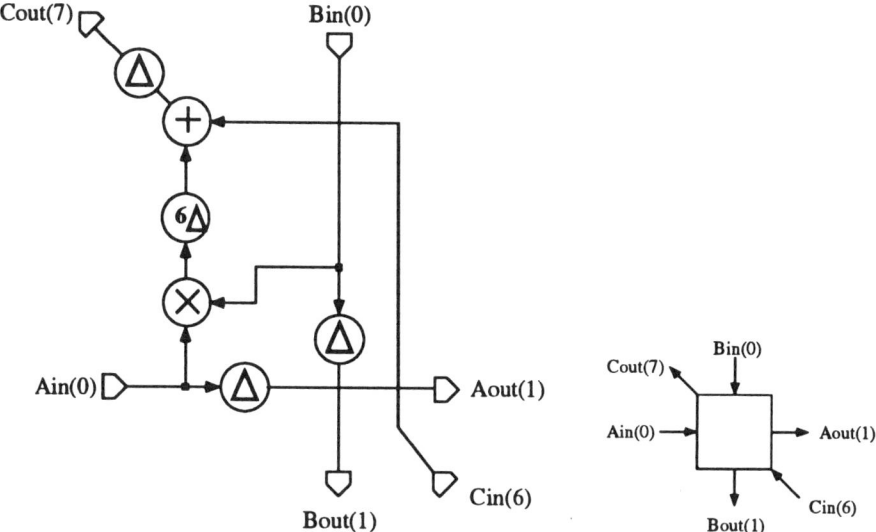

Fig. 11.9: Digit-serial band-matrix multiplier cell and its timing diagram.

assume that the multiplier will have latency of 6, and the adder will have latency of 1. Fig. 11.9 shows the basic cell with timing for digit-serial operators. Fig. 11.9 shows the complete array and the input values for the digit-serial array. This example will be looked at more closely later.

Throughput is not decreased in the digit-serial case. In fact, for a twelve bit word, the net throughput may be increased by a factor of approximately 1.67 (using estimation methods as in Chapter 7) since the digit-serial array can be clocked faster than the fully parallel array. Similarly, we can approximate the change in array area for a twelve bit word-length as a 2 times reduction. Routing area in the array will also be reduced by a factor approaching 1/3. The overall improvement in AT is therefore about $3^1/_3$.

11.4 The General Case - Theory

11.4.1 Cell Specification

A systolic array cell may generally be described by two sets of data, the timing of the input and output signals and a description of the computations carried out in that cell.

Functional specification: The description of the function of the cell can normally be represented by a set of assignments. For instance, for convolution

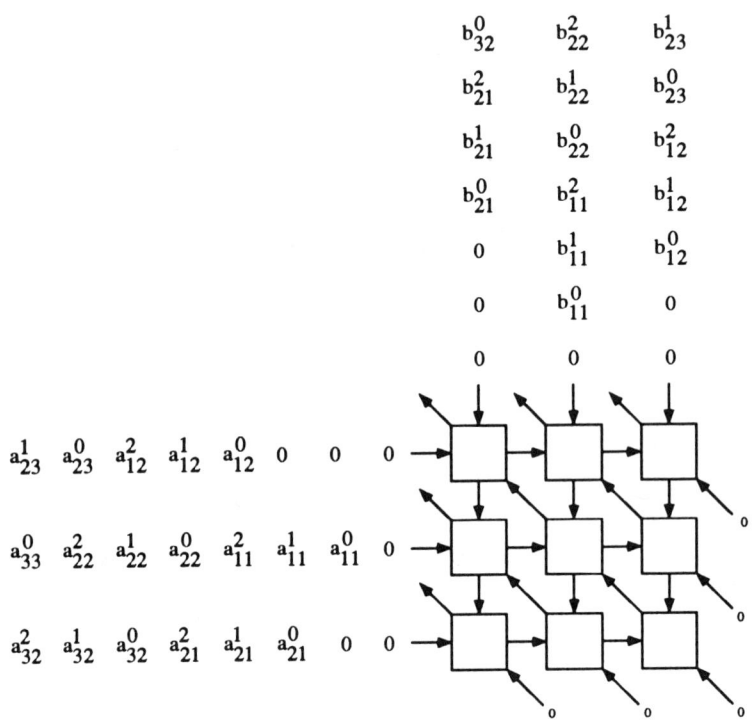

Fig. 11.10: Digit-serial band-matrix multiplier array.

11.4. THE GENERAL CASE - THEORY

array, the equations are

LatchOut ← LatchIn
Xout ← Xin
Wout ← Win
sum ← Sumin + Xin * Win
Sumout ← If (LatchIn) then 0 Else sum
Yout ← If (LatchIn) then sum Else Yin

Here for completeness, we show the complete function of the cell including the means of resetting the accumulated convolution value and for latching the output into the output shift register. LatchIn ↔ LatchOut pair is a signal that moves right to left along with the X input, marking the last X value.

For the band-matrix example, parallel case, the assignments are (this time ignoring reset):

Aout ← Ain;
Bout ← Bin;
Cout ← Cin + Ain * Bin;

Timing specification: These functional descriptions of the cells ignore all timing issues. The timing specification of a systolic cell may be represented by a timing diagram similar to the timing diagram of individual bit- or digit-serial operators, as discussed in Chapter 6, Fig. 6.5. The timing can be expressed by specifying an expected arrival time of the input signals and the corresponding output time of the output signals, all relative to some arbitrary point in time, represented by 0.

For instance, consider Fig. 11.7. The timing numbers in parentheses associated with each input and output denote the timing values. What this means exactly is : If Ain, Bin and Cin arrive at time 0, then the values Aout, Bout and Cout as determined by the functional specification of the cell will be available at time 1.

In arrays using digit-serial computation, the time of arrival of a data value is interpreted to mean the time of arrival of the leading digit of that value. Note that for the digit-serial version of the same cell, (Fig. 11.9) the Cin input is required to arrive 6 cycles later than the Ain and Bin inputs. Cout will then be ready at time 7. Cin must be delayed to make up for the six cycle latency of the multiplier.

11.4.2 Timing of the Whole Array

We make the assumption throughout this discussion that a systolic cell has no feed back or feed-forward loops, which means that each output value can be expressed as a function of the inputs to the cell, and does not depend on any stored state of the cell. Such a cell will be called a combinational cell. Note that this does not mean that the cell does not contain delay latches. Remark that the cell in Fig. 11.1 for the convolution array is not combinational, whereas the cell in Fig. 11.3 is.

Now, individual cells in a systolic array are connected together in a graph, by connecting inputs to outputs of cells. In order to understand the timing in the array, we need to consider loops in the graph. A loop is a sequence of edges of the graph forming a closed path. We do not consider the orientation of edges imposed by the direction of data flow. Instead, we assign an orientation to the edges in a loop in such a way that the edges forming the loop are ordered head to tail. Define the length of a loop to be

$$\sum_{E}(t(\text{tail of } E) - t(\text{head of } E)) \tag{11.1}$$

where t is the timing specification of the head or tail of an edge and the sum is taken over all the edges E in the loop. Note that the length of the loop depends only on the timing of the input and output pins of the systolic cells, and not on the internal timing in the cell. It turns out that the length of the loops in a systolic array is a fundamental property of the array, which determines the computation carried out by the array. In fact, the following theorem states that two arrays for which corresponding cells have the same functional specification, and for which corresponding loops have the same length carry out essentially the same computation.

Theorem 11.1. *Let A1 and A2 be two systolic arrays made up of cells which have the same functional specification, (but not necessarily the same timing of their input and output). Suppose that the cells in the two arrays are connected together in the same topological arrangement. If the lengths of the corresponding loops in the two arrays are the same, then the two arrays carry out the same computation, except that the timing of the inputs and outputs of the two arrays may differ.*

In saying that the two arrays carry out the same computation, we mean that the sequences of values appearing at any pair of corresponding cell IO ports in the two arrays are the same, except that they may be offset in time. It is assumed here that the sequence of input values are the same in both cases (once more modulo timing). This theorem is proven in [58]. A brief sketch is given here to show only the general idea behind the proof.

Proof. Select a pair of corresponding input ports x_0 and x_0' in the two arrays. Since these are input ports to the array, we insist that the sequence of values appearing at these ports in the two arrays are the same. Now let x_1 and x_1' be another pair of corresponding ports in the two arrays, and let p and p' be corresponding paths through the array leading from x_0 to x_1 and from x_0' to x_1'. Let s and s' be the lengths of these paths defined according to formula (11.1). Let v_i and v_i' be the sequences of values appearing at nodes x_1 and x_1' during

11.4. THE GENERAL CASE - THEORY

a computation (where i ranges over all time instances). Then $v_{i+s} = v'_{i+s'}$ for all i. In fact, we insist on this for the input ports of the array, and it may then be verified for the other nodes of the array. □

As an example of this theorem, suppose the band-matrix multiplier array were implemented using two different types of cells, as shown in Figs 11.7 and 11.9. Suppose that both cells were implemented using parallel arithmetic. The timing diagrams for the two cells are different, due to the insertion of 6 cycles of delay after the multiplication in Fig. 11.9. Note, however, that this does not change the total length of the triangular loops in the two arrays. In both cases, the loops have length 3. Thus, two arrays implemented using these cells would give equivalent results. The only difference is that the output Cout (also the input Cin) would be delayed by 6 cycles using the cell of Fig. 11.9. A similar analysis holds for the convolution array implemented using the cells of Figures 11.3 and 11.5.

11.4.3 Generating Digit-Serial Arrays

We define the period of a systolic array to be the highest common factor of the lengths of all loops in the graph. The period of the array is equal to the number of independent computations that may take place on the array, as enunciated in the following theorem.

Theorem 11.2. *If the period, P of a systolic array is greater than 1, then the array actually carries out P identical independent computations on P sets of independent interleaved data.*

Indeed the data may be divided into P independent sets that never meet each other. The next theorem is basic to our investigation of retiming of arrays, and the effect that it has on the computations that are being carried out on the array. If the period P of the array is greater than one, then the array may be implemented using digit-serial arithmetic, as stated next.

Theorem 11.3. *Given any systolic array of period P containing only combinational cells. By replacing the parallel computational elements by digit-serial computational elements satisfying the same timing requirements as the parallel array and working on words with P digits, a new systolic array is obtained which will carry out the same computation as the original.*

Note that this theorem replaces the P independent computations of the original array by one digit-serial computation. In interpreting this theorem, two points are to be noted. First, regarding the timing diagram of a cell, whereas the timing specifications on the input and output ports of a cell using bit-parallel arithmetic indicate the time of arrival of the complete parallel word, in digit-serial cells, the timing specifications indicate the time of arrival of the first digit of the words. It is assumed and required that the succeeding digits of each word arrive in succeeding time instants. Secondly, in interpreting the statement that the arrays carry out identical computations, it must be taken into account that the first array does its computations in bit-parallel arithmetic whereas the derived array uses digit-serial arithmetic.

As an example, we consider again the band matrix multiplication array. As shown in the preceding section, if the two cells in Figs 11.7 and 11.9 were implemented using parallel arithmetic and used in the array of Fig. 11.8, then the computation carried out would be the same in each case. However, the cell in 11.9 can be implemented using digit-serial arithmetic, assuming that multiplication has a latency of 6 cycles. According to Theorem 11.3, the computation carried out by the digit-serial array would be the same as the parallel array, and hence the same as the original array using the cell of Fig. 11.7.

Similarly, the convolution array of Fig. 11.4 has two loops of length 2, and its period is 2. This is true if either the cell in Fig. 11.3 or Fig. 11.5 is used. Consequently, by Theorem 11.2, the array carries out 2 independent interleaved computations. When the bit-parallel operators are replaced by digit-serial operators with 2 digits per word, the timing in Fig. 11.5 will apply. The resulting digit-serial array will carry out the same computation as the original parallel array.

11.4.4 Expanding the Period

Theorem 11.1 may be extended to arrays for which the loop lengths are not the same, but ratios of corresponding loops are the same.

Theorem 11.4. *Let A1 and A2 be two systolic arrays made up of cells which have the same functional specification, but possibly different timing. Further, suppose that the timing in the two arrays is such that the lengths of loops in A1 are some constant, h, (not necessarily integral) times the lengths of the corresponding loops in array A2. Then the period of A1 is h times the period of A2. Furthermore, the computations carried out by the two arrays will be the same, except that they will interleave different numbers of independent computations.*

This theorem may be used to increase the number of digits in a digit-serial implementation of the array, for instance, consider once more the band-matrix multiplication example of Fig. 11.7.

Suppose that the basic band matrix multiplication cell is changed by the addition of an extra pipe-line stage on the Cout output. The resulting cell and its timing diagram are now shown in Fig. 11.11. The duration of the loop is now 4, and the period of the array is 4. The array now carries out 4 overlapping computations, with related data separated by 4 clock cycles, as may be verified. According to Theorem 11.4, however, the basic computation is the same as in

11.4. THE GENERAL CASE - THEORY

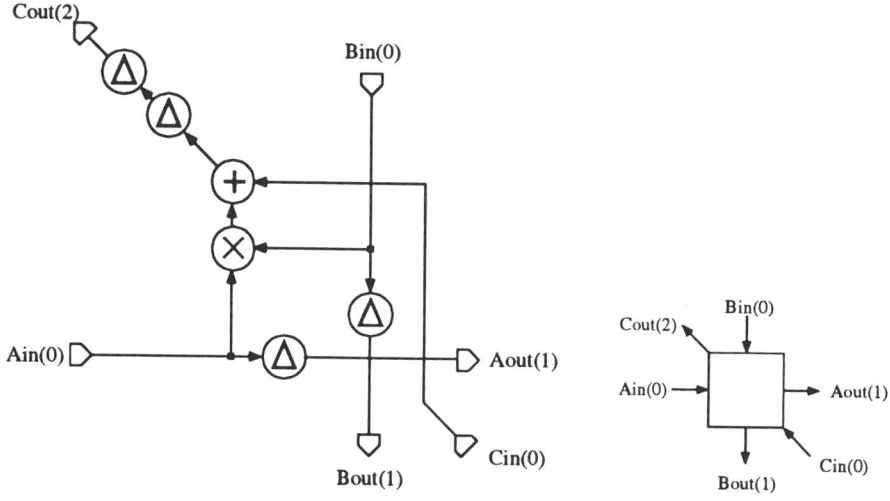

Fig. 11.11: Band-matrix multiplier cell for period 4 array.

the original array. Once again, replacing the arithmetic operators with digit-serial ones will result in an array in which the loops are of the same length. In this way, the array may be implemented with 4 digits in each word.

Similarly, the convolution cell may be modified by the insertion of an extra delay in each loop as in Fig. 11.12. Here, as before we show the timing for the case where multiplication has latency 6 and addition latency 1. Now, each loop has length 3 and the word may be divided into three digits and implemented using digit-serial operators. It is interesting to note how the correct digits of data are synchronized to arrive at the right place at the right time. This is shown in Fig. 11.13.

Although it will give no extra advantage in speed over systolic arrays implemented with the minimal number of digits per word, there are still advantages to be gained from being able to change the digit size of an array. Most notably, these include the ability to trade off area for speed to match the application. By decreasing the digit-size, the size of the circuit is decreased at the cost of a loss in throughput.

Furthermore, it was shown in Chapter 7 that some range of digit-sizes make for more efficient circuits (according to an Area-Time measure) than others. Giving the ability to vary the digit-size allows the designer to choose the most efficient digit-size for the application.

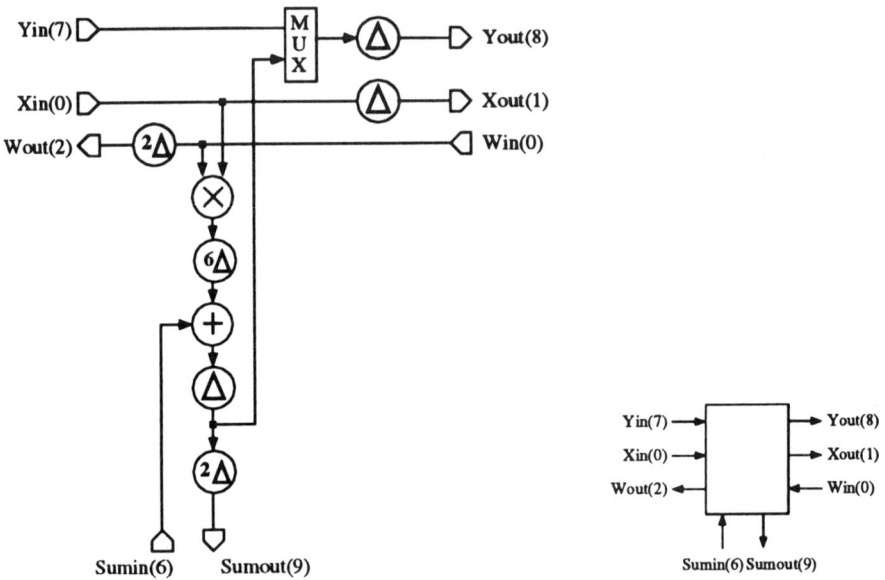

Fig. 11.12: Convolution cell for array with period 3.

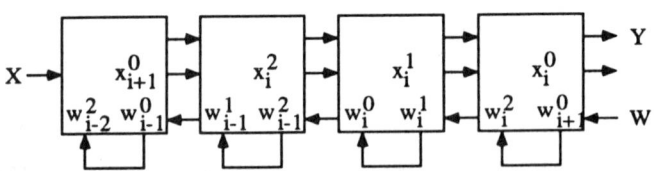

Fig. 11.13: Convolution array of period 3, showing data positions.

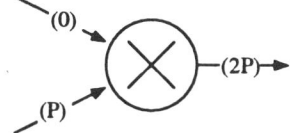

Fig. 11.14: Timing diagram of a digit-serial multiplier.

11.5 Finding the Minimum Digit-Size

This section contains a more detailed investigation of how a digit-size may be chosen to allow a given bit-parallel array to be implemented in digit-serial arithmetic. We can use Theorem 11.4 to formulate a general method of implementing bit-parallel arrays in digit-serial arithmetic. As the examples above show, it is often the case that a simple substitution of digit-serial operators for parallel ones allows the array to be implemented in digit-serial arithmetic with a digits per word, where a is the period of the array, that is, the degree of under-utilization. Similarly, as shown in the previous section, simply increasing the delay on one or more of the outputs of a cell provides an array with period equal to any arbitrary value exceeding a. Theorem 11.3 can then be applied to implement the array with digit-serial format containing some other chosen number of digits per word. It is possible, however, that the substitution of digit-serial for parallel operators changes the timing of the array in such a way as to alter the period of the array. In this case, it is desirable to have a formal method of determining a digit-size for digit-serial implementation of the array. This is discussed in the present section. We assume for simplicity that the array is made up of a single type of cell arranged in a regular array. Arrays which contain more than one type of cell can be treated similarly.

For some types of digit-serial operator the latency is constant, independent of the number of digits in a word. An example of such a cell is a digit-serial adder, in which both inputs may be assumed to arrive at time $t = 0$, and the single output is produced at time $t = 1$. For other operators, the latency of the cell depends on the number of digits in each word. A digit-serial multiplier is an example of such an operator. The digit-serial multiplier described in Chapter 3 has a timing diagram shown in Fig. 11.14, where P represents the number of digits in a word.

Generally, however, the latency or timing specification of any digit-serial operator is expressed in terms of linear functions of P. For instance, in the example of the multiplier shown above, the latency of the output is $2P$, a linear function of P. We are not aware of any digit-serial operator which does not conform to this rule. For digit-serial operators using a most-significant digit first (on-line arithmetic) approach, the latencies of operators are generally small constant values independent of P.

A digit-serial implementation of the bit-parallel cell may be produced by replacing each operator in the signal flow graph by its digit-serial equivalent, and then by scheduling the flow graph and introducing "shimming delays" in the manner described next. This method is similar to the scheduling algorithm of Chapter 6, but has some different features, as we shall see.

First of all, we introduce dummy operators on each input and output of the complete systolic cell. The dummy input operators have no inputs themselves and one output. The output operators, on the other hand have themselves no output, but a single input. As described in Chapter 6, we now attempt to schedule the graph by assigning a "scheduled time" to each operator in the graph. As in Chapter 6, the restriction on scheduling is that any operator must be scheduled at a time after all of its inputs are ready, that is, produced by the previous operator. For each pair of operators as shown in Fig. 6.6 we obtain an inequality of the form

$$S_i + t_i \leq S_j + t_j \ . \tag{11.2}$$

Note that in the case where the timing specifiers are not constants, the variable P (number of digits in a word) may appear linearly in this inequality, which will also be linear in the scheduled times, S_i.

Inequalities (11.2) are concerned with the internal timing of each cell. There will be a further set of equalities concerned with the timing of the complete array. Consider a loop in the array. The length of this loop is equal to $h_\ell P$ where P is the period of the array (the g.c.d. of all loop lengths) and h_ℓ is an integer. According to Theorem 11.4, the length of the corresponding loop in the digit-serial array with P digits per word must be $h_\ell P$. We represent the length of a loop ℓ by $h_\ell P$. The integers h_ℓ are determined by the functionality of the array (see Theorem 11.1), and are therefore assumed known. According to the definition of loop length (11.1), each loop (indexed by ℓ) leads to an equality of the form

$$\sum_E (t(\text{tail of } E) - t(\text{head of } E)) = h_\ell P \tag{11.3}$$

where E ranges over all the edges in the loop. We assume that all cells in the array identical, and hence that the timing specifications of all cells are the same. Thus, we may replace an expression $t(\text{tail of } E)$ or $t(\text{head of } E)$ by S_i, the scheduled time of one of the dummy input or output operands of the cell. This leads to an equality

$$\sum_{k=1}^{n/2} a_k.(S_{ik} - S_{jk}) = h_\ell P \tag{11.4}$$

where S_{ik} and S_{jk} are the scheduled times of a matching pair of input and output ports of the cell, a_k is an integer (usually ± 1 or 0) representing the number of times an edge appears in the loop, and n is the number of input and output ports of the cell. An inequality of type (11.4) exists for each loop in the

11.5. FINDING THE MINIMUM DIGIT-SIZE

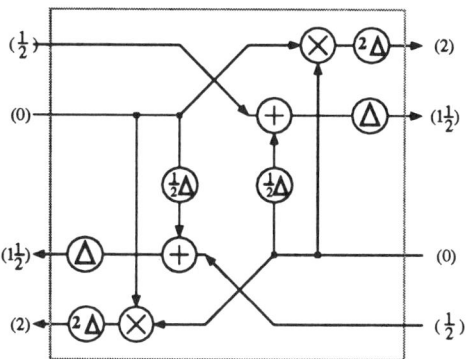

Fig. 11.15: Array cell with minimum non-integral digit-size.

array. Clearly, however, there are at most $n/2$ linearly independent equations of this type. In fact, we need only consider loops starting at a given fixed cell in the array, since all similar loops starting at other cells give rise to the same equality. Further, it is sufficient to consider only a basic set of loops from which all other loops may be derived. For instance, in the case of a planar array, it is sufficient to consider only "inner loops" which contain no other loops in their interior.

The two sets of inequalities of type (11.2) and (11.4) together form a complete set of constraints for scheduling a cell in order for it to be implemented in digit-serial with P digits per word. Any integer solution to these set of inequalities will give a valid schedule. Once a schedule is obtained, it is necessary to add "shimming delays" to the cell to equalize delay paths and ensure that operator inputs arrive at the correct time (see Chapter 6). The least possible number of digits per word is obtained by minimizing P subject to the given constraints. We seek solutions in integers, so it is an integer programming problem that must be solved. Since solving integer programming problems is in general more difficult than the corresponding linear programming problem it is natural to ask whether the solution to the linear programming problem may turn out to be a solution in integers, or easily transformed into one. If this is the case, then we may simply solve the linear programming problem and obtain the minimum solution in integers. Such a favorable situation arises in the problem of minimizing shimming delays in synchronous circuit (see Chapter 6), which is a closely related problem. Unfortunately it is not true in this case, as the example of Fig. 11.15 shows.

In this example we assume a latency of two cycles for multiplication and one for addition. The minimum length of loops is 3. However, the scheduling that gives loops of length 3 has inputs and outputs scheduled at fractional times as shown. It is not intended that the cell of Fig. 11.15 should be realizable, since

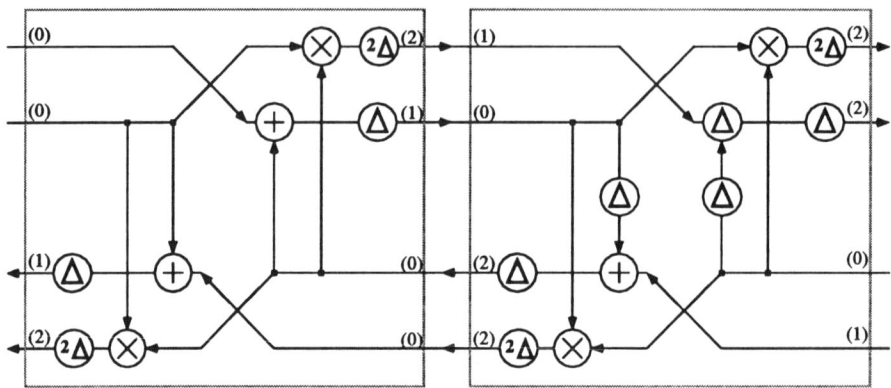

Fig. 11.16: Minimum real implementation using two different cells.

fractional delays have no real meaning. The significance of Fig. 11.15 is that it is the solution of the linear programming problem. It may easily be verified that with integer scheduling it is impossible to attain this minimum loop length. It is interesting to note that if two different variants of the basic cell are allowed, then it is possible to attain the minimum loop length with integer scheduling. This is shown in Fig. 11.16. In general, one may conjecture that with a finite number of different cell types, each with the same functionality but different timing, it is possible to implement the systolic array with the minimum number of digits per word.

In many cases, however, a solution to the constraint set that minimizes P may easily be transformed into a solution in integers with the same value of P. Commonly, this may be done simply by replacing all the scheduled times by the nearest integer value. To give an example of how this may be done we consider any array with the same topology of interconnections as the band multiplier array and assume that it is to be implemented with digit-serial operators in such a way that the scheduling inequalities (11.2) do not depend on P, the number of digits in a word. This would be the case if the array were implemented with MSD first (on-line) digit-serial operators. We show that the solution found to the linear programming problem may be transformed into an integer solution with the same value of P by rounding each scheduled time to the nearest integer value, or alternatively by truncating each scheduled time to the next smaller integer value.

The relevant feature of the band-matrix multiplication array of Fig. 11.8 is that the set of equalities imposed by the array topology reduces to a single equality. In particular, there are just two inner loops in the array and these both give rise to the same equality. With the scheduled times as shown on Fig. 11.17, the equality becomes $(T1 - S1) + (T2 - S2) + (T3 - S3) = P$.

Consider more generally the case where there is a single equality to be

11.5. FINDING THE MINIMUM DIGIT-SIZE

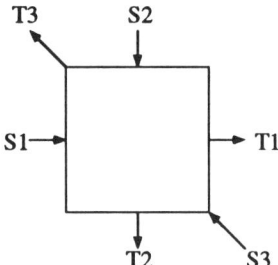

Fig. 11.17: Timing diagram for band-matrix topology.

satisfied :

$$\sum_{k=1}^{n/2} a_k.(T_k - S_k)) = P \tag{11.5}$$

where T_k and S_k are the scheduled time of matching output and input ports, and the a_k are any integers. The linear programming problem may be stated as that of minimizing the expression (11.5) subject to the constraints (11.2). Suppose that we have solved this problem and found a minimum value, P_{min}. Theorem 6.2 now applies, and we can replace all scheduled times S_i of each operator in the cell, (including dummy input and output operators) by $\lfloor S_i - r \rfloor$, for any fixed real number, r to give an optimal valid integral scheduling.

Minimizing the shimming delays: For any chosen digit size and number of digits per word, there may still be several schedules for the systolic cell. As with scheduling complete digit-serial circuits, it will normally be best to use a scheduling that minimizes the number of shimming delays needed, as explained in Chapter 6. The problem is essentially the same as that discussed in Chapter 6, except that one has the extra constraints represented by the loop length equalities, (11.4).

Chapter 12

Canonic Signed Digit Arithmetic

In this chapter we deal with the subject of constant multiplication through the use of shifts and adds. In this manner, an expensive multiplication is replaced by a number of cheap shifts and adds. Encoding the multiplier in the so-called canonic signed digit format leads to the most economical design in terms of conservation of area and power consumption. Power minimization is important to increase battery life in portable applications and to reduce the amount of heat generated [77]. This technique is applicable to bit- and digit-serial designs as well as to pipelined parallel designs. We will therefore extend the domain of interest to include techniques that apply also to pipelined parallel arithmetic.

12.1 Canonic Signed Digit Format

Although multiplication by a fixed constant value may be carried out using a variable-by-variable multiplier, it is often advantageous to use a different technique based on shifts and adds for multiplication with constant numbers. Multiplication with constant numbers is often needed in digital signal processing. For example, the constant coefficients in time-invariant digital filters require multiplication with constant numbers. In these multiplication operations, it is necessary to perform the multiplication with non-zero bits only and this can reduce the implementation area as well as multiplication time or latency. In this section, we consider the representation of numbers as a succession of bits, where each bit is in the set $\{0, 1, -1\}$.

Notation. If a_i is in the set $\{0, 1, -1\}$ for each i, we denote by $a_{N-1}.a_{N-2}\cdots a_1 a_0$ the value $\sum_{i=0}^{N-1} a_i 2^{N-1-i}$. Each a_i will be referred to as a bit, and the number $a_{N-1}.a_{N-2}\cdots a_1 a_0$ will be referred to as an N-bit number.

Two's complement format may be considered as a special case in which the digit a_{N-1} is equal to 0 or -1, whereas $a_i = 0$ or 1 for $i < N-1$. This reflects the special status of the sign bit, a_{N-1}, which has negative weight. Now, suppose that a variable X is to be multiplied by a constant value $A = a_{N-1}.a_{N-2}...a_1 a_0$. The product $A.X$ can be expressed as

$$A.X = \sum_{i=0}^{N-1} a_i.X.2^{N-1-i} = \sum_{i=0}^{N-1} a_i.X >> N-1-i$$

where $X >> i$ represents X shifted right i places. The number of terms to be added together is equal to the number of non-zero bits in the representation of A. Since in a general two's complement number, it is possible for all bits to be non-zero, in the worst case, this multiplication may require as many as N terms to be added. A natural question is what is the minimum number of terms that need to be added or subtracted in order to carry out a given constant multiplication. The answer is given by the canonic signed digit (CSD) number representation.

CSD Format: A number $a_{N-1}.a_{N-2}\cdots a_1 a_0$ is said to be in CSD format if each a_i is 0, +1 or −1 and no two consecutive a_i's are non-zero. Note that whereas two's complement N-bit numbers are restricted to the range $[-1, 1)$, N-bit CSD format numbers cover a wider range of values. For instance the CSD value 1.01010101 is greater than 1. It can be shown in fact that CSD numbers cover the range $(-4/3, 4/3)$. Of greatest interest, however, are the values in the range $[-1, 1)$. The properties of CSD number representations have been considered in [78]. Here is a summary of some of their properties.

Proposition 12.1.

(i) *Any N-bit number in the range $(-4/3, 4/3)$ may be represented in N-bit CSD format.*

(ii) *The representation of a number in CSD format is unique, i.e., a given number may be represented in only one way in CSD format.*

(iii) *The CSD representation of a number contains the minimum possible number of non-zero bits.*

Though it is fairly easy to invent an algorithm to represent any N-bit number in CSD format, a very neat algorithm has been presented in [78] and is repeated here.

12.1. CANONIC SIGNED DIGIT FORMAT

Computing the CSD format for an N-bit number: Let $A = \hat{a}_{N-1}.\hat{a}_{N-2}...\hat{a}_1\hat{a}_0$ represent the two's complement representation of the number A and $A = a_{N-1}.a_{N-2}...a_1a_0$ represent its CSD representation. The CSD representation can be obtained from the two's complement representation using the following iterative algorithm.

$$\hat{a}_{-1} = 0$$
$$\gamma_{-1} = 0$$
$$\hat{a}_N = \hat{a}_{N-1}$$
$$\text{for } (i = 0 \text{ to } N - 1)$$
$$\{$$
$$\quad \theta_i = \hat{a}_i \oplus \hat{a}_{i+1}$$
$$\quad \gamma_i = \overline{\gamma_{i+1}}\,\theta_i$$
$$\quad a_i = (1 - 2\hat{a}_{i-1})\gamma_i$$
$$\}$$

Here, the symbol $\oplus$ denotes exclusive OR, and the overbar indicates complementation.

Here is an example. The input is the number $\hat{a}_{N-1}.\hat{a}_{N-2}...\hat{a}_1\hat{a}_0 =$ 1.01110011.

i	-1	0				...				$N-1$	N
$\hat{a}_i$	1	1	0	1	1	1	0	0	1	1	0
θ_i		1	1	0	0	1	0	1	0	1	
γ_i	0	1	0	0	1	0	1	0	1	0	
$1 - 2\hat{a}_{i-1}$	-1	-1	1	-1	-1	-1	1	1	-1		
a_i		0	-1	0	0	-1	0	1	0	-1	

Number of non-zero bits in a CSD number: Since adjacent bits in a CSD number may not both be non-zero, it is easy to see that the maximum number of non-zero bits is equal to $\lceil N/2 \rceil$. A more interesting question is what is the expected number of non-zero bits in an N-bit CSD number. Here is the answer (see [78]).

Proposition 12.2. *Among those N-bit CSD numbers in the range $[-1, 1)$, the average number of non-zero bits is*

$$E[\# \text{ non-zero bits}] = N/3 + 1/9 + O(2^{-N}) \ .$$

Thus, on the average, approximately one-third of the bits in a CSD number are non-zero. On the average, CSD numbers contain about 33% fewer non-zero bits than two's complement numbers.

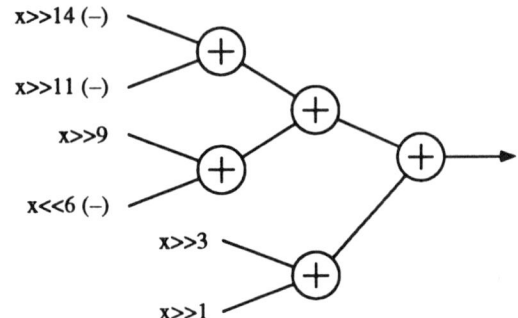

Fig. 12.1: CSD Multiplier.

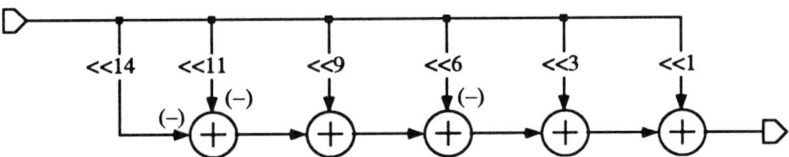

Fig. 12.2: CSD multiplier using linear arrangement of adders.

12.2 CSD Multiplication

As indicated above, constant multiplication may be carried out by combining a number of partial product terms by subtraction or addition in a tree of adders. The tree of adders may range from a balanced tree to a linear arrangement of adders according to the circumstances. Fig. 12.1 shows an arrangement of adders implementing the product $x * 0.10100\bar{1}0010\bar{1}00\bar{1}$ using a balanced tree of adders. The same multiplication can be carried out using a linear arrangement of adders as shown in Fig. 12.2. Which arrangement of adders is preferable will be discussed later.

12.2.1 Pushing Subtractions Towards the Root

For CSD constant multiplication, both adders and subtractors are assumed to be available. The subtractor is used when one of the two inputs has a negative weight. In the topmost addition of Fig. 12.1, however, both the inputs have negative weight. This is easily handled by a technique of pushing subtractions towards the root of the tree. The identity $(-a) + (-b) = -(a + b)$ is used here. Graphically, this corresponds to a transformation:

12.2. CSD MULTIPLICATION

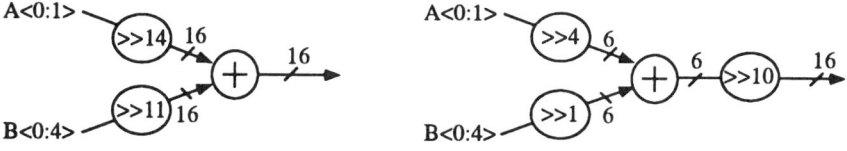

Fig. 12.3: Equivalent circuits.

Adders with only one negative weighted input are left unchanged. Continuing in this way, all subtractions are pushed as far towards the root of the tree as possible. If the original multiplier has at least one +1 bit in its CSD format, the final output of the tree will have positive weight. Otherwise, a final negation will be necessary. This transformation technique applies to all arrangements of adders.

Pushing shifts towards the root: Like subtractions, shifts can also be pushed towards the root of the tree. The basic operation of pushing one shift relies on the identity

$$(a >> 1) + (b >> 1) = (a + b) >> 1 \ .$$

Graphically, this corresponds to the operation:

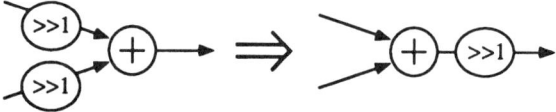

Generally it is desirable to defer shifts as much as possible by pushing them towards the root, because it leads to more efficient use of hardware. For instance, the two circuits in Fig. 12.3 give the same result, though one of them uses a 16-bit adder and the other uses a 6-bit adder. In the circuit on the left, the top 10 bits of the adder are not used productively. The symbol $B<0:4>$ denotes the 5 high order bits of B. If A and B are 16-bit quantities, neither of the two circuits above will give the full precision correct result for $A >> 14 + B >> 11$. To compute the result to full precision will require a 30-bit adder in the circuit on the left, but only a 20-bit adder in the circuit on the right. In some instances, a fixed layout style may constrain all the adders to be of a given size. In such a case, although no hardware saving will be achieved by deferring shifts, nevertheless greater accuracy will be achieved by this technique.

In carrying out a modification of the circuit by deferring shifts, it is important to ensure that addition results in the computation do not overflow. This

CHAPTER 12. CANONIC SIGNED DIGIT ARITHMETIC

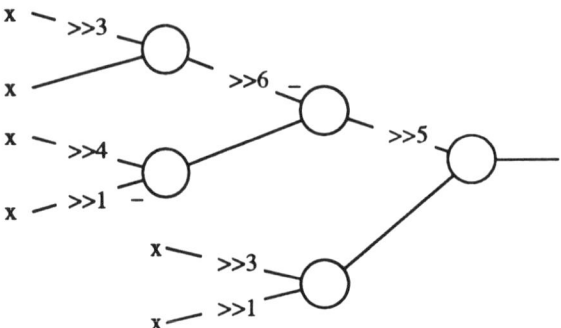

Fig. 12.4: CSD multiplier after shift optimization.

is the reason that in the right hand of Fig. 12.3, the operand B is shifted 1 place. To explain this point, consider the computation of $x >> 5 + x >> 7$. This value may be computed as written, or alternatively as $(x + x >> 2) >> 5$. Suppose that x is a 16-bit quantity. Then $x >> 5$ and $x >> 7$ have 21 and 23 bits. To form their full precision sum, a 23 bit adder is required. On the other hand, to compute $x + x >> 2$ requires only an 18- bit adder. If smaller adders are used, loss of precision through truncation will occur. The truncation will be more severe in the computation of $x >> 5 + x >> 7$. It seems best, therefore to compute $(x + x >> 2) >> 5$. However, if $x > 0.8$, the result of $x + x >> 2$ will overflow and the result will be wrong. Therefore, the correct way is to compute $(x >> 1 + x >> 3) >> 4$. This will require a 19-bit adder and will avoid overflow. In the case of the computation of $x >> 5 - x >> 7$, on the other hand, it is not possible for computation of $x - x >> 2$ to overflow. Hence, the correct way to perform this computation is as $(x - x >> 2) >> 5$. The difference is that in the first case, $x + x >> 2$ equals $1.25 * x$, whereas $x - x >> 2 = 0.75 * x$. Multiplication by a constant greater than 1.0 can cause overflow. A correct approach is as follows :

- Start with a tree of adders as shown in Fig. 12.1.

- Starting with the leaves, push as many shifts towards the root of the tree as possible. Keep track at each node (i.e., adder) in the tree of the value that will occur at the output of that node assuming an input multiplicand $x = 1$. Only push as many shifts as needed to ensure that this value remains in the range $[-1, 1)$.

Fig. 12.4 shows the result of this operation on the tree of Fig. 12.1.

In the case where there are several different variables $x_1, ... x_n$ input to the tree (including different versions of the same variable), at each node, a linear combination $\sum_{i=0}^{N-1} a_i x_i$ will be computed. The inputs at each node should be

12.2. CSD MULTIPLICATION

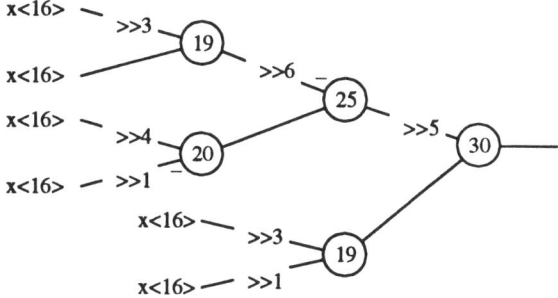

Fig. 12.5: Full precision CSD multiplier.

shifted the minimum amount consistent with the requirement that $\sum_{i=0}^{N-1} |a_i| < 1$.

12.2.2 Size of Adders

First of all suppose that the result is to be computed in full resolution. To determine the required size for all the adders in the tree, start by assuming n-bit values input at all the leaf nodes. Then proceed up the tree towards the root setting the size of each adder. The size of each adder will be equal to the maximum size of its two inputs. For this purpose, an n-bit value shifted s-bits right has size $n+s$ bits. In the case where an adder has inputs of different length (at the LSB end), the shorter addend should be extended with 0s. Fig. 12.5 is the same multiplication as before with adder sizes now shown.

As an alternative to passing the shorter addend with zeros at the LSB end, the adder can match the length of the shorter addend and the overlapping bits can bypass the adder. This method can be used to reduce the size of adders, but it does not work so well for subtractors. We will not explore this possibility further here.

Typically, a designer wishes to multiply two n bit numbers together and obtain fewer than n bits of result. The output value will be obtained by truncation, or rounding. If this is the case, then it is usually not necessary to compute the full-precision result before truncation. Truncation of data causes errors in the final result. Each adder of length shorter than its two inputs causes truncation, so errors are introduced at all levels in the adder tree. The amount of error can be decreased by adding extra hardware in the form of extra length of adders. In general, truncation errors introduced at different parts of the computation tree should be equal. It is inefficient to use extra adder bits to ensure a truncation error of (for example) 2^{-20} at one point in the tree when truncation at another point introduces an error of 2^{-15}. A more efficient use of hardware results from taking bits away from the first adder and giving them to

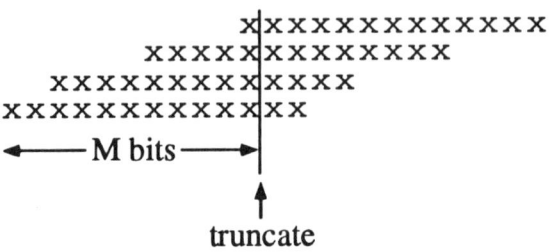

Fig. 12.6: Summing of partial products.

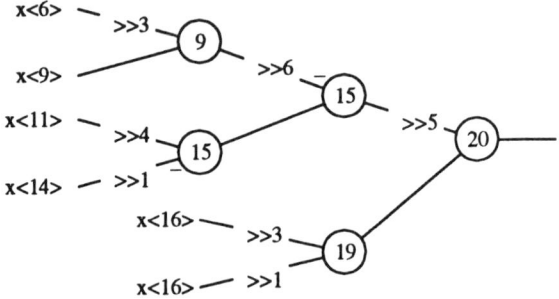

Fig. 12.7: Summation of truncated partial products.

the second, resulting in errors of 2^{-18} in each adder. Our general strategy is to truncate all partial product terms equally. Consider Fig. 12.6 which represents a set of appropriately shifted partial products to be summed.

The vertical line represents a truncation level. All bits to the right will be truncated. Suppose M bits of precision are to be used in the computation. Each input to the tree represents a separate partial product. Suppose $N =$ word-length and M is the number of bits of precision used in the computation. Then, as shown in Fig. 12.7, the number of bits used for a partial product shifted s places is $\min(N, M - s)$. Starting with these values at the leaves of the tree and propagating forward towards the root allows the width of each adder to be computed. Fig. 12.7 shows the resulting adders, assuming a value $M = 20$. The notation $x < 6 >$ (for instance) denotes the 6 high order bits of x.

12.2.3 Error Computation

Using this technique it is possible to compute the average and maximum errors exactly. Assume that each bit in the input value of x is a random variable with equal probability of being a 0 or 1. Then we can compute the contri-

12.2. CSD MULTIPLICATION

bution of each bit to the error. Suppose that the CSD multiplier constant is $b_{K-1}.b_{K-2}...b_1b_0$, where $b_i = 0, 1$ or -1. Suppose that the two's complement multiplicand is $a_{N-1}.a_{N-2}...a_1a_0$, where $a_{N-1} = 0$ or -1 and the other a_i's are non-negative. The product is

$$\sum_{i=0}^{N-1} \sum_{j=0}^{K-1} a_i b_j . 2^{N+K-(i+j)-2} .$$

If we truncate after M bits, then the error will be

$$\sum_{i+j \geq M} |a_i| b_j . 2^{N+K-(i+j)-2} .$$

This equals

$$\sum_{i=0}^{N-1} |a_i| \sum_{j=M-i}^{K-1} b_j 2^{N+K-(i+j)-2} .$$

Denoting $B_i = \sum_{j=M-i}^{K-1} b_j 2^{N+K-(i+j)-2}$ we get

$$\text{Error} = \sum_{i=0}^{N-1} |a_i| B_i .$$

Since $E[|a_i|] = 1/2$, we see that the mean error is given by

$$E[\text{Error}] = \sum_{i=0}^{N-1} B_i/2 .$$

The range of possible errors is from S_- to S_+, which are the sum of the negative and positive B_i respectively. The error distribution is symmetric about the mean, since $|a_i|$ is equally likely to be 0 or 1. The error variance is also easily calculated, and is equal to

$$\sigma^2 = \sum_{i=0}^{N-1} B_i^2/4 .$$

These computations allow for an optimal choice of M for a required error value. For instance, the truncation error in Fig. 12.7 is $(-0.928 \pm 1.584).2^{-18}$.

12.2.4 Bypassing Adders

In the case where an adder has inputs of different length (at the LSB end), the shorter addend should be extended with 0's. Alternatively, the adder can match the length of the shorter addend and the overlapping bits can bypass the adder. This works well when the overlapping operator is to be passed in to an adder, or the positive side of a carry-propagate subtractor. It can also be applied to carry-save operators. Fig. 12.8 shows this optimization applied to the full-precision multiplier of Fig. 12.5.

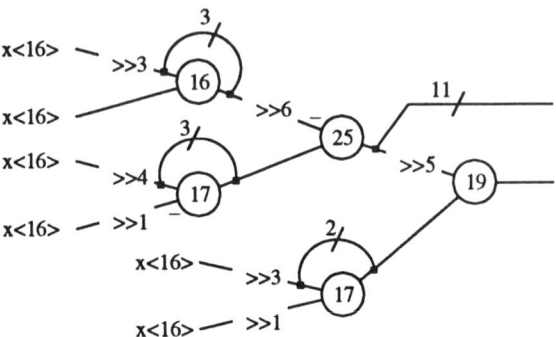

Fig. 12.8: Bypassing bits of the adders.

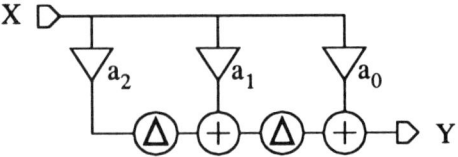

Fig. 12.9: 3-tap FIR filter.

12.3 Sub-Expression Sharing

Next, we consider the savings that may be achieved by sharing common sub-expressions in circuits with one or many constant multiplication operators. We will concentrate particularly on filter designs using CSD multiplication. Constant multiplication by CSD encoded constants has been used in the design of FIR filters in a number of design systems [6], [79]. Fig. 12.9 shows a general FIR filter with 3 taps. Fig. 12.10 shows how this filter may be implemented using CSD constant multipliers, for a choice of filter coefficients

$$\begin{aligned} a_0 &= 0.1000100\bar{1} , \\ a_1 &= 0.0\bar{1}00010\bar{1} , \\ a_2 &= 0.100\bar{1}00\bar{1} . \end{aligned}$$

A M-tap FIR filter is described by the equation

$$y_n = \sum_{i=0}^{M-1} a_i \, x_{n-i} .$$

If each coefficient is expressed in CSD format as $a_i = a_{iN-1}.a_{iN-2}\ldots a_{i1}a_{i0}$

12.3. SUB-EXPRESSION SHARING

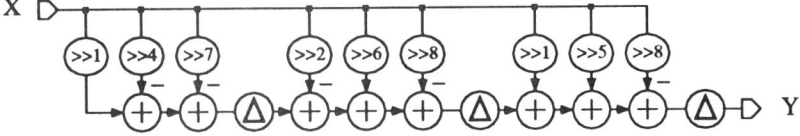

Fig. 12.10: 3-tap FIR filter using CSD multipliers.

then the filter equation becomes

$$y_n = \sum_{i=0}^{M-1} \sum_{j=0}^{N-1} a_{ij}.(x_{n-i}) >> (N-1-j) \ .$$

Denoting $X_{ij} = x_{n-i} >> (N-1-j)$, then

$$y_n = \sum_{i,j} a_{ij} X_{ij} \ . \tag{12.1}$$

12.3.1 Finding Common Sub-expressions

A method of reducing the number of addition operators in an FIR filter was discussed in [80]. This method has not so far been used for the actual design and fabrication of filter chips. However, it has the promise of reducing the number of operators by up to 33% without any adverse effect. It will be particularly useful in filters with many taps and large word-length. The method described in [80] depends on the identification and exploitation of common sub-expressions in (12.1), a technique also used in compiler optimization [81]. Similar ideas have been used as the basis for minimizing the number of addition operators in a single multiplication in [82]. Furthermore, work by Chatterjee et al. ([83, 84]) has considered the gains obtained by sub-expression sharing in linear systems involving multipliers. However, the problem considered and approach used in [83, 84] are rather different from those described here. In addition, sharing sub-expressions in signed-digit multipliers has been considered recently by Potkonjak et al. ([85]) from a somewhat different viewpoint.

An example of common sub-expressions within a single multiplication is

$$y = 0.10\bar{1}00010\bar{1} * x \ .$$

This may be implemented as

$$y = (x >> 1) - (x >> 3) + (x >> 7) - (x >> 9) \ .$$

Alternatively, it can be implemented as

$$\begin{aligned} x_2 &= x - x >> 2 \\ y &= (x_2 >> 1) + (x_2 >> 7) \end{aligned}$$

228   CHAPTER 12. CANONIC SIGNED DIGIT ARITHMETIC

which requires one fewer addition. For more about saving additions in single multiplications, see [82].

Notation: For the present, we write x_1 instead of x, and $x_1[-i]$ to represent x_{n-i}. The notation of $[-i]$ in general represents i sample delays.

A filter can be represented naturally by a table X_{ij} of terms, where the rows are indexed by delay and the columns by shift. The entries are 0, 1 or -1. Row and column indexing starts at 0 and an entry 1 in row i, column j represents $x_1[-i] >> j$. This term is to be added or subtracted according to whether the entry is $+1$ or -1. For example, consider the filter:

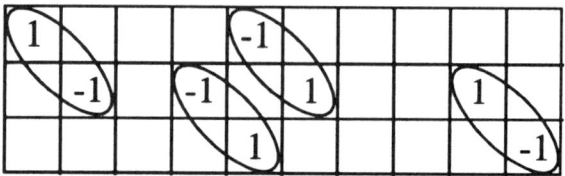

This filter has 8 non-zero terms and would normally require 7 additions. However, we find that the sub-expressions $x_1 + x_1[-1] >> 1$ occurs 4 times in shifted and delayed forms by various amounts as circled in the above table. Therefore, the filter may be written with 4 additions as

$$x_2 = x_1 - x_1[-1] >> 1$$
$$y = x_2 - (x_2 >> 4) - (x_2[-1] >> 3) + (x_2[-1] >> 8) \ .$$

An alternative method of computation is

$$x_2 = x_1 - (x_1 >> 4) - (x_1[-1] >> 3) + (x_1[-1] >> 8)$$
$$y = x_2 - (x_2[-1] >> 1) \ .$$

This last example illustrates a novel feature.

Remark 12.3. *A sub-expression of size m occurring n times is the same as a sub-expressions of size n occurring m times.*

An algorithm for sub-expression matching now follows, illustrated with a 4-tap FIR filter.

1. Starting with a table of terms, identify the "best" sub-expression of size 2. The best sub-expression is chosen according to a benefit function (discussed below) in which the number of occurrences is an important factor.

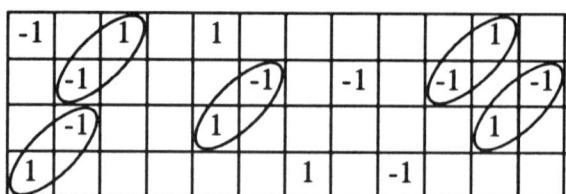

12.3. SUB-EXPRESSION SHARING

2. Remove each occurrence of each sub-expression and replace it by a value 2 or -2 in place of the first of the two terms making up the sub-expression. By "first" is meant first in row-major order on the table entries.

-1		2	1				2	
				-2	-1			-2
	-2							
					1	-1		

3. Record the definition of the sub-expression. This may require a negative value for shift, but do not worry about that right now.

$$x_2 = x_1 - x_1[-1] >> (-1) \ .$$

4. Continue by finding more sub-expressions until done.

-1		3					2	
				-3				-2
	-2							
					1	-1		

$$x_3 = x_2 + x_1 >> 2 \ .$$

5. Write out the complete definition of the filter.

$$\begin{aligned} x_2 &= x_1 - x_1[-1] >> (-1) \\ x_3 &= x_2 + x_1 >> 2 \\ y &= -x_1 + x_3 >> 2 + x_2 >> 10 - x_3[-1] >> 5 - x_2[-1] >> 11 \\ &\quad - x_2[-2] >> 1 + x_1[-3] >> 6 - x_1[-3] >> 8 \ . \end{aligned}$$

6. If any sub-expression definition involves negative shift, then modify the definition and subsequent uses of that variable to remove the negative shift as shown below.

$$\begin{aligned} x_2 &= x_1 >> 1 - x_1[-1] \\ x_3 &= x_2 + x_1 >> 3 \\ y &= -x_1 + x_3 >> 1 + x_2 >> 9 - x_3[-1] >> 4 - x_2[-1] >> 10 \\ &\quad - x_2[-2] + x_1[-3] >> 6 - x_1[-3] >> 8 \ . \end{aligned}$$

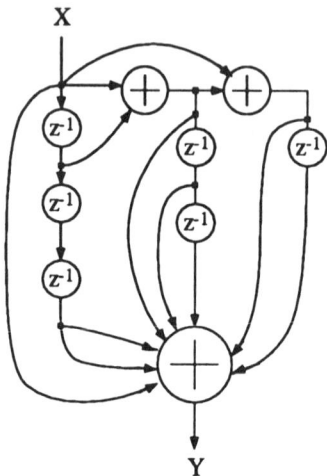

Fig. 12.11: Abstract representation of FIR filter.

12.3.2 Choice of Sub-expression

The choice of which sub-expressions to choose at each iteration is not dependent only on the number of times an expression occurs, but rather on the expected reduction in number of operators. There are two types of operators in the circuit, adders and delay latches.

Estimation of the number of adders: The number of adders that will be eliminated by a common sub-expression occurring N times is $N-1$. The only trick in making sure that the count is accurate is to take care that overlapping sub-expressions are not counted twice. In other words, if a given term occurs twice in sub-expressions of the same type, (once as first member and once as second member) then it should not be counted twice. The number of times each sub-expression occurs is counted by an exhaustive search. This requires a time complexity of $O(n^2)$ where n is the total number of non-zero terms in the table.

Estimation of latch count. At one level of abstraction, an FIR filter implemented as a network of adders may be represented as a graph of multiple input adders and z^{-1}-delays representing the passage from one data value to the previous value. For the present, we ignore shift operations, since they may be easily implemented by appropriately hard-wiring the inputs to each adder. Furthermore, we will ignore the type of each of the operands (addition or subtraction), assuming that all operators are adders. For instance, the circuit described by the example above is represented by Fig. 12.11.

12.3. SUB-EXPRESSION SHARING

In reality, this circuit will be implemented using two-input adders and delay latches. We may use the iterative splitting algorithm described in sections 6.7 and 6.8 to split the multiple-input adders into two-input adders. At the same time, this algorithm schedules the circuit. Next, delay latches must be inserted for two reasons – to realize the z^{-1} delays and to pipeline the circuit to prevent excessive cascading of adders circuits. To get a good estimate of the cost of such a realization, it is necessary to make an estimate of latch count. The number of latches needed will in general depend on the maximum number of adders that may be cascaded between two latches. In order to estimate accurately the number of latches needed, it is necessary to carry out a complete scheduling of the circuit, followed by latch insertion.

The method of delay insertion for the circuits considered in this section differs from that proposed in chapter 6 because we are allowing several operators to be cascaded between latches. In the following discussion, we consider the problem of adding pipelining latches in a pipelined circuite. We investigate particularly how the number of latches needed can be estimated. We start by considering a network of adders without any z^{-1} operators. Such a network is shown in Fig. 12.12(a). We define the time taken by an addition operator to be one time unit. Unlike in chapter 6, this "latency" of the adder does not represent the number of clock cycles taken by the adder, but rather the signal propogation delay from input to output of the adder. One can then schedule the circuit as shown, where the vertical lines indicate the time steps. The time when a given addition operation takes place is called as in chapter 6 the "scheduled time" of the operator. The output of the operator will appear at the output one time unit later. This is called the "ready time" of the operand. Thus the ready time of the output of an adder is one greater than the scheduled time of the adder. For circuit input operands, we may assign a ready time $t = 0$. The scheduling problem is to assign scheduled times to all operators so that no operator is scheduled before the ready times of its inputs.

We may define the "lifetime" of an operand value as being the time elapsed between the moment is was produced as the output of some operator (in this case an adder) and the time it is last used. This may be written formally as an interval

$$[t_{\text{ready}}, \max_i(t_{\text{sink}_i})]$$

where t_{sink_i} represents the scheduled time of an operators where the signal is consumed as input, and i runs over all such operators. For example, the lifetime of the signal x_2 in Fig. 12.12 is the interval $[1, 3]$.

If we wish to limit the number of adders that may be cascaded together, then it is necessary to introduce delay latches. For instance, if we wish to allow no more than two consecutive adders between latches, then we must introduce latches as shown in Fig. 12.12(b). In general, the number of latches that must be inserted can be expressed in terms of the time the signal is created (t_{ready}) and the time it is consumed by the input of another operator, (t_{sink}). By

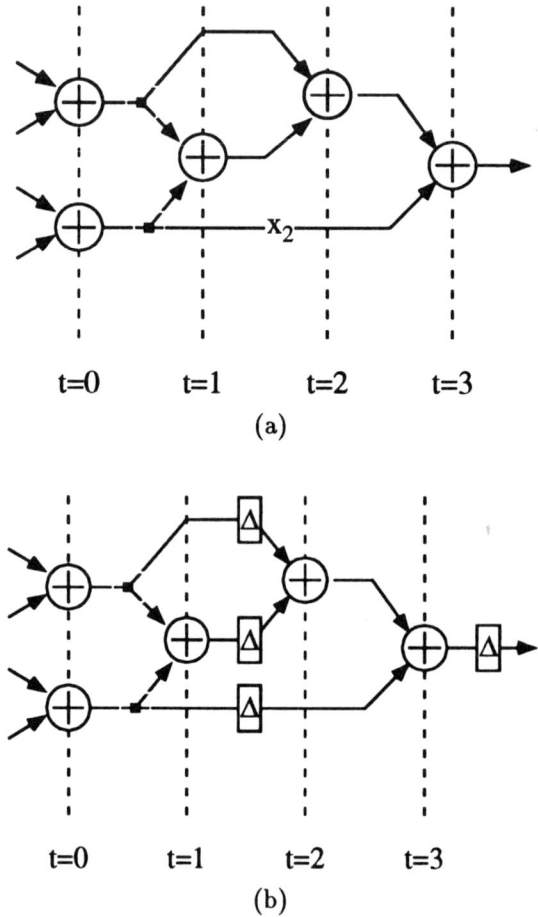

Fig. 12.12: (a) A network of adders. The dotted vertical lines indicate the scheduled time of the adders. (b) Insertion of latch delays to limit the number of cascaded adders to 2.

12.3. SUB-EXPRESSION SHARING

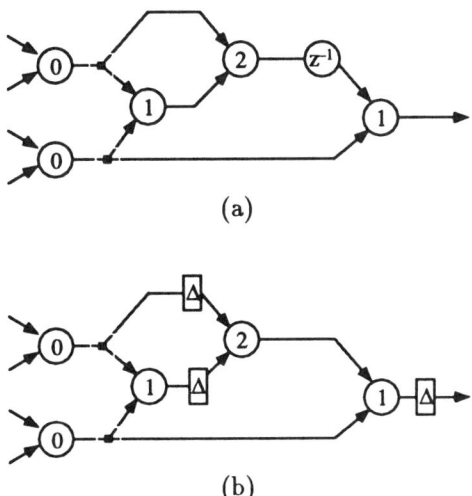

Fig. 12.13: (a) A scheduled newtork of adders. The numbers in the circles represent the scheduled times of the adder operators. (b) The same circuit after removal of the z^{-1} operator and insertion of latches. Note how the position of the latches does not correspond with the location of the z^{-1} operator.

inspection, one determines that the number of latches necessary is equal to

$$\lfloor t_{\text{sink}}/L \rfloor - \lfloor (t_{\text{ready}} - 1)/L \rfloor \tag{12.2}$$

In general, each signal serves as the input of several operators, known as its fanout operators. For a signal with multiple fanout, the total number of latches that must be applied to the signal is equal to the maximum value of (12.2) over all fanout operators for the signal. This is because one may create a sequence of delays and tap the appropriately delayed signal for use as input to each fanout operator. This is illustrated in Fig. 6.8. Consequently, one may write the total number of latches applied to a signal in terms of its lifetime. For a signal with lifetime $[a, b]$, the total number of latches to apply is equal to

$$\text{Total \# latches} = \lfloor b/L \rfloor - \lfloor (a-1)/L \rfloor . \tag{12.3}$$

Now, we turn to networks including z^{-1} operators. Such circuits may be correctly scheduled by assigning a latency of $-L$ time units to the z^{-1} operator. Thus the previous value of a signal (the output of the z^{-1} operator) is available L time units before the present value (the input of the z^{-1} operator. The circuit may as before more be scheduled using the algorithms of chapter 6. Such a circuit is shown in Fig.12.13. Since an FIR filter has no feedback loops, an earliest-possible scheduling may be carried out in a single pass in the direction from input towards output of the graph. The z^{-1} operator does not itself correspond to any actual circuitry. After scheduling is complete, the z^{-1} operator is removed, and delay latches are inserted. This is the same approach as was used for fully synchronous digit-serial circuits in Chapter 6.

To determine the total number of latches required in a circuit, we define a "signal" to correspond to each of the (hyper)-edges in the network after the z^{-1} operators have been removed. Thus, different delayed versions, $x, x[-1], x[-2], \ldots$ of an operand x are all considered part of the same signal and are referred to as different versions of that signal. The number of latches that need to be attached to any signal is equal to the maximum number that will be required for any of the fanout operator inputs attached to that signal. Consider a signal produced at time t_{ready}. If $t_{\text{sink}}(i)$ is the scheduled time of the i-th fanout operator of the signal, and if version $[-k(i)]$ of the signal is used at that operator input, then the number of latches that must be applied to the signal is given by

$$\max_i (k(i) + \lfloor t_{\text{sink}}(i)/L \rfloor - \lfloor (t_{\text{ready}} - 1)/L \rfloor) \qquad (12.4)$$

where i runs over all the fanout operator inputs of the signal.

This formula may be expressed in terms of the lifetime of each signal. We consider how the lifetime of a signal should be defined in circuits containing different versions of signals. If version $x[-k(i)]$ of a signal x is used at time $t_{\text{sink}}(i)$, then $x[-k(i)]$ must survive until time $t_{\text{sink}}(i)$. Since $x[-k(i)]$ represents the $k(i)$-th previous value of x, it follows that x itself must survive until time $t_{\text{sink}}(i) + Lk(i)$. Thus, the lifetime of the signal x is the interval

$$\text{lifetime}(x) = [t_{\text{ready}}, \ \max_i(Lk(i) + t_{\text{sink}}(i)] \ . \qquad (12.5)$$

Applying this definition to equation (12.4) we see that if the lifetime of a signal is the interval $[a, b]$, then as before, the number of latches that must be applied to a signal is given by the formula (12.3). The total number of latches in a circuit may be computed by summing over all signals in the circuit.

Estimating change in latch count. In estimating the change in latch count that will occur when a given sub-expression is chosen, one possible way is to do a complete rescheduling of the circuit assuming the sub-expression in question will be chosen. It is feasible to carry this out for a set of attractive sub-expressions chosen on the basis of numbers of adders saved. A weighted sum of the number of adders and latches saved by the choice of each of the sub-expressions may be used to choose the best sub-expression.

Since completely rescheduling the graph is a lengthy process to carry out repeatedly, an alternative method has been used in computing examples. This heuristic method relies on the remark that all other factors being equal, it is best to combine signals with overlapping lifetimes in common sub-expressions,

12.3. SUB-EXPRESSION SHARING

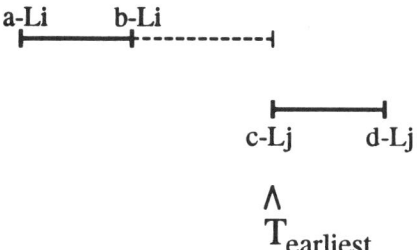

Fig. 12.14: Combining two non-overlapping lifetimes. The lifetime of signal x_1 must be extended to $c - Lj$ in order to be able to sum $x_1[-i]$ and $x_2[-j]$. The earliest time at which the two signals may be added is marked $T_{earliest}$

since otherwise, the lifetime of one of the signals must be extended in order for the operation to be correctly scheduled. Such an extension of lifetime of a signal has a cost in terms of extra latches.

The situation may be analyzed more precisely as follows, with reference to Fig. 12.14. Suppose that two terms $x_1[-i]$ and $x_2[-j]$ are to be combined. Normally, either i or j will be zero according to our algorithm, but we may ignore this fact. Suppose that the lifetime of x_1 is $[a, b]$, and the lifetime of x_2 is $[c, d]$. It may be reasoned that the lifetimes of $x_1[-i]$ and $x_2[-j]$ are $[a-Li, b-Li]$ and $[c-Lj, d-Lj]$ respectively. The earliest possible scheduled time of the new operator combining $x_1[-i]$ and $x_2[-j]$ will be at time $\min(a-Li, c-Lj)$. If the two lifetime intervals overlap, then the scheduled time of the new operator will lie within the lifetime of the two variables, and neither lifetime needs to be extended for the new operator to be scheduled. Suppose, on the other hand that the two lifetime intervals do not overlap, and assume without loss of generality that $c - Lj > b - Li$. The new operator will be scheduled at time $c - Lj$ at the earliest. The lifetime of signal x_2 will not need to be extended, but the lifetime of $x_1[-i]$ will need to be extended to $[a - Li, c - Lj]$. Equivalently, the lifetime of x_1 will be extended to $[a, c - Lj + Li] = [a, c + L(i - j)]$. Expressing numbers of latches in terms of the lifetimes, using (12.3), the original number of latches on the signal x_1 was $\lfloor b/L \rfloor - \lfloor (a - 1)/L \rfloor$, whereas the new number of latches is $\lfloor (c + L(i - j))/L \rfloor - \lfloor (a - 1)/L \rfloor$. The increase in the number of latches is therefore equal to

$$\lfloor (c + L(i - j))/L \rfloor - \lfloor b/L \rfloor = \lfloor c/L \rfloor - \lfloor b/L \rfloor + (i - j) \;.$$

Taking account of the other possibility that $a - Li > d - Lj$, we see that in the general case, the additional number of latches required to combine $x_1[-i]$ and $x_2[-j]$ will be

$$\max(0, \lfloor c/L \rfloor - \lfloor b/L \rfloor + (i - j), \lfloor a/L \rfloor - \lfloor d/L \rfloor + (j - i)) \qquad (12.6)$$

It should be emphasized that this number represents the number of additional latches that must be added to the presently scheduled circuit simply to allow the terms $x_1[-i]$ and $x_2[-j]$ to be added together, and does not take account in any way various complicating factors such as :

- The new term will be used in place of x_1 and x_2 in some places where they are used. This will change the lifetime of x_1 and x_2.

- Saving adders by combining these terms will cause a saving in the number of latches needed.

- Substituting the common sub-expression will lead to a totally different topology and scheduling of the graph.

Therefore, it is clear that expression (12.6) does not in any way accurately represent the actual change in latch count resulting from the substitution of a sub-expression, nor is it intended to. In fact, as will be seen from the examples the latch count actually decreases when sub-expression are used. Rather expression (12.6) represents an approximation to the extra latch count that may result from selecting the given pair of terms, compared with selecting a pair of terms (if available) with overlapping lifetimes, assuming that all other factors are equal for the candidate pairs.

Cost Function : The cost function used to choose the best common subexpression will be a weighted difference of the number of adders saved and the additional latch count as specified by formula (12.6).

This estimate of latch-count discriminates quite strongly against choosing sub-expressions involving operators with widely different delay. In particular, suppose that a sub-expression $x + x[-i]$ is a candidate. Let $[a, b]$ be the lifetime of the signal x. According to (12.6), the additional latch count is equal to $i - (\lfloor b/L \rfloor - \lfloor a/L \rfloor)$, assuming that this quantity is non-negative. For large i, this may be quite significant. For instance, in an example discussed below, during one iteration, a sub-expression $x_1 + x_1[-11]$ was found to occur 14 times, easily the most common sub-expression at that point. Nevertheless, it was never chosen, since the presence of the $x[-11]$ term increased the lifetime of the variable x_1 by 10 cycles, thereby making the choice less attractive.

In fact, because of the latch-count factor, it is rare for delayed versions of variables to be involved in chosen sub-expressions. In the example below, out of 20 sub-expressions chosen, only two involved variables with different amounts of delay.

12.3.3 An Example

This algorithm was run on a filter containing 23 coefficients of 32 bits chosen at random. The results are shown in the following table.

12.3. SUB-EXPRESSION SHARING

# operators between latches	Before finding common sub-expressions		After finding common sub-expressions	
	# adders	# latches	# adders	# latches
1	206	208	100	104
2	206	78	98	57
3	206	42	99	39
4	206	22	98	25
5	206	21	98	22
10	206	21	98	22

The filter was designed with differing values for the maximum number of adders that may be cascaded in series without a pipelining latch as shown in first column of above table. Fig. 12.10 has the minimum number of latches necessary to implement the circuit. In that figure, adders are cascaded together without intervening latches. In a digit-serial or bit-serial design, however, it is usually desirable to include pipelining latches to break up long chains of cascaded adders. At the other extreme, a latch may be inserted after each adder. As the number of adders that may be cascaded together increases, the number of latches decreases, but not below the minimum number required to implement the z^{-1} operators in the filter.

As shown, the extraction of common sub-expressions cuts the amount of hardware by about 50%. An interesting point is that increasing the number of operators cascaded between latches beyond about 3 operators rarely decreases the total number of latches in the circuit, and so decreases the performance with no benefit. The reason is that a minimum number of latches are necessary to implement the sample delays. As soon as the point is reached where all additions necessary to implement one multiplication are accomplished in one sample delay, no further benefit is derived from increased cascading. The necessity of including the latch cost when choosing sub-expressions is demonstrated by comparing with the results obtained without considering latch cost. In this case, with 3 operators between latches the result is 93 adders and 70 latches. This is certainly inferior to the 99 adders and 39 latches shown in the above table. This example requires under 1 minute to run on a Sun 4.

12.3.4 Routability

The above discussion does not consider routability. As the number of sub-expressions grows, it becomes more difficult to maintain a structured layout style for the filter such as the linear layout style described previously. In the example above, a total of about 23 common sub-expressions were found. Most of the last of these were expressions occurring only twice or three times. If parallel arithmetic is used, it is probably counter-productive to calculate and reuse

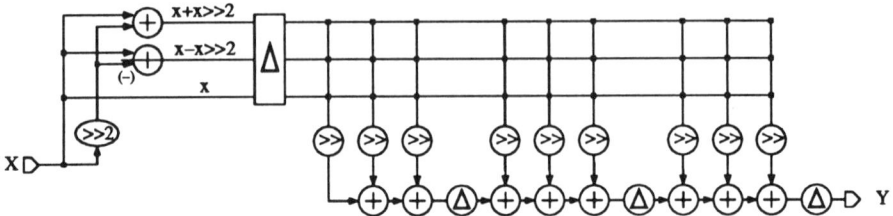

Fig. 12.15: FIR filter using term sharing.

such common expressions, since one is then faced with the task of laying out a complicated data-flow graph. On the other hand, the two most common sub-expressions in the graph occurred 42 and 21 times respectively. As remarked, these may alternatively be considered as sub-expressions of sizes 42 and 21 being reused just once each. This will clearly have little adverse effect on routability, while decreasing the amount of hardware by 30% – a very substantial gain.

12.3.5 Using the Two Most Common Sub-expressions

In our example, the two most common sub-expressions are $x - x \gg 2$ and $x + x \gg 2$, and statistically, this will always be so. Substantial gains in circuit area can be achieved very easily by finding the occurrences of only these two sub-expressions. An expression $x - x \gg 2$ corresponds to a sequence $10\bar{1}$ in one of the filter coefficients, and the expression $x + x \gg 2$ corresponds to a sequence 101. It can be shown that asymptotically an N-bit CSD number can be broken down into $N/18 + O(1)$ pairs of type $10\bar{1}$, $N/18 + O(1)$ pairs of type 101, and $N/9 + O(1)$ isolated 1 or $\bar{1}$ bits. Referring to pair $10\bar{1}$, a pair 101, or an otherwise isolated non-zero bit as a term, it can be shown that the asymptotic expected number of terms in an N-bit CSD number is $2(N+1)/9$. This represents a 33% saving compared with the total number of non-zero bits, which is equal to $(3N+1)/9$. Furthermore, taking advantage of these common sub-expressions can be done without a major increase in routing cost. Fig. 12.15 shows a possible linear layout for an FIR filter designed using term sharing. In this figure the two terms $x + x \gg 2$ and $x - x \gg 2$ are pre-computed. The three terms are then fed along the row of cells. Each of the adders taps its inputs off one of the three buses (all three connections are shown in the diagram, but connection is made to one of the buses only). Although the three buses are shown above the data path, in reality, the best plan is to have them running right over the adder cells, or else with one bit between each pair of bit-slices in the adders.

Comparing this design style with that of Fig. 12.10, which does not use term sharing, four differences may be detected. (Note that the filters in the two examples are not the same.)

12.3. SUB-EXPRESSION SHARING

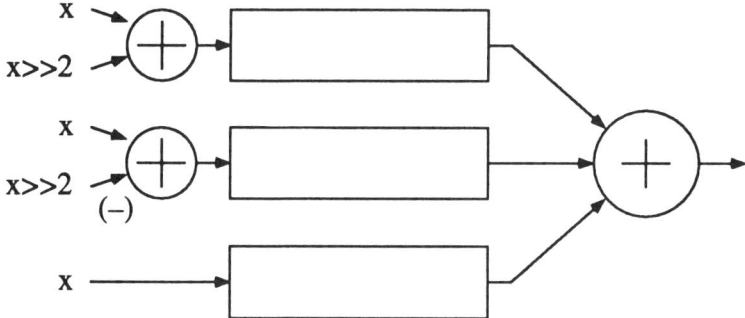

Fig. 12.16: Alternative layout for carry-save FIR filter.

1. The design of Fig. 12.15 requires the two extra complete adders to produce the shared terms. However, this is a once only cost and is relatively negligible for large filters.

2. The design of Fig. 12.15 requires that three buses run the length of the data path, instead of only one. This may require a slight increase in the pitch of the bit-slices.

3. The design of Fig. 12.15 may be expected to have about 30% fewer adders, which should far outweigh the increased operator size due to possibly greater pitch.

4. In the design of Fig. 12.15, there will in general be fewer adders between the delay stages. This is because the number of adders in one stage is equal to the number of terms in the corresponding filter coefficient. The maximum number of non-zero bits in a CSD number is equal to $\lceil N/2 \rceil$, in a number such as 101010101. On the other hand, if pairs are counted as a single term, the maximum number of terms which may occur is $\lceil N/3 \rceil$. If the number of terms in a single stage is too high, then it may be the limiting factor in determining the maximum clock speed. In such a case, the design of Fig. 12.15 will run at a higher clock speed.

Alternative layout style: An alternative layout shown in Fig. 12.16 is possible, in which all the terms of each type are combined in a separate data-path. Then the three partial results are summed to obtain the final output.

In this layout, best results will be achieved if the three data-paths are to some extent balanced, that is, contain the same number of terms. One benefit of this may be layout convenience. Another important benefit relates to the clock rate. The maximum clock rate of the circuit is dependent on the maximum number of adders in one stage of the pipeline. In order to keep the number of

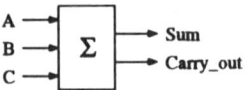

Fig. 12.17: One bit full adder circuit.

adders in each stage at a minimum, it is desirable to keep the number of adders balanced in each of the three data paths. As remarked above, however, the number of terms in the x data path is on the average twice as many as in the $x+x \gg 2$ and $x-x \gg 2$ paths. This inequality can be redressed by swapping x terms for $x+x \gg 2$ and $x-x \gg 2$ terms. This is done by departing from a strict adherence to CSD number representation. For instance, a bit sequence 1001 containing two isolated 1-bits and leading to terms $x + x \gg 3$ can be rewritten as $101\bar{1}$. This gives rise to terms $(x + x \gg 2) - x \gg 3$. Thus, one of the x-terms has been replaced by an $x + x \gg 2$ term. Similarly, a sequence 10001 can be written as $10\bar{1}00 + 00101$. In this way, two isolated 1-bits have been replaced by two pairs.

12.4 Carry-Save Arithmetic

In applications in which speed is particularly important and the goal is to achieve the maximum possible throughput, carry-save arithmetic has been used with success. This technique is applicable mainly to designs using pipelined parallel arithmetic [1][2][86]. The idea behind carry-save arithmetic is to reduce propagation delay to a minimum by avoiding any carry-propagation from bit to bit. Carry-save arithmetic is the basis of the well-known Wallace-tree multiplier [87]. The concept of carry-save arithmetic is easily illustrated by considering carry-save addition. A one-bit full adder circuit, as shown in Fig. 12.17, takes three input bits of equal weight, denoted A, B and C, and produces two output bits, denoted **Sum** and **Carry_Out** according to the formula

$$A + B + C = \text{Sum} + 2.\text{Carry\_Out} .$$

In other words, the **Carry_Out** bit has weight 2, whereas all the other bits of input and output have weight 1. Note that there is no essential difference between the three input bits. Normally, in a ripple-carry adder, the carry-out from one bit is connected to the carry-in of the bit of next-highest order. In a carry-save adder, on the other hand, the **Carry_Out** bit is not connected to the next highest-order adder within the same addition operation, but rather is "saved" and connected to the carry-in of an adder in the next addition operation, hence the name. Fig. 12.18 shows an array of full adders configured as an N-bit carry-save adder. As may be seen, the effect of a carry-save adder is to reduce three input words to two. Normally, the one-bit adder used

12.4. CARRY-SAVE ARITHMETIC

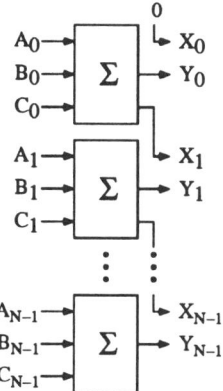

Fig. 12.18: N-bit carry-save adder.

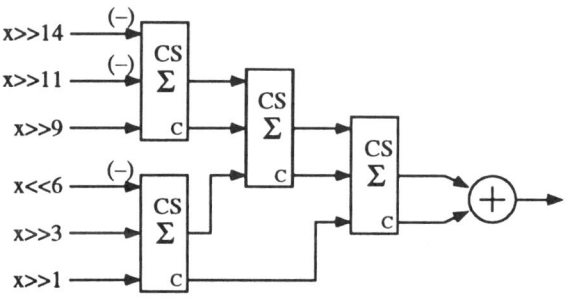

Fig. 12.19: A CSD multiplier using carry-save arithmetic.

in carry-save arithmetic will be designed differently from the adder used in a ripple-carry adder. In a ripple-carry adder design, the one-bit adder cell will be optimized for maximum speed of propagation from **CarryIn** to **CarryOut**, possibly at the cost of increasing the time used to produce the **Sum** output bit. In a design of an adder used in carry-save arithmetic, the time taken for the production of the **Sum** and **Carry_Out** output bits should be balanced.

Carry-save arithmetic may be used in CSD constant multiplication. The resulting circuit will be a sort of fixed coefficient Wallace tree multiplier. Fig. 12.19 shows a possible topology for the same multiplication shown in Fig. 12.1. As shown in the figure, a final full addition is necessary to combine the final two terms. This is a standard feature when using carry-save arithmetic. The final addition should be a fast adder such as a carry-select or carry-look-ahead adder in order to maintain a fast clock rate. It is also possible to use a linear arrangement of carry-save adders much as in Fig. 12.2. The advantages of such an arrangement will be a much simpler layout task.

242   CHAPTER 12. CANONIC SIGNED DIGIT ARITHMETIC

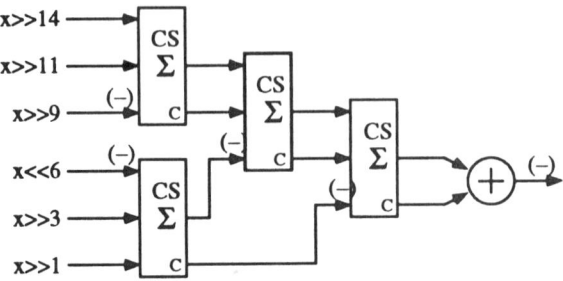

Fig. 12.20: CSD multiplication using carry-save adders.

12.4.1 Carry-Save Subtraction

In carrying out CSD multiplication using carry-save arithmetic, it is necessary to have available a carry-save subtractor. Consider first the case where one of the inputs is to be subtracted and the other two are to be added. It is easy to modify the CS-adder in Fig. 12.17 to a subtractor. Each bit of the input to be subtracted is inverted and a 1 value is fed in at the X_0 output position (see Fig. 12.18). This corresponds to the usual way of negating a number in two's complement format by inverting the bits and carrying in a 1 at the least significant bit position. The technique of pushing subtractions towards the root can be used to avoid the situation where more than one of the inputs to the CS-adder must be subtracted. In the case of a CS-adder/subtractor with two inputs to be subtracted the circuit modification relies on the identity $a - b - c = -(-a + b + c)$. The corresponding circuit modification is as shown :

Fig. 12.20 shows the multiplier circuit of Fig. 12.19 with subtractions pushed towards the root.

As may be seen, the subtraction of the $x \gg 14$ and $x \gg 11$ terms propagate all the way to the output, necessitating a final negation. This may be avoided by modifying the subtractor such that the **Carry_Out** word is negated, and hence has negative weight. In other words, we design a circuit to carry out the operation

$$\textbf{Sum} - 2.\textbf{Carry\_Out} = A + B - C$$

Since the least significant bit of the **Carry_Out** output (X_0) equals 1, the **Carry_Out** word may be negated by inverting all the bits except X_0.

Using this circuit, the appropriate transforms for propagating subtractions towards the root are shown in Fig. 12.21, where the operators on the right represent the new negative-carry circuit. Fig. 12.22 shows the same multiplier

12.4. CARRY-SAVE ARITHMETIC

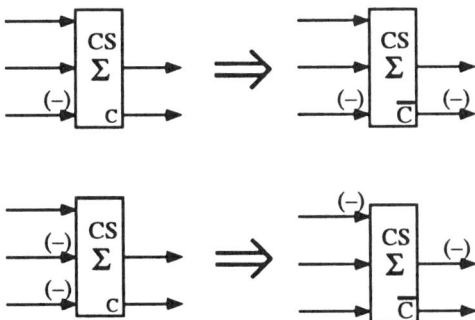

Fig. 12.21: Propagating subtractions towards the root.

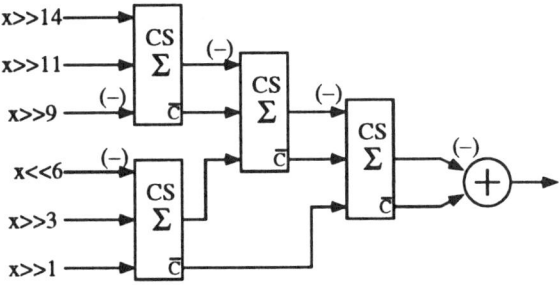

Fig. 12.22: Multiplier using modified CS subtractor.

as before using this modified subtractor. In this case, the final result can never be negative unless all partial product terms are negative, since a positive input operand will propagate all the way to the output. The technique of deferring shifts may be applied to multiplier trees using carry-save adders in a manner analogous to its use in Fig. 12.3.

12.4.2 Speed of a Carry-Save Multiplier Circuit

Assume that there are no latches in the circuits. It may appear that the circuit of Fig. 12.19 will be almost 3 times as fast as the circuit in Fig. 12.1, since the time for a carry-save addition is almost negligible compared with that of the final full addition. However, the advantage will not in fact be as great as it may appear. The reason is that the time taken by N cascaded ripple-carry adders is not equal to N times the time taken by one adder. Since all the ripples are in the same direction, a subsequent addition may begin computation of its low-order outputs before the previous one has completed the correct computation of its high order bits. In fact, if an ordinary ripple-carry adder is used for the full adders in each circuit, then the circuit of Fig. 12.19 may even be slower than the

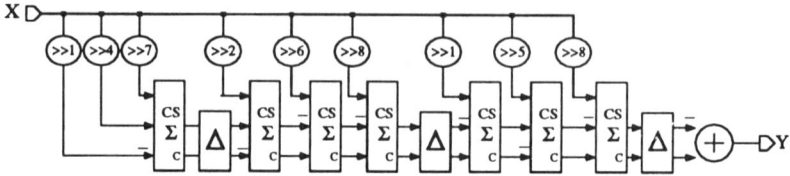

Fig. 12.23: 3-tap FIR filter using carry-save multipliers.

one in Fig. 12.1, since it has one more level. The advantage of Fig. 12.19 is that a fast (expensive) adder may be used for the final full addition. To use a fast adder for each of the adders in Fig. 12.1 would be excessively costly. Carry-save arithmetic really becomes attractive in cases similar to the FIR filters discussed in section 12.3, where both sum and carry words are carried through several latch stages with only one final addition at the very end.

12.4.3 Filters using Carry-Save Arithmetic

Filters may also be implemented using carry-save adders to carry out the multiplications using CSD representation of the filter coefficients. Fig. 12.23 shows the FIR filter of Fig. 12.10, this time implemented using carry-save arithmetic. The final carry-save adder produces two output words, which must therefore be combined in a final full adder. In order to match the speed of the rest of the circuit, this final adder should be a fast adder, such as a carry-select or carry-look-ahead adder. This final full-word addition is a characteristic of all carry-save arithmetic circuits.

The FIR filter design style using carry-save arithmetic and CSD representation of the filter coefficient has been incorporated into an automatic design tool called FIRGEN [79] and laid out using gate arrays. In this system, very high clock rates were achieved due to the use of carry-save addition and a high degree of pipelining. The maximum clock rate is limited by the maximum of the propagation times of the carry-save adders within each latch stage and the time taken for the final full addition. If the total number of adders in a given stage is excessive (corresponding to a filter coefficient with many non-zero bits) then it may be necessary to add extra pipeline stages in order to maintain a high clock rate. This would normally result in a longer latency for the filter output. Operation rates of the order of 2 Gops (Giga operations per second) were reported in [88]. This count is obtained by multiplying the clock rate by the number of multiply or add operations in the filter equation. In Fig. 12.23, the adders constituting a single multiplication are combined in a linear array. The layout for such an arrangement of adders is extremely simple. The data path consists of a linear array of carry-save adder/subtractor and delay cells. A single bus carrying the input X word runs along the length of the array of

12.4. CARRY-SAVE ARITHMETIC

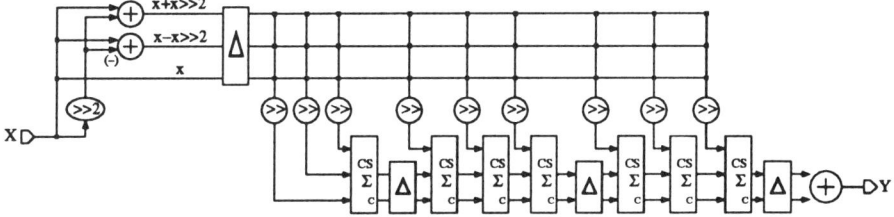

Fig. 12.24: Carry-Save FIR filter using term sharing.

cells, preferably right over the top and is shifted up or down by the appropriate number of places to obtain a hard-wired implementation of the shift operation. By contrast, an arrangement of the adders in a tree arrangement will make the routing of signals much more difficult. In particular, consider the second stage of the filter of Fig. 12.23. This stage contains three additions, and could be carried out by a tree of adders of height two. This would possibly increase the maximum clock rate of the circuit, but would be at the cost of layout complexity. In particular, there would be no way to lay out such a non-linear tree of adders in such a way that signals are passed only to adjacent cells.

One point needing care is the load on the X input bus. In a filter topology such as the one used here, the fan-out of the X input is quite large, and this is especially the case for filters with many taps. The solution used in FIRGEN is to lay the cells of the filter out in several columns and to have separate drivers for the X bus along each column.

Term sharing with carry-save arithmetic: The method of term sharing to cut down the number of additions may be applied to filter designs based on carry-save arithmetic as well. The filter design shown in Fig. 12.15 may easily be adapted to carry-save arithmetic, as shown in Fig. 12.24 Note that the initial additions in this diagram must be full adders (as opposed to carry-save adders), since otherwise there will be twice as many terms to combine.

The alternative layout shown in Fig. 12.16 may also be implemented in carry-save arithmetic. Once again the initial additions must be carried out using full adders. With carry-save arithmetic, the outputs of each of the three data paths consist of Sum and CarryOut components. The summation of the six remaining terms will be accomplished using a carry-save tree.

Finally, a layout that avoids the expensive initial full additions (see Fig. 12.16) may be obtained by deferring the computation of the two shared terms. This transformation is related to the remark 12.3. This layout is shown in Fig. 12.25.

Once again, in reality, each of the values being summed by the final adder has both Sum and CarryOut components. The ten values will normally be summed using a carry-save adder network. The advantage of this layout is that

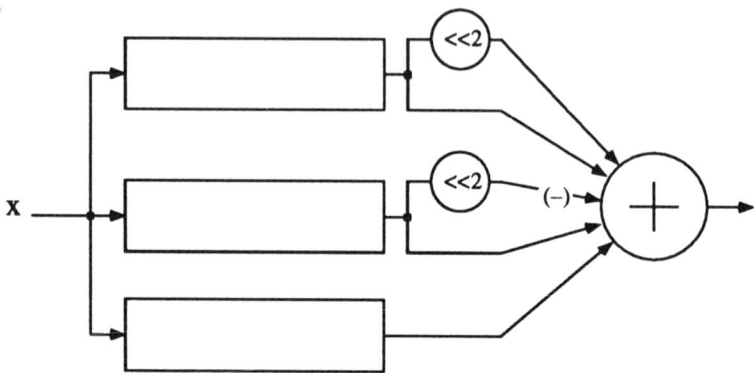

Fig. 12.25: Deferring the expensive full addition to last.

only one full addition (the final addition) is necessary.

12.5 Asymptotic Occurrence of Pairs.

In this last section of the chapter, we derive formulae for the number of non-zero bits and pairs in CSD numbers. This section is somewhat mathematical in nature, and can be skipped if desired. Previous expressions deriving formulae for the number of non-zero bits have been published in [78] using different methods. On the way to our goal of deriving formulae for the frequency of occurrence of various bit combinations, the formulae of Reitwiesner will appear as preliminary results. The general approach used here is to cast the problem as recurrence equations and to solve them using generating functions (otherwise known as the z-transform). In doing this, the symbolic algebra program Macsyma [89] was used to derive and verify results, especially for the computation of partial fraction expansions of rational functions. Readers are asked to take the correctness of these expansions on faith, though they may offer a recreational opportunity for those with a penchant for formal algebraic manipulation.

12.5.1 Number of n-bit CSD integers.

Denote by f_n the number of possible CSD values with n bits. It is easily seen that all possible numerical values between the minimum value, $101010\ldots$ and the minimum value, $\bar{1}0\bar{1}0\bar{1}0\ldots$ may be expressed as n-bit CSD values. This is approximately $\frac{4}{3}2^n$ values. We could easily deduce the correct value directly. Instead, it will be computed by difference equations. The set of all n-bit CSD integers can be divided up into those that start with 0 and those that start with

12.5. ASYMPTOTIC OCCURRENCE OF PAIRS.

1. Consequently, we have a recurrence formula

$$\begin{align} f_1 &= 3 \\ f_2 &= 5 \\ f_n &= f_{n-1} + 2f_{n-2} \end{align} \quad (12.7)$$

The two terms in the recurrence formula correspond to the number of terms starting with 0 and 1 or $\bar{1}$ respectively. In order to satisfy the recurrence, it is most natural to define $f_0 = 1$, which we will do. A common method of solving recurrence equations is to use generating functions. Consequently, we define

$$F(z) = \sum_{i=0}^{\infty} f_n z^n$$

The recurrence may now be expressed by

$$F(z)(1 - z - 2z^2) = 1 + 2z$$

or

$$F(z) = \frac{1 + 2z}{1 - z - 2z^2}$$

Expressing this in terms of partial fractions gives

$$F(z) = \frac{4}{3} \frac{1}{1 - 2z} + \frac{-1}{3} \frac{1}{1 + z} \quad (12.8)$$

Since for any a

$$\frac{1}{1 - az} = \sum_{n=0}^{\infty} a^n z^n$$

we deduce from (12.8) that

$$f_n = \frac{4}{3} 2^n - \frac{1}{3}(-1)^n \quad (12.9)$$

12.5.2 Number of non-zero bits in CSD integers.

Next, we denote by g_n the total number of all 1 or $\bar{1}$ bits in all n-bit CSD integers. As before, this can be divided up into the number of bits in numbers starting with 0 and the number of bits in numbers starting with 1 or $\bar{1}$. These two contributions are easily seen to be equal to g_{n-1} and $2(g_{n-2} + f_n)$ respectively. Therefore

$$\begin{align} g_0 &= 0 \\ g_1 &= 2 \\ g_{n+2} &= g_{n+1} + 2g_n + 2f_n \end{align} \quad (12.10)$$

Multiplying equation (12.10) by z^n for each n and summing gives

$$z^{-2}(G(z) - 2z) = z^{-1}G(z) + 2G(z) + 2F(z) \ .$$

Multiplying by z^2 and substituting formula (12.8) for $F(z)$ leads to

$$(1 - z - 2z^2)G(z) = 2z + \frac{2z^2(1 + 2z)}{1 - z - 2z^2} \ .$$

So

$$G(z) = \frac{2z}{(1 - 2z)^2(1 + z)^2} \ . \tag{12.11}$$

The partial fraction expansion of this is

$$G(z) = \frac{1}{27}\left(\frac{-4}{1 - 2z} + \frac{12}{(1 - 2z)^2} + \frac{-2}{1 + z} + \frac{-6}{(1 + z)^2}\right) \tag{12.12}$$

In order to handle the terms with the square in the denominator, we note that in general,

$$\frac{1}{(1 - az)^2} = \sum_{n=0}^{\infty}(n + 1)a^n z^n \ .$$

Therefore, we derive from (12.12) the general formula for g_n :

$$g_n = \frac{1}{27}\left((12n + 8)2^n - (6n + 8)(-1)^n\right) \ . \tag{12.13}$$

12.5.3 Restricting the range of CSD integers

We are particularly interested in the CSD integers that are obtained from n-bit twos' complement integers, that is, those that lie in the range $[-2^{n-1}, 2^{n-1})$. It is easily seen that the number $1000\ldots$ lies outside this range, as do all numbers starting with $10\ldots$ or $\bar{1}0\cdots$ in which the next non-zero bit is of the same sign as the first bit. All other numbers lie in the desired range.

We denote by f'_n and g'_n the quantities analogous to f_n and g_n for integers in the restricted range. Since every integers can be expressed in CSD form, it is clear that $f'_n = 2^n$, but we will derive this fact separately. It can be seen that all CSD integers starting with 0 are in the desired range. On the other hand, exactly half of the values starting with 1 or $\bar{1}$ lie in the range. This is because for each CSD integer with two leading non-zero bits of the same sign (hence out of range) there is one identical except in the sign of the second non-zero bit, which therefore lies in range. Further, the value $\bar{1}00\cdots$ lies in range, whereas $10\cdots$ does not. Now, the number of values starting with 0 is f_{n-1}, and so the number not starting with 0 is $f_n - f_{n-1}$. Therefore,

$$\begin{aligned} f'_n &= f_{n-1} + (f_n - f_{n-1})/2 \\ &= (f_n + f_{n-1})/2 \end{aligned} \tag{12.14}$$

12.5. ASYMPTOTIC OCCURRENCE OF PAIRS.

From this it may easily be computed that $f'_n = 2^n$. The computation may be carried out directly, or in the z-domain.

To compute g'_n it may be observed that the same reasoning holds that was applied to f'_n. In particular, (12.14) holds with g replacing f. Direct computation using (12.13) yields

$$g'_n = \frac{1}{9}\left((3n+1)2^n - (-1)^n\right) \qquad (12.15)$$

It is interesting to note how this could have been derived by computations in the z-domain. Indeed, from (12.14) applied to g_n may be derived

$$\begin{aligned} G'(z) &= G(z)(1+z)/2 \\ &= \frac{z}{(1+z)(1-2z)^2} \end{aligned} \qquad (12.16)$$

Expressing (12.16) in terms of partial fractions will lead to (12.15). The absence of the term $(1+z)^2$ in the denominator of (12.16) explains the absence of the $n(-1)^n$ term in (12.15) compared to (12.13).

12.5.4 Asymptotic frequency of non-zero bits.

The expected frequency of non-zero bits in an n-bit CSD integer is our goal. This is equal to g_n/f_n and is easily computed by dividing formula (12.13) by formula (12.9). As $n \to \infty$ the asymptotic value is easily computed to be

$$\lim_{n \to \infty} (g_n/f_n) = n/3 + 2/9 \ . \qquad (12.17)$$

For numbers in the restricted range $[-2^{n-1}, 2^{n-1})$, the answer is slightly different:

$$\lim_{n \to \infty} (g'_n/f'_n) = n/3 + 1/9 \ . \qquad (12.18)$$

12.5.5 Frequency of pairs.

We now turn to the number of pairs in n-bit CSD integers. A pair is defined as a pair of non-zero bits separated by a single 0 bit. Thus 101, 10$\bar{1}$, $\bar{1}$01 and $\bar{1}$0$\bar{1}$ are the four possible pairs. We are interested in the occurrence of pairs wherever they appear in the word. In counting pairs, we do not allow a single bit to occur in more than one pair. Thus, a sequence 001010100 contains one pair, not two, whereas 00101010100 contains two pairs. Before determining the number of pairs, we need to determine the total number of n-bit CSD values starting with an even or an odd length string of ± 1s separated by single 0 bits. Thus, the word 10$\bar{1}$0100$\cdots$ starts with a string of an odd number (3) of ± 1s, whereas 0100$\cdots$ starts with a string of an even number (0) of ± 1s. Denote by b_n the number of values starting with a string of even length, and by c_n the number of values starting with a string of odd length. Since every value starts with an even or odd string of ± 1s, we have $f_n = b_n + c_n$, with f_n defined as before. Various recurrence relations hold between b_n and c_n. First, if a word starts with a string of odd length, then truncating the first two bits gives a word starting with an even string. Therefore, since there are two choices (1 or $\bar{1}$) for the leading bit of the original word

$$c_n = 2b_{n-2} \ .$$

Similarly, a word starting with a string of even length must either start with 0 or be a continuation of a word starting with a string of odd length. Consequently,

$$b_n = 2c_{n-2} + f_{n-1} \ .$$

Since $f_{n-1} = b_{n-1} + c_{n-1}$, this becomes

$$b_n = b_{n-1} + c_{n-1} + 2c_{n-2} \ .$$

Passing to the z-domain, and taking account of the starting conditions $c_0 = 0$, $c_1 = 2$ and $b_0 = b_1 = 1$, we get

$$\begin{aligned} B(z) &= 1 + zB(z) + (z + 2z^2)C(z) \\ C(z) &= 2z + 2z^2 B(z) \ . \end{aligned}$$

From these two equations, it is easy to derive

$$C(z) = \frac{2z}{(1+z)(1-2z)(1+2z^2)} \ . \tag{12.19}$$

It is not necessary for our purposes to compute the formula for c_n, so we remain in the z-domain. Next we turn to the computation of the number of pairs. Denote by p_n the total number of pairs in all n-bit CSD integers. The recurrence relation is

$$p_n = p_{n-1} + 2(p_{n-2} + c_{n-2}) \ . \tag{12.20}$$

The term p_{n-1} represents the number of those pairs occurring in words starting with 0, and $2(p_{n-2} + c_{n-2})$ is the number of pairs in words starting with 1 or $\bar{1}$. The pairs in words starting with ± 1 are of two types – those already present in the words of length $n - 2$ obtained by truncating the two leading bits (represented here by the p_{n-2} term) and those formed by the leading ± 1 bit and ± 1 in the third bit position. However, such a pair can only be counted if the truncated word of length $n - 2$ starts with an odd-length string of ± 1s. This explains the c_{n-2} term. Switching to the z-domain, taking account of the fact that $p_0 = p_1 = p_2 = 0$, gives

$$P(z)(1 - z - 2z^2) = 2z^2 C(z) \ .$$

12.5. ASYMPTOTIC OCCURRENCE OF PAIRS.

Using formula (12.19), we obtain

$$P(z) = \frac{4z^3}{(1-2z)^2(1+z)^2(1+2z^2)}$$

Once again using partial fraction expansion, we obtain

$$P(z) = \frac{1}{81}\left(\frac{-20}{1-2z} + \frac{12}{(1-2z)^2} + \frac{4}{1+z} + \frac{-12}{(1+z)^2} + \frac{16-28z}{1+2z^2}\right).$$

Hence, we may derive a formula for p_n :

$$\begin{aligned}p_n &= \frac{1}{81}\left((12n-8)2^n - (12n+8) + 16(-2)^{n/2}\right) \quad \text{for } n \text{ even} \\ &= \frac{1}{81}\left((12n-8)2^n + (12n+8) + 14(-2)^{(n+1)/2}\right) \quad \text{for } n \text{ odd} \quad (12.21)\end{aligned}$$

We are also interested in finding the number of pairs in the set of CSD integers in the range $[-2^{n-1}, 2^{n-1})$. This time it is easier to work in the z-domain (supposing computer aid in computing partial fraction expansions). As with the derivation of equation (12.14), we may deduce $p'_n = (p_n + p_{n-1})/2$, and hence (since $p_0 = p'_0 = 0$)

$$P'(z) = \frac{(1+z)P(z)}{2} = \frac{2z^3}{(1-2z)^2(1+z)(1+2z^2)}$$

The partial fraction expansion is

$$P'(z) = \frac{1}{27}\left(\frac{-6}{1-2z} + \frac{3}{(1-2z)^2} + \frac{-2}{1+z} + \frac{5-2z}{1+2z^2}\right).$$

This leads to

$$\begin{aligned}p'_n &= \frac{1}{27}\left((3n-3)2^n - 2 + 5(-2)^{n/2}\right) \text{ for } n \text{ even} \\ &= \frac{1}{27}\left((3n-3)2^n + 2 + (-2)^{(n+1)/2}\right) \text{ for } n \text{ odd} \quad (12.22)\end{aligned}$$

12.5.6 Asymptotic frequency of pairs.

The frequency of pairs is equal to p_n/f_n. In the limit, this value is

$$\lim_{n\to\infty} p_n/f_n = \frac{n}{9} - \frac{2}{27}.$$

For values restricted to the range $[-2^{n-1}, 2^{n-1})$ the value is slightly different :

$$\lim_{n\to\infty} p_n/f_n = (n-1)/9.$$

Now, considering that the asymptotic number of non-zero bits in a CSD integer is approximately $n/3$ according to equation (12.18), we see that the frequency of pairs is approximately one third the frequency of ± 1s. However, each pair accounts for two non-zero bits. Therefore, it follows that asymptotically two-thirds of the non-zero bits occur in pairs.

Chapter 13

Online Arithmetic

13.1 Redundant Data Formats

The topic of most-significant-digit-first (msd-first) serial arithmetic has been discussed in the literature of computer arithmetic for many years starting with the paper of Avizienis in 1961 ([90]). The basic idea can also be applied to the design of parallel arithmetic cells for avoiding ripple-carry propagation. It relies on the use of a redundant format number representation to allow serial operations to proceed in a most-significant-bit first mode. The particular number representation used is also called signed-digit number representation and computation using signed-digit number representations and msd-first serial arithmetic has come to be known as online arithmetic.

The carry-free property of the redundant number system overcomes the speed limitation imposed by the carry-ripple operation in two's complement architectures. The carry-free property allows the redundant format architectures to be operated at a speed which is independent of the word-length. In this chapter we present radix-2 and higher radix (including radix-4) architectures for online addition and multiplication [91] [92]. Redundant number based architectures for division, square-root and rotation operations have been reported in several references and are not studied in this book.

In signed-digit representation, numbers are represented by sequences of digits A_i where the sequence $A_{m-1}A_{m-2}...A_0$ represents the number $K \sum_{i=0}^{m-1} A_i r^i$. The base r is called the radix and K is some constant that specifies the range. The length of the sequence, m will be termed the *word-length*. In this chapter we will be concerned exclusively with radices that are powers of 2. Furthermore, the range specifier, K will also be a power of 2. For simplicity, we will assume throughout the chapter that $K = 1$ and so the numbers represented are integers. This will rarely matter except for multiplication.

Notation. Digits will be represented in this chapter by upper case letters, such as A or A_i. Usually digits are represented in two's complement format. Thus if the individual bits of a digit A need to be considered it will be denoted as $<a_n a_{n-1} \ldots a_0>$ where the a_i's are the individual bits. Thus, the binary bit-string $<a_n a_{n-1} \ldots a_0>$ represents the number $-a_n 2^n + \sum_{j=0}^{n-1} a_j 2^j$. Negative digits will often be represented by using overbars. For instance the notation $\bar{1}$ represents the digit value -1.

It will always be assumed that the digits belong to a consecutive set of integers containing at least $r+1$ distinct values. This has the effect that numbers have more than one representation as a signed-digit string. For this reason, we distinguish semantically between a *number*, which is a member of the set of arithmetic values being represented (that is, a real number, usually an integer) and the signed-digit string that represents it. The latter string will be called a *signed-digit string*. It is usual to make two further assumptions about the radix set.

(13.1) (Range of the digit set) Any digit A satisfies $|A| < r$.

(13.2) (Symmetry of the digit set) If A is a possible digit value, then so is $-A$.

Under these assumptions some simple properties are easily observed. From (13.1) follows two useful properties.

(13.3) The arithmetic sign of number represented as a signed-digit string is equal to the sign of the leading non-zero digit.

(13.4) The arithmetic value 0 has a unique signed-digit representation, in which all digits are zero. Other values have more than one representation.

The assumption (13.2) makes it simple to take the negative of a number.

(13.5) The negative of a number represented as a signed-digit string is obtained by negating each digit.

For radix $r = 2^n$, the set $\{-(r-1), \ldots, r-1\}$ is called the maximally redundant digit set, whereas $\{-r/2, \ldots, r/2\}$ is called the minimally redundant set. In either case, each digit may be represented using $n + 1$ bits. For radix $r = 4$, the maximally redundant set is $\{\bar{3}, \bar{2}, \bar{1}, 0, 1, 2, 3\}$ and the minimally redundant set is $\{\bar{2}, \bar{1}, 0, 1, 2\}$. For radix $r = 2$, each digit can be $\{\bar{1}, 0, 1\}$ and radix-2 redundant number representation is also referred to as signed-binary number representation. Consider the representation of a number -5 for a word-length of 4. In radix-2, this number can be represented as $0\bar{1}0\bar{1}$ or $\bar{1}10\bar{1}$ or $\bar{1}1\bar{1}1$ or $0\bar{1}\bar{1}\bar{1}$ or $\bar{1}011$. This non-unique or redundant representation leads to the carry-free property. Usually, digits are represented in two's complement format; however other schemes have also been suggested ([93][94] [95][67]) and are used later in this chapter.

The redundant number representation leads to the desirable carry-free property, but makes the sign checking of a number difficult. While the most significant bit in a two's complement number provides sign information, the leading non-zero digit in a redundant number provides the sign information. The leading non-zero digit can be in any position from most significant to least significant and in the worst case all digits need to be checked to determine the sign of the digit. Thus, implementation of the comparison operation is less efficient in most significant digit (msd) first or online implementations. In redundant arithmetic, the carry-free arithmetic operations can be implemented either in least-significant-digit (lsd) first or msd-first mode. However, the redundant arithmetic architectures are attractive for msd-first implementations in division, square-root, and compare operations which are inherently msd-first. These architectures are also suitable in implementation of DSP operations which involve loops such as recursive and adaptive digital filters where the use of msd-first redundant arithmetic makes the loop latency independent of word-length [94].

13.2 Reduction of the Range of Digits

Before we begin to consider the familiar arithmetic operations, a basic operation on signed-digit strings will be discussed – the operation of reduction of the range of digits, or simply *reduction*. Let the radix be $r = 2^n$. In the maximally or minimally redundant digit set, each digit can be represented in $n + 1$ bits. We consider however a signed-digit string in which each digit lies in an extended range and is represented by $n + q$ bits with $q > 1$. The goal is to reduce the range of each digit, and hence the number of bits used in its representation. The reduction takes place in two stages, namely

(13.6) *Generation of carry and remainder.* Each digit A_i is split into two parts, a *carry* and a *remainder* such that

$$A_i = r.\text{carry}_i + \text{remainder}_i .$$

(13.7) *Summing carry and remainder.* The new digit A'_i is formed according to the formula

$$A'_i = \text{carry}_{i-1} + \text{remainder}_i .$$

There is a certain amount of freedom in how one divides the digit up into carry and remainder. A simple choice is as follows.

(13.8) *Range of remainder.* The remainder lies in the range $-r/2 \le$ remainder $< r/2$.

This choice is made because it allows each digit to be split up into carry and remainder in a particularly simple way. In fact it may be seen that for a radix $r = 2^n$, and a digit $A = <a_{n+q-1} a_{n+q-2} \ldots a_0>$ of length $n+q$, the remainder has two's complement representation

$$\text{remainder} = <a_{n-1} a_{n-2} \ldots a_0>, \tag{13.1}$$

whereas the carry is the number

$$\text{carry} = <a_{n+q-1} a_{n+q-2} \ldots a_n> + a_{n-1}. \tag{13.2}$$

To verify this, observe that the two's complement number $<a_{n-1} \ldots a_0>$ lies in the range $[-2^{n-1}, 2^{n-1} - 1]$, satisfying step (13.8) of the reduction algorithm. Furthermore

$$\text{remainder} + r.\text{carry} = -a_{n-1} 2^{n-1} + \sum_{i=0}^{n-2} a_i 2^i$$

$$+ 2^n \left(-a_{n+q-1} 2^{q-1} + \sum_{i=0}^{q-2} a_{n+i} 2^i + a_{n-1} \right)$$

$$= -a_{n+q-1} 2^{n+q-1} + \sum_{i=0}^{n+q-2} a_i 2^i$$

$$= A$$

as required by step (13.6) of the reduction algorithm.

As seen, the *carry* value is either $<a_{n+q-1} a_{n+q-2} \ldots a_n>$ or one more than this value, depending on whether a_{n-1} is 0 or 1. It is not necessary to do the addition to form this correct carry value right away. It is more efficient to wait until the carry is added to the remainder of the previous digit in step 13.7 of the reduction algorithm. Then the bit a_{n-1} serves as the low-order *carry – in* to the adder computing the final digit value.

Fig. 13.1 shows the choice of remainder and carry bits for radix $r = 4$ for digits in the range $-6, \ldots, 6$.

Example 13.9. Consider an example in which $r = 4$ and each digit lies in the range $-6 \leq A_i \leq 6$, and hence is representable in 4 bits. The goal is to reduce this to a signed-digit string in which the digits lie in the range $-3 \leq A'_i \leq 3$ and hence may be represented using 3 bits. Consider the signed-digit string $\bar{2}64\bar{6}$.

Step 1: Generation of carry and remainder:

$$\begin{array}{cccc} \bar{2} & 6 & 4 & \bar{6} \\ \hline \bar{2} & \bar{2} & 0 & \bar{2} \\ 0 & 2 & 1 & \bar{1} \end{array}$$

13.2. REDUCTION OF THE RANGE OF DIGITS

y_j	carry	remainder	y_j	carry	remainder
−6	−1	−2	6	2	−2
−5	−1	−1	5	1	1
−4	−1	0	4	1	0
−3	−1	1	3	1	−1
−2	0	−2	2	1	−2
−1	0	−1	1	0	1
0	0	0			

Fig. 13.1: Carry-sum generation for radix 4 addition.

The first of the two bottom lines represents the remainder digits, whereas the bottom line represents the carry digits.

Step 2: Summation of carry and remainder: The carry digits are shifted one place to the left and added to the remainder digits.

$$
\begin{array}{ccccc}
 & \bar{2} & \bar{2} & 0 & \bar{2} \\
0 & 2 & 1 & \bar{1} & \\
\hline
0 & \bar{1} & \bar{1} & \bar{2} &
\end{array}
$$

It may be verified that the original and final signed digit strings represent the same number −22.

For a general radix $r = 2^n$ it may be seen that digits represented by $2n$ bits (that is in the range $-2^{2n-1} \leq A_i \leq 2^{2n-1} - 1$) split up into carry and remainder each of n bits in step 1 of the above algorithm. Hence the summation of carry and remainder, plus a low-order carry-in bit equal to a_{n-1} produces an $n+1$-bit result. In short, one step of reduction can reduce $2n$-bit digits to $n+1$-bit digits.

If $q > n$, then the remainder is n bits and the carry is q-bits. In this case, summation of remainder and carry words results in a $q+1$-bit digit. Thus, one step of reduction can reduce the $n+q$-bit digits to $q+1$-bit digits. This process can be repeated to reduce the size of the digit eventually to $n+1$ bits.

Inhibiting overflow: In the example above the signed-digit string produced after normalization represents the same number as the original digit string. This is because the carry produced in the high order digit position was zero. Any carry value produced in the high order digit position will be lost unless the number of digits in the output is to be increased. For this reason, it is a good idea to treat the high order digit differently by inhibiting any carry-out from the most-significant digit. Instead, the final digit should simply be truncated to the number of bits used to represent the digit in the final output word.

Example 13.10. Consider another example, the reduction of the string $\bar{4}66\bar{6}$, where radix $r = 4$, and we wish to reduce from the 4-bit range $-8, \ldots, 7$ to a range $-4, \ldots, 3$ requiring just 3 bits.

Step 1: Generation of carry and remainder:

$$
\begin{array}{cccc}
\bar{4} & 6 & 6 & \bar{6} \\
\hline
\bar{4} & \bar{2} & \bar{2} & \bar{2} \\
0 & 2 & 2 & \bar{1}
\end{array}
$$

Step 2: Summation of carry and remainder: The carry digits are shifted one place to the left and added to the remainder digits.

$$
\begin{array}{ccccc}
 & \bar{4} & \bar{2} & \bar{2} & \bar{2} \\
0 & 2 & 2 & \bar{1} & \\
\hline
\bar{2} & 0 & \bar{3} & \bar{2} &
\end{array}
$$

The result is equal to decimal -142. Note how carry-out is inhibited from the most-significant digit. Using the method of Example 13.9 the computation would overflow from the top digit and an incorrect result would be achieved.

Supposing that radix is $r = 2^n$ and that the result of the reduction is to be represented in $n+1$ bits, truncation of a digit A to $n+1$ bits produces a remainder in the range $-r \leq$ remainder $< r$ satisfying the congruence remainder $\equiv A$ (mod $2r$).

Example 13.11. A final example of reduction shows that as long as the mathematically correct final digit value can be represented correctly in $n+1$ bits (or $q+1$ bits if $q > n$), the result will be correct despite the truncation. Consider the reduction of the string $\bar{5}66\bar{6}$, where radix $r = 4$, and we wish to reduce from the 4-bit range $\{-8, \ldots, 7\}$ to a range $\{-3, \ldots, 3\}$ requiring just 3 bits. First of all, we carry out the operation without truncation or carry (that is, retaining full precision) in the high-order digit.

Step 1: Generation of carry and remainder:

$$
\begin{array}{cccc}
\bar{5} & 6 & 6 & \bar{6} \\
\hline
\bar{5} & \bar{2} & \bar{2} & \bar{2} \\
0 & 2 & 2 & \bar{1}
\end{array}
$$

13.2. REDUCTION OF THE RANGE OF DIGITS

Step 2: Summation of carry and remainder: The carry digits are shifted one place to the left and added to the remainder digits.

$$\begin{array}{ccccc} & \bar{5} & \bar{2} & \bar{2} & \bar{2} \\ 0 & 2 & 2 & \bar{1} & \\ \hline & \bar{3} & 0 & \bar{3} & \bar{2} \end{array}$$

This gives the correct result, decimal -206. Next, we consider what happens when the last digit is truncated to $n+1 = 3$ bits in the last digit. In this case, only values in the range $\{-4, \ldots, 3\}$ can be represented, and any value that falls outside this range will be reduced, modulo 8 to a number in this range. The computation is then

Step 1: Generation of carry and remainder:

$$\begin{array}{cccc} \bar{5} & 6 & 6 & \bar{6} \\ \hline 3 & \bar{2} & \bar{2} & \bar{2} \\ 0 & 2 & 2 & \bar{1} \end{array}$$

In this step, the value $\bar{5}$ is reduced to 3 by truncation.

Step 2: Summation of carry and remainder: The carry digits are shifted one place to the left and added to the remainder digits.

$$\begin{array}{ccccc} & 3 & \bar{2} & \bar{2} & \bar{2} \\ 0 & 2 & 2 & \bar{1} & \\ \hline & \bar{3} & 0 & \bar{3} & \bar{2} \end{array}$$

In the second step, because of overflow, the summation of 3 and 2 gives a result of $\bar{3}$. The result of the computation is correct, despite the truncation and overflow.

This verifies the principle.

> The result obtained by reduction, with overflow inhibition will be correct, as long as the mathematically correct high-order digit may be correctly represented in the available number of bits.

Because each digit of output depends only on the input digit and the next less significant digit, reduction may be carried out in a parallel ripple-free manner. A circuit that carries out parallel digit reduction is shown in Fig. 13.2.

Reduction may easily be carried out on most-significant-digit first serial data. The circuit is shown in Fig. 13.3.

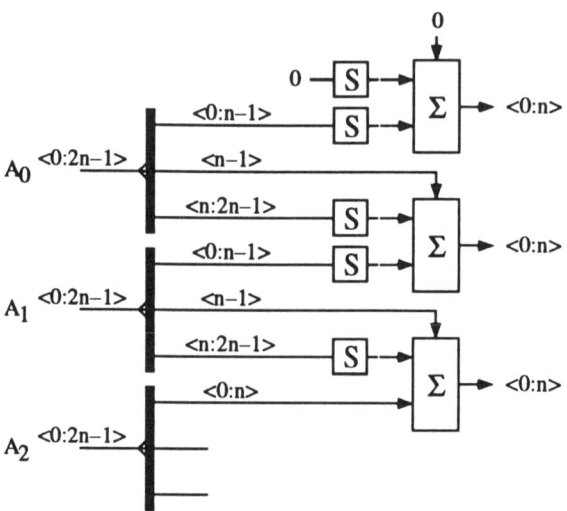

Fig. 13.2: Parallel Reduction Circuit.

The boxed symbol S denotes sign extension of the input to $n+1$ bits. Note that in the high order digit, the $n+1$-th bit is not obtained by sign extension, but by truncation of the input digit. Bits $n-1$ and $<n:2n-1>$ of the input word are added to give the carry word according to (13.2).Note that they are added by providing bit $n-1$ as the carry-in to the final addition.

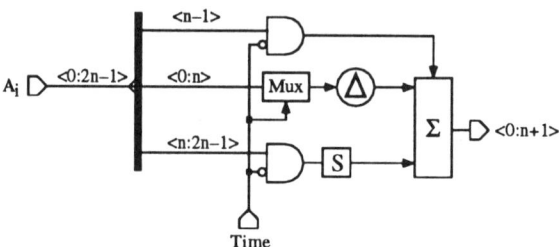

Fig. 13.3: Online Reduction Circuit.

The periodic control signal TIME is used to mark (high) the most-significant digit (the first digit) of each word. It is used to set the carry word and the adder carry-in to zero to prevent them from interfering with the computation (still not complete) of the previous word. The TIME signal also controls the MUX circuit, which accepts an $n+1$-bit input word, and produces an $n+1$-bit output word either by overwriting the $n+1$-th bit of the input with the duplicated n-th bit (when TIME is low), or by passing the input unchanged to the output (when TIME is high). The remainder word (bits $<0:n>$ must be delayed for one clock cycle to synchronize it with the carry word from the next, less significant digit.

13.3 Addition

Given the reduction algorithm, it is easy to see how addition may be carried out on numbers represented in signed-digit format. We assume that the radix is $r = 2^n$ and each digit is represented in $n + 1$ bits. Addition is carried out in two steps:

13.12 Addition of corresponding pairs of $n + 1$-bit digits to produce $n + 2$-bit results.

13.13 Reduction of the resulting digits to $n + 1$ bits.

It was seen that the process of reduction can reduce digits of $2n$ bits to digits of size $n + 1$. Hence, the addition will succeed as long as $2n \geq n + 2$, that is as long as $n \geq 2$, or $r \geq 4$, and as long as the resulting high-order digit is representable in $n + 1$ bits.

It is important also to notice that if the digits are in the maximally redundant set $\{-3, \ldots, 3\}$ for radix 4, then the result of summing two digits will be in the range $\{-6, \ldots, 6\}$. In this case, the generated remainder and carry digits (see table in Fig. 13.1) will be such that when added, the result will again lie in the range $\{-3, \ldots, 3\}$. Thus, the maximal redundant set is preserved by this operation. The same is true of higher radices.

One may also notice that the digit set $\{-3, \ldots, 2\}$ is preserved by the addition operation, though this does not seem to have any particular application.

Example 13.14. Consider the addition of the numbers represented by signed-digit strings $\bar{3}33\bar{3} = -135$ (decimal) and $\bar{2}33\bar{3} = -71$ (decimal). In the summation step 13.12, pairs of corresponding digits are added, producing a string $\bar{5}66\bar{6}$. Example 13.11 shows how this is reduced to the string $\bar{2}0\bar{4}1$ representing the decimal number -206.

Addition of signed-digit numbers may either be carried out on digit-parallel words, or on most-significant-digit first digit-serial data. (Least-significant-digit first addition is also possible but is not of much interest since this can be performed in two's complement number system using less hardware.) The adder circuit is constructed by placing a digit-adder ahead of the reduction circuit shown in Fig. 13.2 or Fig. 13.3.

Addition using the reduction circuit of Fig. 13.3 has a latency of one cycle, though an additional delay on the output (not shown) is mandated by the pipelined nature of serial architectures as described in chapter 6. In describing the latency of online operations, the term essential latency may be used to describe the delay associated with the operator excluding all pipeline delays not necessary for the logically correct function of the operator. Fig. 13.3 shows that the essential latency of online addition is one cycle. The exception is radix 2 addition which has an essential latency of 2 cycles (see section 13.6). (It is

y_j	carry	remainder	y_j	carry	remainder
−6	−1	−2	6	1	2
−5	−1	−1	5	1	1
−4	−1	0	4	1	0
−3	−1	1	3	1	−1
−2	0	−2	2	0	2
−1	0	−1	1	0	1
0	0	0			

Fig. 13.4: Carry-sum generation for radix 4 addition.

shown in [67], however, that with a different representation of the digits, one cycle latency is sufficient).

13.4 An Alternative Implementation

The choice or the method of division of a digit into remainder and carry, as determined by rules (13.6), (13.7) and (13.8) has a significant influence on the computational properties of the redundant format system. Other choices are possible. A common choice ([96]) is to replace the rule 13.8 by the rule

(13.15) *Range of remainder.* The absolute value of the remainder is as small as possible, subject to the conditions 13.6 and 13.7.

With this policy, the choice of remainder and carry words for digits in the range $\{-6,\ldots,6\}$ is shown in Fig. 13.4.

One observes that with this method of carry/remainder generation the maximally redundant set ($\{-3,\ldots,3\}$ in radix 4) is preserved by addition. The disadvantage of this system of carry/remainder generation is that it requires more circuitry. This is especially true in higher radix cases.

Now we describe design of radix-4 redundant adder architectures in digit-parallel, msd-first and lsd-first implementation styles [97]. Fig. 13.5(a) shows the digit-parallel implementation of this carry-free adder. Fig. 13.5(b) shows the msd-first and Fig. 13.5(c) shows the lsd-first implementation of this carry-free adder. Note that in the msd-first adder, the delayed w or sum bit of A cell is input to the B cell whereas in the lsd-first adder the delayed t or transfer-out digit (which is similar to the carry-bit) of the A cell is input to the B cell. The multiplexors associated with the t bits ignore the most significant transfer bit of the current word and set $t_{-1} = 0$ for the next word in the same clock cycle, once every word-length number of cycles. In this adder, the latency of the add operation is one clock cycle (where each clock cycle represents the delay of the digit-adder cell A) which is independent of the word-length. Note that the cell B does not perform any computation but obtains the sum digit by concatenating the two input digits appropriately.

13.4. AN ALTERNATIVE IMPLEMENTATION

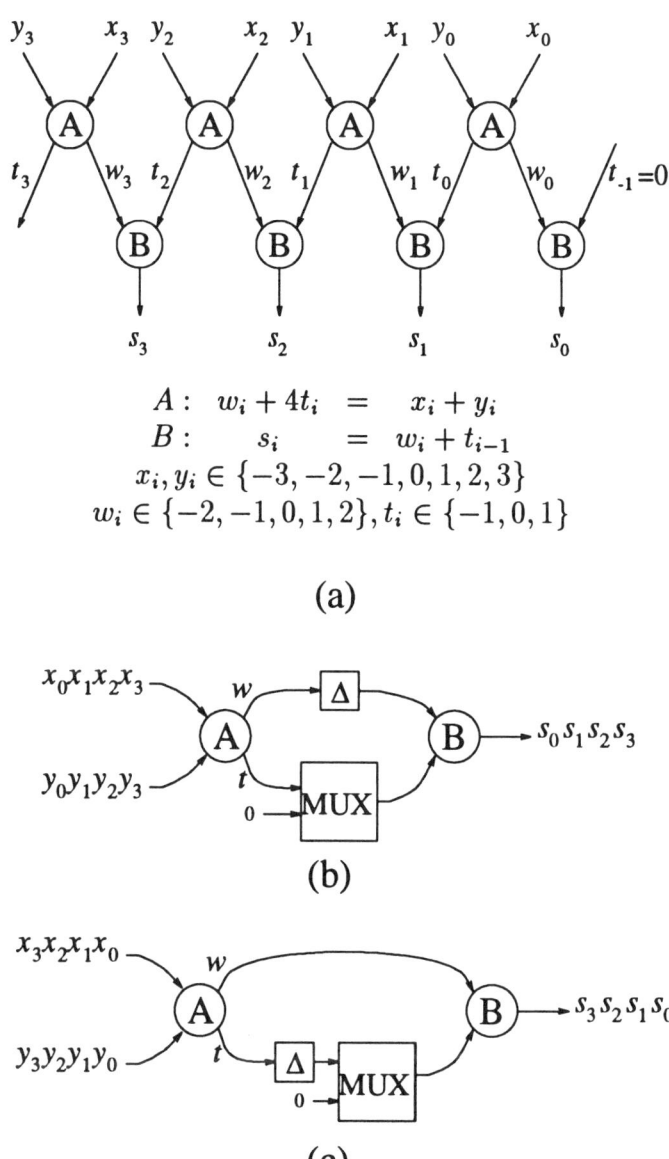

Fig. 13.5: Radix-4 maximally redundant adder.

Consider the adder implementation in the minimally-redundant radix-4 representation as shown in Fig. 13.6. The latency of this architecture is greater than that of the radix-4 maximally-redundant adder. The shorter latency of the radix-4 maximally redundant adder makes it preferable to the radix-4 minimally redundant adder.

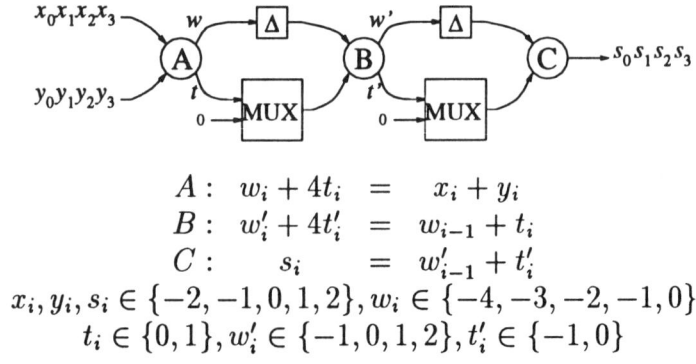

$$A: \quad w_i + 4t_i = x_i + y_i$$
$$B: \quad w'_i + 4t'_i = w_{i-1} + t_i$$
$$C: \quad s_i = w'_{i-1} + t'_i$$
$$x_i, y_i, s_i \in \{-2, -1, 0, 1, 2\}, w_i \in \{-4, -3, -2, -1, 0\}$$
$$t_i \in \{0, 1\}, w'_i \in \{-1, 0, 1, 2\}, t'_i \in \{-1, 0\}$$

Fig. 13.6: Radix-4 minimally redundant adder.

13.5 Data Range and Overflow

In the case of two's complement number representation the properties of the number range is sufficiently well understood to need little explanation. For instance, with n-bit two's complement number representation any number in the range $[-2^{n-1}, 2^{n-1} - 1]$ can be represented. Furthermore, for all the common operations like addition and multiplication, if both the input and output operands lie in the representable range, then the standard computation algorithms will give the correct arithmetic result, free from overflow problems. The same statement can also be made regarding lsd-first digit-serial computation as described in this book. For the redundant signed-digit format used in on-line arithmetic on the other hand, it is necessary to be far more careful about overflow problems and correctness of computational methods.

In a digit-serial architecture, either lsd-first or msd-first, the number of digits input to an operator and the number of digits output must be equal. Consequently, the result of an operator must be representable in the same number of digits as the inputs, or else overflow will occur. (The only alternative is to represent the output in double or multiple-precision format, as may be done with digit-serial multiplication as described in chapter 3.) Although temporary overflow may sometimes be tolerated, it is generally to be avoided, since it may

13.5. DATA RANGE AND OVERFLOW

lead to incorrect arithmetic results.

We begin by identifying the source of the difficulty with redundant signed-digit number representation. Non-redundant data formats such as two's complement binary or higher radix representations have the following pleasing property. If a number a is representable in n digits as $a_{n-1}a_{n-2}\ldots a_0$, then it may also be representable in more than n digits as well (by sign extension). However the truncation of the longer representation to n digits will result in the original number, and hence will be a correct representation of the number. This is not true of redundant signed-digit number representations. For instance, consider radix 4 and a word-length of 4 digits. The smallest representable value is 0001. This can also be represented as $\bar{1}3333$ in which the number has overflowed into a fifth digit. The truncation to 4 digits is not a correct representation of the number. This difference means that it is not possible to tell simply from the numerical value of a signed-digit number whether it is safe to truncate it to n digits.

Consequently, without knowledge of the particular implementation of operations which will be carried out on data, there is no way of knowing, simply from the arithmetic value of the result, that the result will be correct without overflow when truncated to n digits. *With* knowledge of the operations that will be carried out on data, the situation is more promising, however. For instance, one might hope that a range R of numbers might be found such that if the inputs and the result of a given arithmetical operation falls within the given range, then the result of the operation is correct, without overflow. This condition may be stated formally as follows.

Definition 13.16. Let R be an interval of real numbers and **op** an arithmetic operation with an online (or some other) implementation, **OP**. The operator **OP** is said to be *overflow-safe on* R if the following condition holds :

> Whenever A and B are signed-digit strings representing arithmetic values a and b in R such that a **op** b is also in R, then A **OP** B is a representation of the number a **op** b.

A set of online operators is called overflow-safe on R if each operator is individually overflow-safe on R.

Thus, the overflow-safe range for an operator is the range of numbers for which the online digital operation **OP** is a valid representation of the abstract arithmetic operation **op**. For instance, addition as defined abstractly for integers does not overflow, whereas digital addition on words of finite length does overflow. Only if the numbers being added lie in a restricted range, R can be regarded as a specific digital addition algorithm correctly carrying out an abstract addition operation. This concept allows an algorithm designer to evaluate an algorithm (perhaps by simulation) using abstract arithmetic operations. If inputs, outputs and all intermediate results of an algorithm lie inside some range

R of real numbers, then it can be guaranteed that a circuit implementing the algorithm using online arithmetic (or some other digital implementation) will generate the correct results provided all the operators **OP** used are overflow-safe on the range R.

It is clear that the overflow-safe range for addition of two's complement binary or higher-radix non-redundant values is the whole range of numbers which may be represented for the given word-length. The goal of the next few paragraphs is to consider radix-4 redundant format addition operation defined in section 13.3 and to determine a range R on which it is overflow-safe. Thus, the operation **OP** is the signed-digit addition algorithm of section 13.3, in which the output is truncated to the same number of digits as the input words. The operation **op** is the abstract arithmetic addition.

13.5.1 Reduction without Overflow Inhibition

In describing the operation of reduction in section 13.2 we showed that the technique of overflow inhibition in the high-order digit may be used to extend the range of the operation by avoiding data loss from the high order digit. It may be thought that this operation is optional, and is required only if one wishes to use the full possible range of representable numbers. It has the disadvantage that the high-order digit must be treated differently from the others. It will be shown here that in fact this operation of overflow inhibition is essential. Without it, one is never free of the threat of overflow. We will consider in this section radix-4 addition without overflow inhibition.

First a few examples will give some feel for the problem. A word-length of 4 digits is assumed.

Example 13.17. It is possible to obtain overflow simply by adding zero to a number. Consider the sum of 3000 and 0000. Since the digit 3 decomposes into a carry of 1 and a remainder of -1 (see Fig. 13.1), the result is $1\bar{1}000$, which when truncated to four digits is $\bar{1}000$ – not the correct result. Conclusion : Numbers starting with 3 or $\bar{3}$ are certainly not in the overflow-safe range. In fact, only numbers starting with 1, 0, $\bar{1}$ or $\bar{2}$ will not overflow when added to zero.

The problem becomes worse when one considers the next example

Example 13.18. Adding $1\bar{3}\bar{3}\bar{3} + 1\bar{3}\bar{3}\bar{3}$ gives $2\bar{6}\bar{6}\bar{6}$ which when reduced according to Fig. 13.1 and truncated to 4 digits gives $\bar{3}\bar{3}\bar{3}2$ which overflows. The numbers being added, however, are both equal to 1. Thus we see that adding $1+1$ causes overflow.

As a result, we deduce that addition is overflow safe only when the result is no greater than 1. This ludicrous result means that the reduction scheme of the table in Fig. 13.1 is unworkable without a scheme for overflow inhibition from the high-order digit.

13.5. DATA RANGE AND OVERFLOW

13.5.2 Reduction with Overflow Inhibition

With overflow inhibition, however, the result is quite different. It was shown at the end of section 13.2 that the result of reduction will be correct, as long as the result of the computation lies in the range of representable digits in the high digit position. If we are using the maximally redundant digit set $\{-3,\ldots,3\}$, then as long as the high order digit would lie in this range, the result will be correct. Accordingly, we seek the smallest positive number which has a representation with high-order digit outside of this range. The number in question is $4\bar{3}\bar{3}\bar{3} = 3001$. One may verify that the number 3000 or any smaller number cannot be represented with a 4 in the high order bit position and with the other digits in the range $\{-3,\ldots,3\}$. A similar argument holds with negative numbers, and one deduces that if the result of the addition lies in the range $[\bar{3}000, 3000]$, then the addition will be correct.

By definition, therefore, the addition is overflow-safe on the range $[\bar{3}000, 3000]$. This is not quite the largest overflow-safe range for this addition, however. One notices that the result of an addition must have one of the values $\bar{2}$, $\bar{1}$, 0 or 1 in the least-significant digit, since these are the only possible remainder values according to table 13.1. The smallest positive overflowing result then is $4\bar{3}\bar{3}\bar{2} = 3002$, and the largest negative overflowing result is $\bar{4}331 = \bar{3}00\bar{3}$. This means that the largest overflow-safe range for addition is equal to $[\bar{3}00\bar{2}, 3001]$, and any addition operation for which the result falls in this range will give the correct result. The following example shows that this range cannot be extended.

Example 13.19. Adding $2\bar{3}\bar{3}\bar{3} + 21\bar{3}\bar{3}$ gives $4\bar{2}\bar{6}\bar{6}$. When the high-order bit is truncated to 3 bits we get $\bar{4}\bar{2}\bar{6}\bar{6}$ which when reduced according to Fig. 13.1 and truncated to 4 digits gives $\bar{4}3\bar{3}\bar{2}$ - obviously not the correct answer.

The above discussion assumes that we will not accept a digit value $\bar{4}$ in the high order position. One may relax this constraint if one is willing to accept that negation may cause overflow. If one does, then the overflow-safe range for addition is extended even further to be the range $[\bar{4}00\bar{2}, 3001]$.

13.5.3 Overflow with the Alternative Reduction Scheme

If instead of the reduction rule given by the table in Fig. 13.1 we use the reduction rule given in Fig. 13.4, then once again, the overflow-safe range for addition is different. We assume that each digit is treated equally, and there is no overflow inhibition from the high-order digit. In this case, similarly to example 13.17 we find that addition of 0 to numbers with high-order digit equal to 3 or $\bar{3}$ causes overflow. The following example also causes overflow.

Example 13.20. The sum $2\bar{3}\bar{3}\bar{3} + 1\bar{3}\bar{3}\bar{3}$ gives $3\bar{6}\bar{6}\bar{6}$ which when reduced according to Fig. 13.4 gives $1\bar{2}\bar{3}\bar{3}\bar{2}$, which overflows when truncated to 4 digits. Arithmetically, the result of this computation is the same as $1001 + 0001 = 1002$. One concludes that computations whose results exceeds 1001 risk overflow. Conclusion : The overflow-safe region must be contained in the interval $[\bar{1}00\bar{1}, 1001]$.

In fact, it is not difficult to see that addition is overflow-safe on the range $[\bar{1}00\bar{1}, 1001]$, which is therefore the largest overflow-safe range for this implementation of addition.

Summary: These considerations show that seemingly innocent changes to the basic reduction scheme can greatly affect the numerical range on which the operation is reliable. This shows that one must be careful in designing arithmetic operators, and some analysis of the range on which they are reliable is advisable. The reduction scheme summarized by the table in Fig. 13.1 gives a larger overflow-safe range than the one given by Fig. 13.4 as long as an overflow inhibition scheme is used. Otherwise, it is not workable at all.

The analysis given here for radix-4 may easily be extended to higher radices. Using similar examples and proofs for higher radices, the following proposition may be shown :

Proposition 13.21. *For radix r and word-length of m digits, the addition algorithm of section 13.3 gives the correct result provided the result lies in the arithmetic range $R = [\bar{s}00\bar{1}, s001]$, where $s = r - 1$. Hence signed-digit-addition is overflow-safe on this range.*

13.6 Radix-2 Carry-Free Addition

The design of a radix-2 msd-first adder is shown in Fig. 13.7. In this architecture, the latency is the delay of 2 digit-level adder cells (not including the delay of the cell C which simply concatenates the two input digits to form the sum digit). It may be noted that the *carry-save representation* is an alternative to the signed digit or redundant representation [86]. In carry-save representation, each digit can be 0, 1, or 2 and this representation is not symmetric like the signed digit representation where each digit value, which can be -1 or 0 or 1, maintains symmetry with respect to the value 0 [67]. The architecture of Fig. 13.7(a) can be used as a carry-save adder if the ranges of digits as shown in Fig. 13.7(b) are used.

The lower level implementation architectures of the msd-first radix-2 redundant and carry-save adders of Fig. 13.7 are shown in Fig. 13.8(a) and Fig. 13.8(b), respectively. (Note that the bubbles in the cells do not represent inversion operations; these are used to show the negative weight associated with the signals.) In the radix-2 signed-digit redundant representation, the value of x is given by $x_+ - x_-$ where x is encoded using two bits x_+ and x_-. Similarly, the carry-save value of the digit x is given by $x' + x''$. Note that this encoding is different from the two's complement encoding of the digits.

13.6. RADIX-2 CARRY-FREE ADDITION

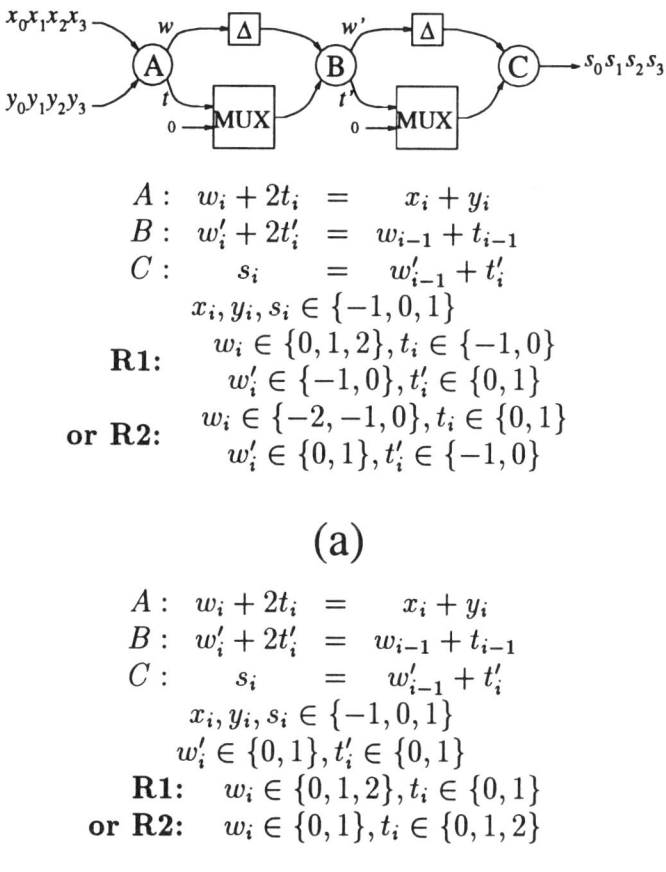

$$A: \quad w_i + 2t_i \quad = \quad x_i + y_i$$
$$B: \quad w'_i + 2t'_i \quad = \quad w_{i-1} + t_{i-1}$$
$$C: \quad s_i \quad = \quad w'_{i-1} + t'_i$$
$$x_i, y_i, s_i \in \{-1, 0, 1\}$$

R1: $\quad w_i \in \{0, 1, 2\}, t_i \in \{-1, 0\}$
$\quad w'_i \in \{-1, 0\}, t'_i \in \{0, 1\}$

or R2: $\quad w_i \in \{-2, -1, 0\}, t_i \in \{0, 1\}$
$\quad w'_i \in \{0, 1\}, t'_i \in \{-1, 0\}$

(a)

$$A: \quad w_i + 2t_i \quad = \quad x_i + y_i$$
$$B: \quad w'_i + 2t'_i \quad = \quad w_{i-1} + t_{i-1}$$
$$C: \quad s_i \quad = \quad w'_{i-1} + t'_i$$
$$x_i, y_i, s_i \in \{-1, 0, 1\}$$
$$w'_i \in \{0, 1\}, t'_i \in \{0, 1\}$$

R1: $\quad w_i \in \{0, 1, 2\}, t_i \in \{0, 1\}$
or R2: $\quad w_i \in \{0, 1\}, t_i \in \{0, 1, 2\}$

(b)

Fig. 13.7: (a) Radix-2 and (b) Carry-Save redundant msd-first adders.

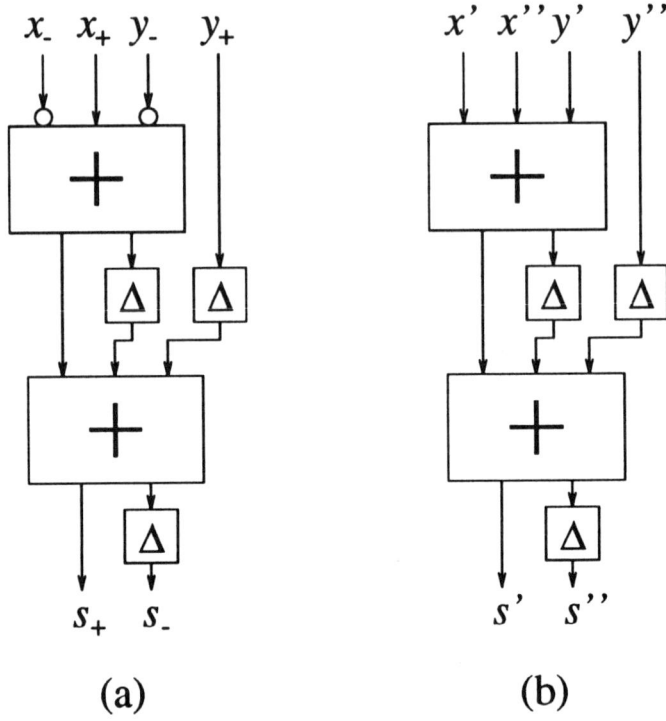

Fig. 13.8: Low-level radix-2 signed-digit and carry-save adders.

The latency of the radix-2 adder can be reduced even further by making use of a hybrid radix-2 representation where one input and the output operands are represented in radix-2 redundant representation and the other input operand is represented using two's complement representation [95][98][99][67]. The hybrid number based architecture is shown in Fig. 13.9 for a msd-first digit-serial operation where the input operand x and the output operand s are in redundant representation and the input operand y is represented using two's complement number system. The redundant numbers x and s are assumed to be represented as signed digit numbers in Fig. 13.9(a) and as carry-save numbers in Fig. 13.9(b).

In the architecture of Fig. 13.9, proper encoding of the digits can make the cell B a dummy cell. The lower level implementation architectures of such hybrid msd-first adders are shown in Fig. 13.10(a) and Fig. 13.10(b) for radix-2 redundant signed-digit and carry-save representations, respectively. In such a representation, the latency is limited by the delay of one binary full adder only. This adder is the most attractive for VLSI implementation due to its short latency and simplicity of the cells.

13.7. DATA FORMAT CONVERSION

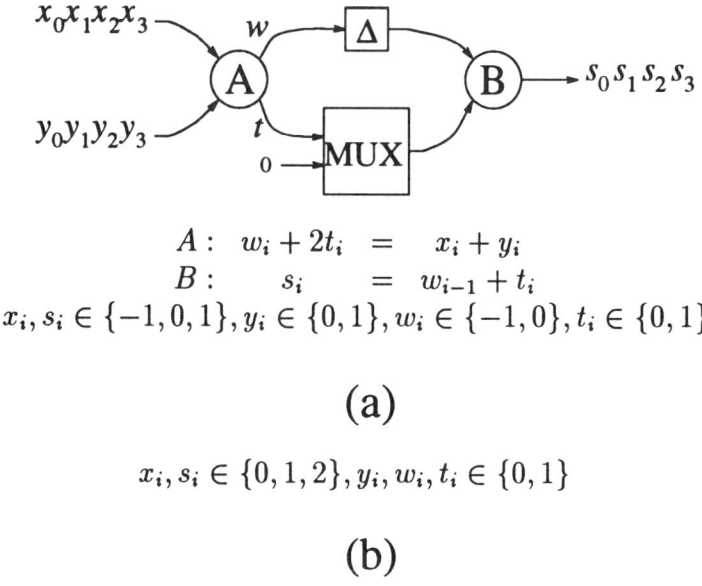

Fig. 13.9: Radix-2 hybrid redundant adders.

13.7 Data Format Conversion

If one is to use redundant msd-first digit-serial format, then there must be some way to convert from this format to two's complement format and vice versa. The conversion from two's complement to redundant format is relatively easy, since transforming from two's complement to redundant format involves simply extending each digit by the addition of one bit, by sign extension in the most significant digit and by zero-extension in other digits. More interesting is the problem of conversion from redundant format to two's complement format. Consider a radix 4 digit in maximally redundant format transmitted msd-first. The digits take on values from the set $\{-3, -2, -1, 0, 1, 2, 3\}$. The problem is to transform this number into a parallel two's complement number. A key observation is that in general it is not possible to know the value of any of the two's complement digits until the least significant redundant format digit becomes available. For instance, consider the redundant format number $10000\bar{1}$. To emphasize that this is in redundant format, it will be denoted $(10000\bar{1})_R$, where the subscript R indicates redundant radix-4 format. In two's complement radix-4 format this is $(033333)_4$. On the other hand, changing the least significant digit from -1 to 1, it is immediately seen that $(100001)_R = (100001)_4$. Notice, therefore, that changing the least significant digit from 1 to -1 causes all of the preceding (more significant) digits to be changed. In other words,

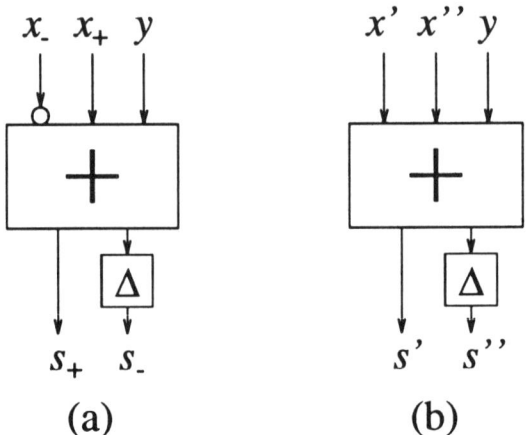

Fig. 13.10: Low-level hybrid adder architectures.

if the number $(10000\bar{1})_R$ is being scanned msd-first and transformed to two's complement radix-4 format, it is impossible to output a single digit of the result until the least significant digit of the input has been processed.

If the digits were to be transmitted in reverse or lsd-first order, however, it would be very simple to convert from redundant to two's complement format. Here is a software algorithm to do the conversion, where a is the input and b is the output.

```
carry = 0;
for (i=0; i<k; i++)
    {
    val = a[i] + carry;
    if (val < 0)
            {
            carry = -1;
            b[i] = val + r;
            }
    else
            {
            carry = 0;
            b[i] = val;
            }
    }
```

In converting this circuit to hardware, it must be noted that for radix-4 if *val* is denoted in two's complement format $(a_2 a_1 a_0)_2$, then $b[i] = (a_1 a_0)_2$ and $-\text{carry} = a_2$. A simple circuit to do the conversion from lsd-first redundant

13.7. DATA FORMAT CONVERSION

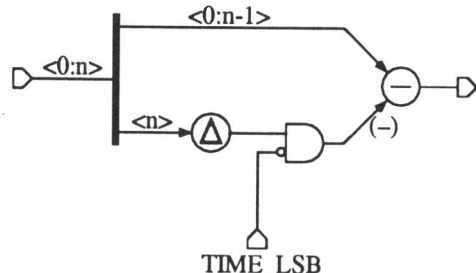

Fig. 13.11: LSD-first redundant to non-redundant format converter

format to lsd-first digit-serial (two's complement) format is shown in Fig. 13.11. Fig. 13.12 illustrates use of this algorithm to convert the radix-2 redundant digit represented by 00$\bar{1}$101 to the two's complement number 111101. This conversion algorithm is an inherently lsd-first operation and is not suitable as an online converter. Two architectures for the hardware implementation of radix-2 lsd-first converter have been presented in [98][100]. An online or msd-first

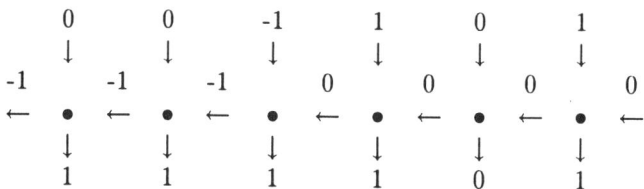

Fig. 13.12: Radix-2 to two's complement conversion example.

conversion algorithm can be designed by using the observation that the carry from one digit to the next most significant digit is either 0 or −1 [101]. Turning now to msd-first transmission, it can be seen that as each redundant digit is viewed, there are only two possibilities for the output digits, one possibility assuming that there is no carry-in from the as-yet unseen digits, and the other assuming that there is a carry-in of −1. Correspondingly, at each stage, we keep the two possible partial output words in shift registers. One shift register contains the digits to be output assuming that there is no further carry-in from the less significant digits: denote the value contained in this shift register at time k by P_k (P = positive). The other shift register contains the digits to be output assuming that there is a carry-in of −1 from the less significant digits: denote the value contained in this shift register by N_k (N = negative). The Table in Fig. 13.7 gives the values to be kept in the registers P and N at time $k+1$ in response to the possible input values for a radix-4 online converter.

Input Digit	P_{k+1}	N_{k+1}
3	$3P_k$	$2P_k$
2	$2P_k$	$1P_k$
1	$1P_k$	$0P_k$
0	$0P_k$	$3N_k$
-1	$3N_k$	$2N_k$
-2	$2N_k$	$1N_k$
-3	$1N_k$	$0N_k$

Fig. 13.13: Operation of Format Converter

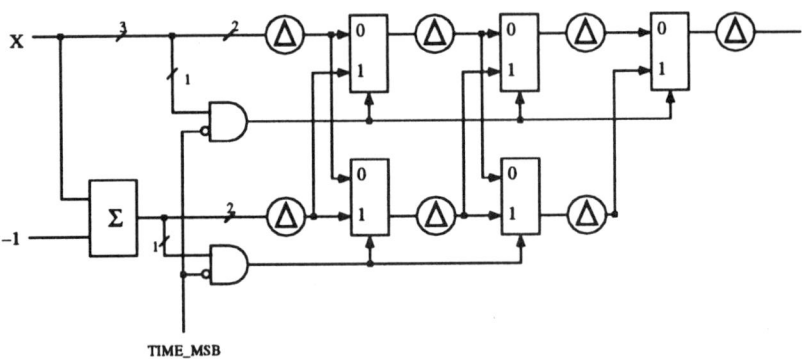

Fig. 13.14: Redundant format to non-redundant format converter.

The notation $3P_k$ denotes a radix-4 value consisting of a 3 digit preceded by the digits of P_k. By way of explanation, suppose that the current input digit is a 3. The value of the digit output in this position will be either a 3 or 2 depending on whether there will be a carry in of -1 or not. In either case, the carry will not propagate further to the more significant places, so the N_k value can now be discarded. The output will be $3P_k$ assuming no carry-in, or $2P_k$ if there is a carry in. The other rows of this table may be similarly explained.

The converter in Fig. 13.14 shows a redundant to non-redundant radix-4 msd-first converter, for the special case of 4 digits per word. If the word consists of more digits, then the length of the shift registers must be increased. In this converter, two shift registers are kept, the P shift register on the top and the N shift register on the bottom. It is assumed that the input digits, X are coded in two's complement format in 3 bits. The P shift register is fed with the two lower bits of the input digit, while the sign bit is used to control the multiplexors. Each input digit is diminished by 1 and used to feed the N shift register, once more the sign bit being used to control the multiplexors. It may be verified that this circuitry corresponds to the functionality expressed in the

13.8. RADIX-2 MULTIPLICATION ARCHITECTURES

table of Fig. 13.13 above. The signal *TIME_MSD* is assumed to be high during the first (most significant) digit of each word, and otherwise low. A high signal, therefore, indicates the start of a new word, and hence no further carry-in from lower order digits is possible. The P_k values are switched into both the N and P shift registers in response to this signal.

The converter as shown is able to handle a continuous stream of new values, producing MSB-first non-redundant digit-serial output. If parallel output is required, the signal *TIME_MSD* may be used to latch the contents of the P register. It is evident that this converter can be generalized to the case of arbitrary radix $r = 2^n$ by replacing signal widths 2 and 3 by n and $n+1$.

Exercise 13.22. Using the ideas explained here, design a MSB-first non-redundant format bit-serial adder.

13.7.1 Radix-2 Format Conversion

Fig. 13.15 shows the table used to update the registers P and N at time step $k+1$ using their values at time step k and the value of the digit received for radix-2 online conversion. The reader can use this msd-first conversion algorithm to verify the conversion of the number represented by $00\bar{1}101$ and verify that the lsd-first and msd-first conversion algorithms output the same two's complement number.

Digit	P_{k+1}	N_{k+1}
1	$1P_k$	$0P_k$
0	$0P_k$	$1N_k$
-1	$1N_k$	$0N_k$

Fig. 13.15: Register update operation in msd-first redundant to two's complement conversion.

13.8 Radix-2 Multiplication Architectures

In this section we derive several architectures for radix-2 redundant number based online multiplication using the mapping or projection technique studied in Chapter 9. These architectures are presented at a higher level and the low level implementations and the architectures which include the effect of encoding of digits to bits are not considered. We present two classes of redundant architectures referred to as *doubly-redundant* and *singly-redundant* [67].

Consider two possible multiplication dependence graphs as shown in Fig. 13.16 and Fig. 13.17 for multiplication of redundant numbers a and b.

In the dependence graph of Fig. 13.16, the carry moves vertically down whereas it moves diagonally in the dependence graph of Fig. 13.17.

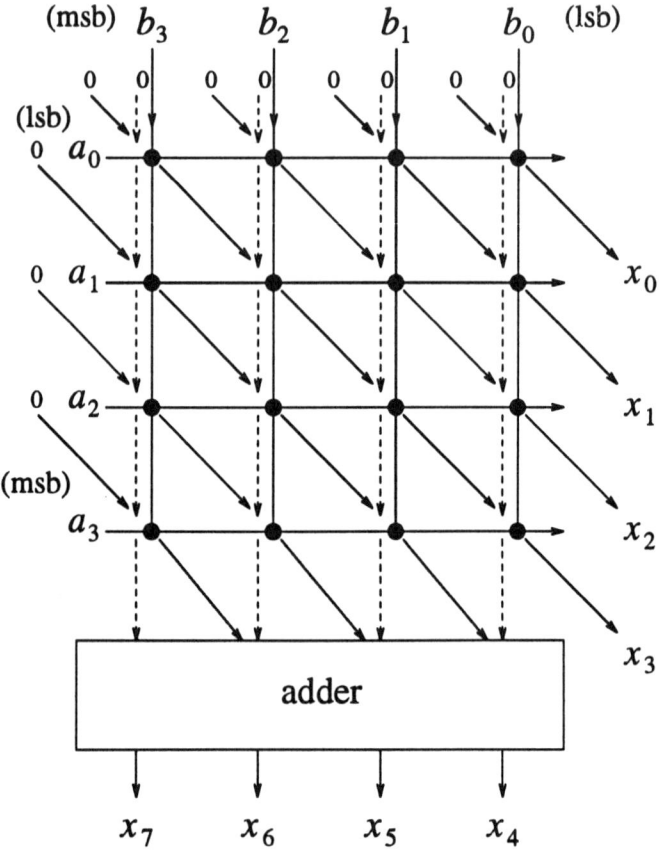

Fig. 13.16: Multiplication dependence graph with vertical carry.

13.8.1 Doubly-Redundant Multiplier Architectures

In doubly-redundant numbers, the partial sum digits belong to the set $\{\bar{2}\ldots 2\}$ and each digit-level adder performs the add operation with the two input operands in the range $\{\bar{2}\ldots 2\}$ and $\{\bar{1}\ldots 1\}$ and the output operand in the range $\{\bar{2}\ldots 2\}$. Although this adder can perform the add operation in one cycle, it requires the digits in the range $\{\bar{2}\ldots 2\}$ to be encoded using 4 bits which is inefficient.

Using projection technique, we can derive two doubly-redundant architectures as shown in Figures 13.18 and 13.19 from the two dependence graphs

13.8. RADIX-2 MULTIPLICATION ARCHITECTURES

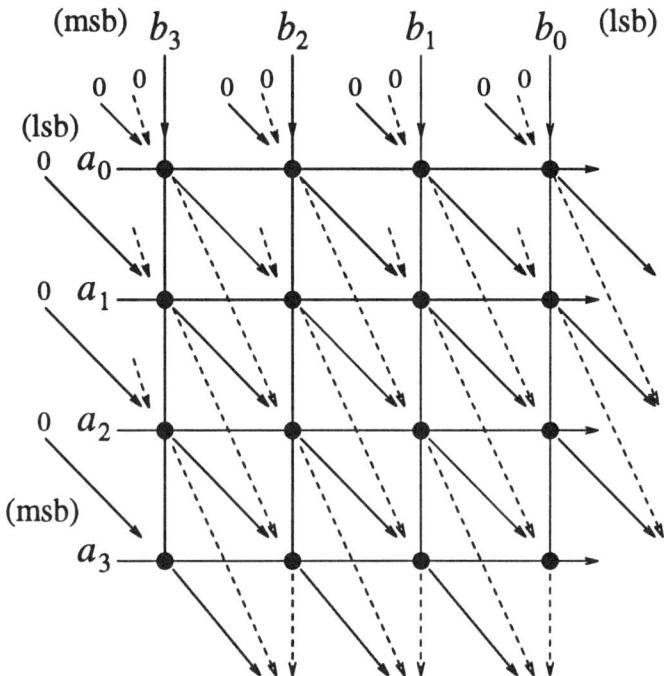

Fig. 13.17: Multiplication dependence graph with diagonal carry.

shown in Fig. 13.16 and Fig. 13.17, respectively. To obtain the architecture of Fig. 13.18 from the dependence graph with vertical carry as shown in Fig. 13.16, we consider the iteration vector, $\mathbf{d}$, to be $[10]^T$. The processor space, $\mathbf{P}^T$, is selected as $[01]$. The msd-first schedule is obtained by $\mathbf{S}^T = [10]$. With this projection, the processor displacements, $\mathbf{P}^T \mathbf{e}$, for the edges (e's) a, b, carry, and result signals are obtained as 0, −1, −1 and −1, and their associated communication delays, $\mathbf{S}^T \mathbf{e}$, are given by 1, 0, 0, and 1, respectively. Note that the x and y axes are assumed to be in standard representation, i.e., the x axis is directed from left to right, and the y axis from bottom to top. The same mapping parameters are also used to derive the architecture of Fig. 13.19 from Fig. 13.17. With this projection, the processor displacements, $\mathbf{P}^T \mathbf{e}$, for the edges (e's) a, b, carry, and result signals are obtained as 0, −1, −2 and −1, and their associated communication delays, $\mathbf{S}^T \mathbf{e}$, are given by 1, 0, 1, and 1, respectively.

The adder cells in these two architectures, shown in Fig. 13.20(a) can be implemented using either the architecture of Fig. 13.20(b) or Fig. 13.20(c). However, the cells of Fig. 13.20(b) and Fig. 13.20(c) cannot be used together

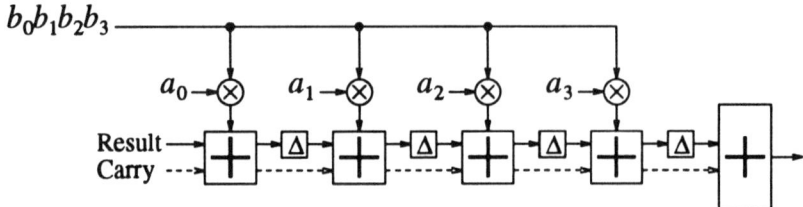

Fig. 13.18: Doubly redundant architecture obtained from the dependence graph of Fig. 13.16.

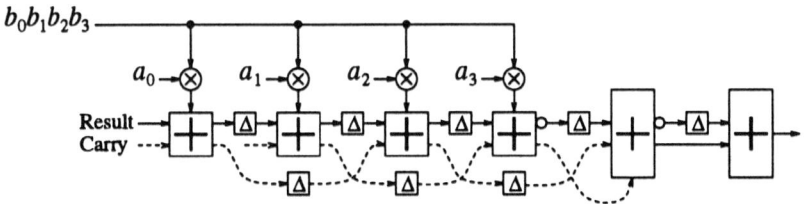

Fig. 13.19: Doubly redundant architecture obtained from the dependence graph of Fig. 13.17.

in an architecture.

In both these doubly-redundant adders, the three inputs and the two outputs are all in redundant representation. The basic cell adds a number in the range $[-2, 2]$ (which corresponds to the sum and carry output combination) and another number in the range $[-1, 1]$ and generates a number in the range $[-2, 2]$. The output carry-assimilation now involves a conversion from doubly-redundant to a simple redundant representation which can be achieved by using a msd-first radix-2 adder as shown in Fig. 13.7. The architecture in Fig. 13.18 contains a critical path and latency proportional to the word-length. On the other hand, the critical path and latency of the architecture in Fig. 13.19 are independent of word-length.

13.8.2 Singly-Redundant Multiplier Architectures

Singly-redundant architectures are efficient since the adder cell outputs (which contains the sum-carry combination) in this architecture belong to the set $\{\bar{1}01\}$ which can be encoded using two bits rather than four bits as was needed for doubly-redundant architectures. The singly-redundant architectures exploit the natural redundancy inherent in the multiplication operation. One input and the output (sum-carry combination) operands belong to the set $\{\bar{1}01\}$ and another input operand belongs to the set $\{01\}$ and the add operation requires a single

13.8. RADIX-2 MULTIPLICATION ARCHITECTURES

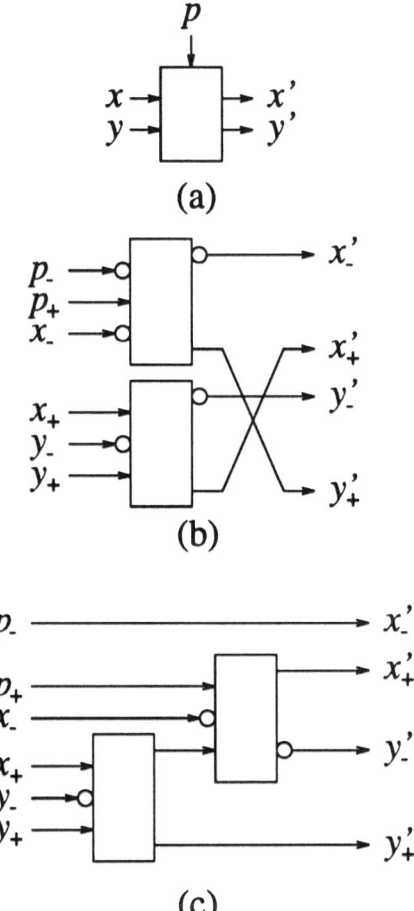

Fig. 13.20: Adder cells used in doubly redundant architectures.

cycle. The key to achieving coding and area efficiency in the singly-redundant architecture lies in the hybrid representation of the input and output signals of the cell. Using the projection technique, we can derive two singly-redundant architectures from the two dependence graphs shown in Fig. 13.16 and Fig. 13.17. These two high-level architectures are similar to the doubly-redundant architectures shown in 13.18 and 13.19; however, the cells used in the singly-redundant architecture are much simpler and are similar to the adder in Fig. 13.9. We conclude that the singly-redundant architecture similar to the architecture in Fig. 13.19 is the most efficient architecture in terms of critical path and computation latency. This architecture has a critical path of one adder delay and has a fixed latency of 3 clock cycles. The msd-first operation was the key to achieve a one-bit adder critical path delay obtained without insertion of any pipeline latches [67].

13.9 Radix-4 and Higher Radix Multiplication

13.9.1 Constant Multiplication

Various methods of constant multiplication have been treated in the literature ([96], [93], [102]). The method of [96] is based on coding the constant multiplier value in redundant format. A constant multiplier will be described here that is based on a two's complement encoding of the constant multiplier value. Since the encoding of the multiplier is done off-line, there is no particular reason for it to be represented in redundant format. It will be assumed that the multiplicand is in digit-serial msd-first redundant format, and each digit is represented in two's complement format.

In a radix-4 multiplication, each digit is represented using 3 bits. At each clock cycle, the two's complement multiplier is multiplied by the current digit of the multiplicand in an array multiplier. Unlike the lsd-first digit-serial multiplier discussed in chapter 3, the array used here is arranged in such a way that the sum bits are propagated straight down and the carry bits are propagated down and to the left. Fig. 13.21 shows the general idea behind an online constant multiplier. The figure shows a complete array used to multiply a redundant format multiplicand, X by a constant value to produce a redundant format result, Y. A digit-serial multiplier is produced (as was seen earlier for lsd-first digit-serial multipliers) by pipelining and folding the multiplier. The particular array shown in Fig. 13.21 multiplies an 8-bit two's complement multiplier value by a 4-digit radix-4 multiplicand. The digits of the multiplicand are represented in two's complement format by 3 bits each. Notice that the array is drawn in such a way that bits of the same significance are in the same vertical column. Because of the redundant format of the multiplicand, the least significant bit of one digit has the same significance as the high-order bit of the next digit, as shown in the figure. Otherwise stated, each block is moved two

13.9. RADIX-4 AND HIGHER RADIX MULTIPLICATION

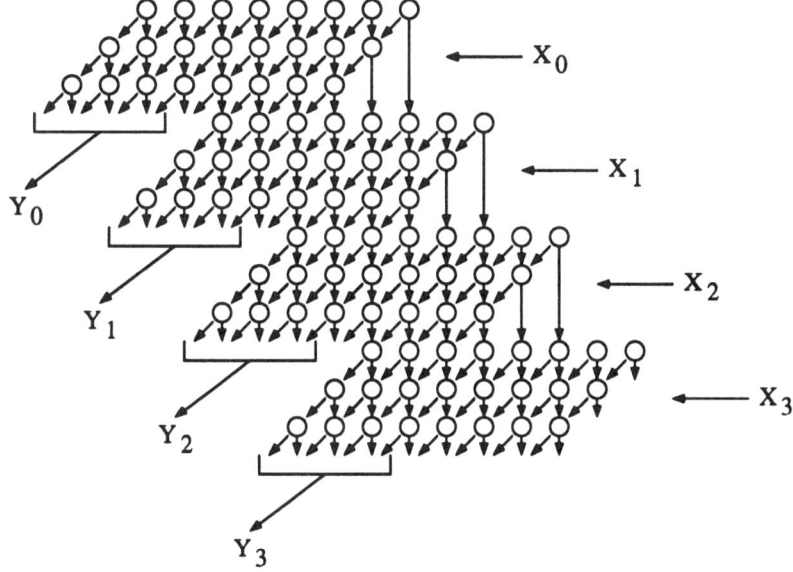

Fig. 13.21: Redundant format Array Multiplier.

places to the right with respect to the previous block.

The three bits of the multiplicand are passed along three rows of the array and are used to generate the appropriate partial products. It must be taken into account that the high order bit of each digit has negative weight. The bit generated at each position of the array is added to the incoming sum and carry bits resulting in output sum and carry bits which are passed straight down and down to the left respectively. Because of the staggering of the rows the overlap at the left end of the array produces carry and sum digits. The sequence of carry and sum output digits is subsequently passed through an online adder to produce the output word.

In order for this method to work correct handling of negative digits in the multiplicand is necessary. Since the high-order bit of each digit has negative weight, the corresponding partial product must be subtracted instead of added. This is done by inverting each bit of the partial product and adding 1 in to the least-significant bit of the partial product. This extra 1 is easily supplied as a carry out bit in the least-significant bit position. This will become clearer in the example to follow.

In order to understand the operation of this multiplier let us consider an example.

Example 13.23. Multiply the redundant format value $2\bar{1}32$ by the constant value 0.11010111 (binary). Expressed in radix-4 this is the product 1332 ×

0.3113, and the correct answer is 1221.3102.

To carry out this product, we extend the multiplier (0.11010111) by an extra bit at the left hand end to give 0011010110 (henceforth ignoring the binary point) and set up a 3 by 10 array of adder cells. This extension is done so that the remainder and carry words carried over from the previous step may be conveniently added. The bits of the multiplier are fed diagonally down the array and partial products are formed by ANDing with the current multiplicand bit. The bits of the last row are inverted and a 1 is carried in at the least-significant bit position in order to negate the bottom row of the array. In the following diagrams, the bit values (1 or 0) of each partial product are shown at each location in the array. Arrows correspond to propagated 1 bit values. The propagated 0 bit values are not shown. In the first clock cycle, the digit of the multiplicand is 2, represented as bits 010.

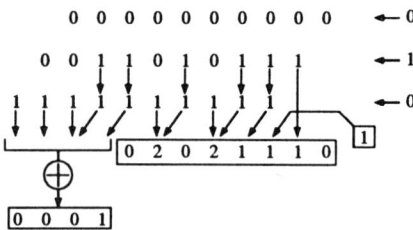

This diagram shows the input digit on the right. Each bit of the multiplicand is fed along the corresponding row and is used to create the partial products in the usual manner. Note where a 1 (in a box) is provided as the carry bit at the right hand end of the bottom row. In this clock cycle, the output is 0001 = 1 (decimal) obtained by summing 1110 and 0011. The sum and carry words combine to give 02021110, though in reality they are not combined, but are carried separately to the next digit.

In the second clock cycle, the multiplicand digit is $\bar{1}$, represented as 111. Note how the sum output word from the previous digit is carried in, suitably shifted at the top of the array.

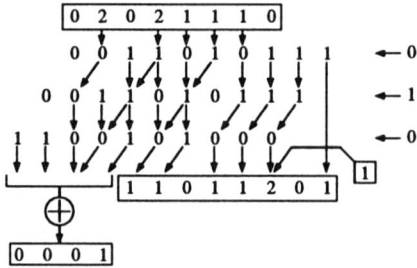

13.9. RADIX-4 AND HIGHER RADIX MULTIPLICATION

The output value is 1 = 0001, formed by summing the binary sum (1110) and carry (0011) output digits. The sum and carry words to be input to the next cycle are 00011101 and 11000100 respectively, combining to give 11011201.

The next two figures show the continuation of the operation for the remaining two digits.

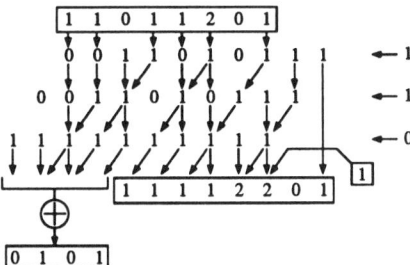

Output is 0101 = 5. Carry is 11112201.

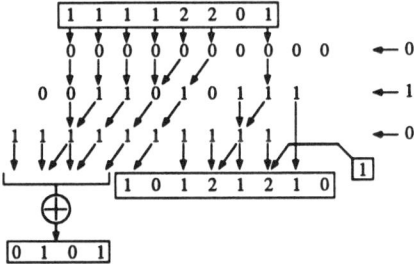

Output is 5. Carry is 10121210.

At this point, the four bits of digit output have been 1155. The least significant four bits may be computed by adding the remaining sum and carry words two bits at a time. The four digits are 2442, which gives a final product of 1155.2442 = 1221.3102, in agreement with the previously computed result.

13.9.2 Generalization to Higher Radices

The above idea for an array multiplier may be extended to higher radices of the form 2^n by adding more rows to the array. Exactly how this is to be done requires a little care, however. For the present let us assume still that the multiplier value is positive. Fig. 13.22 indicates how the design may be extended to radix $32 = 2^5$. From this is will be obvious how the general case works. Fig. 13.22 shows the left hand bits of an array used for radix-32 multiplication. The diagonal line in Fig. 13.22 shows the left end of the array proper, the bits just to the right of the diagonal line being the sign bits of each partial product.

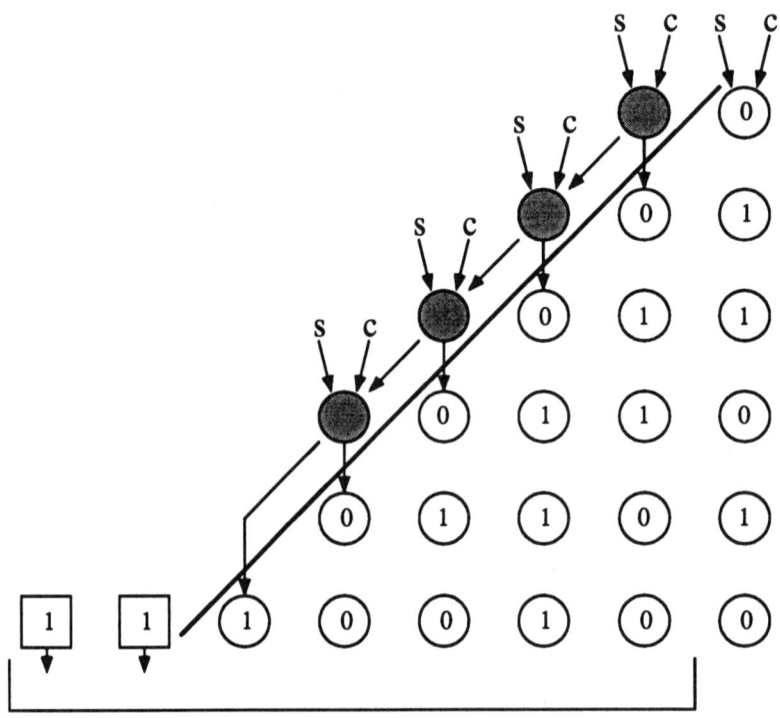

Fig. 13.22: Radix 32 multiplier.

13.9. RADIX-4 AND HIGHER RADIX MULTIPLICATION

The shaded cells to the left of the diagonal line are full-adder cells used to sum the carry and remainder words carried over from the previous step. Note that there is no partial product generated in these cells. Indicated in the figure also are the positions of the bits that form the current output words (at the bottom) and the positions of the carry-in words from the previous digit. It is assumed in this design that the multiplier value is positive, so that the sign bit in each of the partial products is zero, except in the final row, where it is inverted. The two 1-bits (in boxes) at the left end of the array are sign extension bits required to give sufficient dynamic range to avoid overflow.

Next, we consider how this array must be modified in order to handle negative multiplier constants. One approach which can be used, is to use only positive multiplier constants. Where multiplication by a negative constant is required, a multiplication by the corresponding positive constant is substituted and the final result negated. The negation may easily be carried out during the final addition of the sum and carry output digits. However, the design of a multiplier that will multiply by a true negative constant will now nevertheless be presented. Such a multiplier design can be used to carry out serial/parallel multiplication where a parallel two's complement value is multiplied by a digit-serial multiplicand.

The online multiplier relies on a well known trick used in two's complement array multipliers. In this method, it is not necessary to extend partial products on the left by sign extension bits. Instead, the partial product bit generated in the most-significant bit (sign-bit) position in each partial product is inverted. In addition, a carry-in of 1 is applied to the most significant bit position of the least-significant partial product and also to the left of the most significant bit in the array. As explained previously, a third 1 bit must be added in the least-significant bit position of the partial product corresponding to the sign bit of the multiplicand. The following example shows how this is done.

Example 13.24. Multiply 101101 (decimal −19) by 1010 (decimal −6). This example is shown in Fig. 13.23. The figure shows the partial product bit that is generated in each position in the array. This partial product bit is obtained by ANDing the multiplicand bit passed horizontally across the rows with the multiplier bit passed diagonally down the columns. In the circled position in the array, the bit is inverted. These positions are the last row of the array (corresponding to the sign bit of the multiplicand) and the rightmost cells in each row (corresponding to the sign bit of the multiplier). In the cell corresponding to the sign bit of both multiplicand and multiplier the bit value is not inverted. The three 1s in boxes are the three extra 1 bits that must be added to get the correct result. In the first and third lines, the partial product is zero, but the most significant bit is inverted giving a 1. In the second line, the partial product is 101101, but the most significant bit is inverted giving 0. In the final line, the sign bit of the multiplicand value is 1 which causes the partial product word to be inverted bit-by-bit, giving 010010. However, the sign bit is inverted, giving

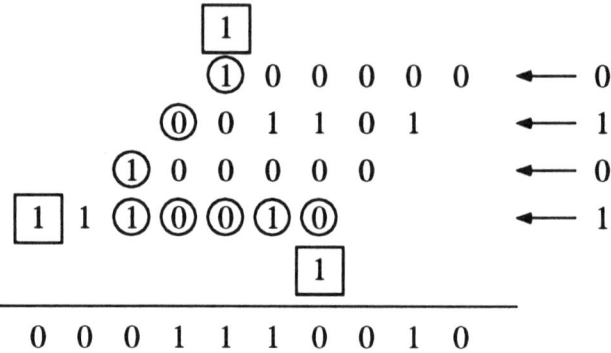

Fig. 13.23: Signed multiplier array.

110010. The correctness of the result may be verified. The user may verify that the array as shown will give the correct result for all choices of multiplier and multiplicand. The left-most bit will be needed in the case of multiplication of 10000 by 1000 (that is -32 by -8) giving a result 0100000000 = 256.

Justification of this method is as follows. Suppose that all the partial products are positive, so that all the sign bits are 0. Then by inverting them to give 1's we are effectively adding a 1 bit in each sign bit position. In the example of Fig. 13.23, these bits add up to a value 0111100000. The two bits in boxes add up to 1000100000. Now adding these two words together gives 0000000000, with a 1 bit being carried out to the controls the MUX circuit, which accepts an $n+1$ bit input word, and controls the MUX circuit, which accepts an $n+1$ bit input word, and controls the MUX circuit, which accepts an $n+1$ bit input word, and left end beyond the range of the computation. Thus, effectively, if all partial products are positive, then a zero value is added to the result. If however one of the partial products is negative, then the sign bit is changed from 1 to 0. This has the effect of subtracting a 1 in the sign bit position, which gives the correct result, since the sign bit has negative weight.

Next, we consider how this method may be applied in the case of the online array multiplier. Fig. 13.24 shows the left end of a radix-32 array multiplier. The cells with diagonal shading are those in which the generated partial product bits must be inverted before being added. The addition of the extra 1 bits is accomplished as shown (the bits in boxes). Because of the extra range forced by the input carry and sum words injected at the top of the array, it is necessary to extend the output word by one extra bit. Consequently, we inject two extra 1 bits at the left end of the array (shown in boxes). The other 1 bit is applied as a carry-in at the least significant bit during the final addition of the output sum and carry words. A further 1 bit must be added in the carry-in position at the right end of the bottom row of the array (not shown).

13.9. RADIX-4 AND HIGHER RADIX MULTIPLICATION

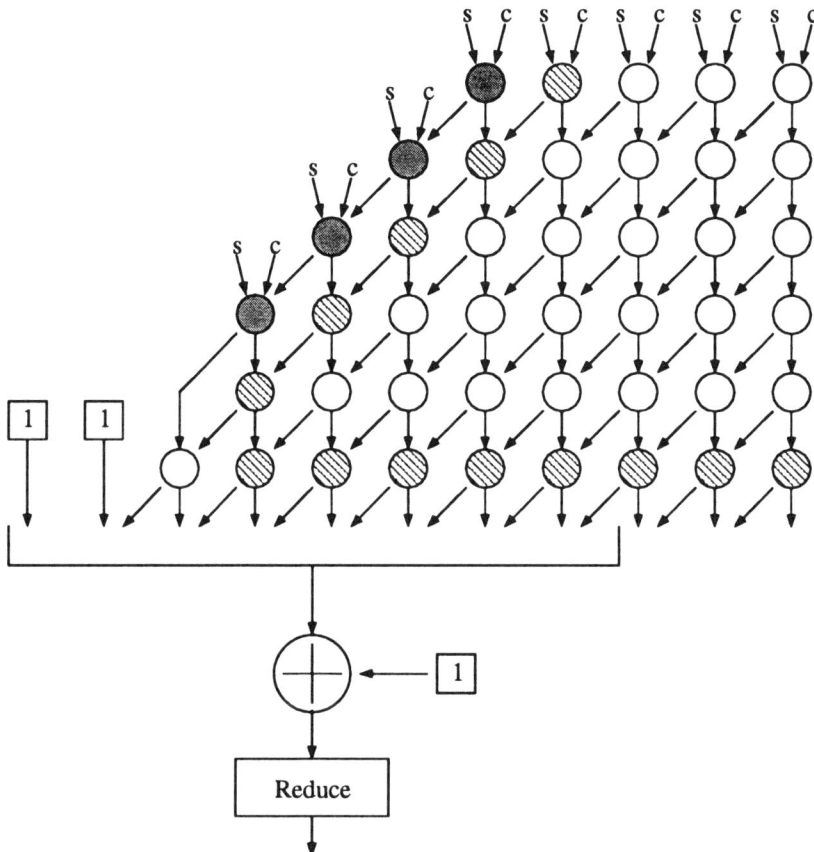

Fig. 13.24: Radix 32 two's complement multiplier.

The output carry and sum words from this array are added together and then reduced. If the radix is equal to $r = 2^n$, then the result of the summation is a digit represented in $n + 3$ bits. This may be reduced to $n + 1$ bits by one or more stages of reduction. The online array constant multiplier described here has a very regular structure. Since each of the rows of the array has the same length, it can be efficiently laid out in a rectangular array (rather than in the parallelogram shape shown in Fig. 13.24). Further refinements are also possible. For instance advantage can be taken of the fact that for bit positions corresponding to 0 bits of the multiplier constant, the generated partial product bit is always zero. This means that the cell in that position can be simplified. In particular, the partial product bit generation circuitry may be deleted and the full-adder replaced by a half-adder.

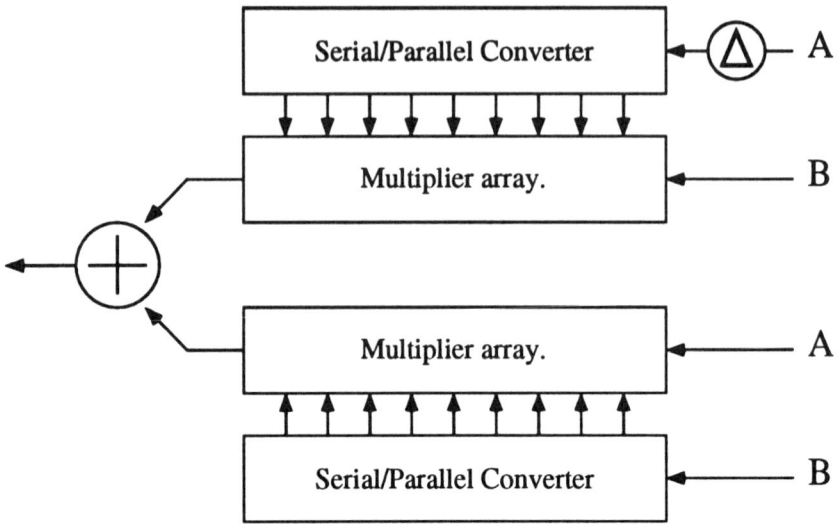

Fig. 13.25: Variable-variable multiplier.

13.9.3 Variable-by-Variable Multiplication

Next we turn to variable-by-variable online multiplication. Rather than attempt multiplication of redundant digits, the approach will be to convert each of the two operands digit-by-digit into parallel two's complement format as they arrive and to apply the ideas described above for digit-serial multiplication with constant two's complement multiplier.

Let A and B represent the values during some given clock cycle of two input values to be multiplied. Suppose that the product AB has been computed and at least certain output digits have been generated and sum and carry words have been computed which when summed would give the complete product. During the next clock cycle, the values of the two operands will be updated in response to the arrival of the next digits. Let the new values be represented by $A+a$ and $B+b$. The product $(A+a).(B+b)$ may be computed as follows.

$$(A+a).(B+b) = A.B + A.b + a.(B+b) \ .$$

Since the product $A.B$ has been evaluated already, it is necessary in computing the new result to update the previous result by adding $A.b + a.(B+b)$. This formula indicates that the new digit, b, must be multiplied by the previous value of the operand A, whereas the digit a must be multiplied by the current value $(B+b)$ of the other operand. This leads to the design (Fig. 13.25) of a variable-variable multiplier. Each of the two multiplier arrays is of the parallel/serial type described in section 13.9.1. The two serial-parallel redundant-format/two's complement converters are of the type described in section 13.7. Initially, they

13.9. RADIX-4 AND HIGHER RADIX MULTIPLICATION

contain a value zero and they are filled up from the left hand end as the digits of the two operands arrive. As seen in the design of the redundant-format to two's complement converter, the value seen at the output of the converter may change entirely in all bits in response to the arrival of a new input digit, but this poses no problems for the correct operation of the multiplier. At any time, the bits of the serial/parallel converter not yet filled are assumed to contain zero values. In order for the multiplier to be able to handle a steady stream of input values, each new input operand word following directly after the previous one, it is necessary to provide for the resetting of the serial/parallel register and also the sum and carry words in the multiplier array. Any more detailed discussion of these points is omitted here. In addition, if both the high and low order parts of the product are required, then it is necessary before the start of a new operation to latch the remaining sum and carry words and to sum them using a separate online adder in much the same way as described for lsd-first digit-serial multipliers.

Many variations on the basic design shown in Fig. 13.25 are possible. For instance, it may be observed that the method of combining the partial products design shown in the figure results in two separate sum and two separate carry words being produced. To compute the output digit of the final result, it is necessary to combine these four terms. Other methods of combining the partial products resulting in fewer carry words are also possible, but these are left for the reader to discover.

Further Reading

Fast multiplication using tree architectures has been presented in [103] [104][105][106]. Bit-serial multiplication using Booth encoding has been considered in [4] and [107].

Division, square-root and CORDIC rotation architectures using redundant arithmetic have not been studied in this book. These have been addressed in several publications including [102],[108] and the recent monograph [32]. SRT division, named after the three persons Sweeny, Robertson and Tocher, was first presented in [109][110]. Various radix-2 and higher radix SRT type division algorithms have been presented in [111] [112][113][114][115]. Another class of division which uses prescaling has been addressed in [116][117][118] [119][120][121] [122]. Architectures for square-root [123][124] [125] using on-line approaches have been addressed in [126][127] [128]. Various architectures for implementation of rotation operations using redundant numbers have been presented in [129] [130] [131].

A good collection of papers on computer arithmetic can be found in [132] [133]. In particular, [133] contains several papers on on-line arithmetic.

Testing of bit- and digit-serial chips has not been considered in this book, but has been discussed in [5] and [134]. The article [134] relates specifically to the architecture of Parsifal.

Bibliography

[1] M. Hatamian and G. L. Cash, "A 70-MHz 8-bit × 8-bit parallel pipelined multiplier in 2.5 μ CMOS," *IEEE Journal of Solid-State Circuits*, vol. 21, no. 4, pp. 505–513, August 1986.

[2] T. G. Noll *et al.*, "A pipelined 330-MHz multiplier," *IEEE Journal of Solid-State Circuits*, vol. 21, no. 3, pp. 411–416, June 1986.

[3] L. B. Jackson, J. F. Kaiser, and H. S. McDonald, "An approach to implementation of digital filters," *IEEE Transactions on Audio and Electroacoustics*, vol. 16, pp. 413–421, September 1968.

[4] R. F. Lyon, "Two's complement pipeline multipliers," *IEEE Transactions on Communications*, vol. 24, pp. 418–425, April 1976.

[5] P. B. Denyer and D. Renshaw, *VLSI Signal Processing: A Bit-Serial Approach*. Reading, MA: Addision-Wesley, 1985.

[6] R. Jain *et al.*, "Custom design of a VLSI PCM-FDM transmultiplexor from system specifications to circuit layout using a computer aided design system," *IEEE Transactions on Circuits and Systems*, vol. 38, no. 2, pp. 183–195, February 1986.

[7] R. Hartley and P. Corbett, "Digit-serial processing techniques," *IEEE Transactions on Circuits and Systems*, vol. 37, no. 6, pp. 707–719, June 1990.

[8] S. G. Smith and P. B. Denyer, *Serial Data Computation*. Boston, MA: Kluwer Academic, 1988.

[9] K. K. Parhi and C. Wang, "Digit-serial DSP architectures," in *Proceedings of Int. Conf. on Applications Specific Array Processors*, (Princeton, NJ), pp. 341–351, September 1990.

[10] K. K. Parhi, "A systematic approach for design of digit-serial signal processing architectures," *IEEE Transactions on Circuits and Systems*, vol. 38, no. 4, pp. 358–375, April 1991.

[11] J. M. Rabaey, S. Pope, and R. W. Brodersen, "An integrated automated layout generation system for dsp circuits," *IEEE Transactions on Computer Aided Design*, pp. 285–296, July 1985.

[12] J. M. Rabaey, C. Chu, P. Hoang, and M. Potkonjak, "Fast prototyping of datapath intensive architectures," *IEEE Design and Test of Computers*, vol. 8, pp. 40–51, June 1991.

[13] H. De Man, J. Rabaey, P. Six, and L. Claesen, "Cathedral-II: A silicon compiler for digital signal processing," *IEEE Design and Test of Computers*, vol. 3, pp. 13–25, December 1986.

[14] J. Vanhoof, K. Van Rompaey, I. Bolsens, G. Goossens, and H. De Man, *High-Level Synthesis for Real-Time Digital Signal Processing*. Norwell, MA: Kluwer Academic, 1993.

[15] M. C. McFarland, A. C. Parker, and R. Composano, "The high-level synthesis of digital systems," *Proceedings of IEEE*, vol. 78, no. 2, pp. 301–318, February 1990.

[16] P. G. Paulin and J. P. Knight, "Force directed scheduling for the behavioral synthesis of ASICs," *IEEE Transactions on Computer Aided Design*, vol. 8, no. 6, pp. 661–679, June 1989.

[17] C.-T. Hwang, J.-H. Lee, and Y.-C. Hsu, "A formal approach to the scheduling problem in high-level synthesis," *IEEE Transactions on Computer Aided Design*, vol. 10, no. 4, pp. 464–475, April 1991.

[18] C. H. Gebotys and M. I. Elmasry, "Optimal synthesis of high-performance architectures," *IEEE Journal of Solid-State Circuits*, vol. 27, no. 3, pp. 389–397, March 1992.

[19] C. H. Gebotys, "Synthesizing embedded speed-optimized architectures," *IEEE Journal of Solid-State Circuits*, vol. 28, no. 3, pp. 242–252, March 1993.

[20] C.-Y. Wang and K. K. Parhi, "High-level synthesis using concurrent transformations, scheduling, and allocation," *IEEE Transactions on Computer Aided Design*, March 1995 (to appear).

[21] K. Ito, L. E. Lucke, and K. K. Parhi, "Module selection and data format conversion for cost-optimal DSP synthesis," in *Proceedings of IEEE Int. Conf. on Computer Aided Design*, (San Jose, CA), pp. 322–329, November 1994.

[22] R. I. Hartley and J. R. Jasica, "Behavioral to structural translation in a bit-serial silicon compiler," *IEEE Transactions on Computer-Aided Design*, vol. 7, no. 8, pp. 877–886, August 1988.

[23] N. Weste and K. Eshraghian, *Principles of CMOS VLSI Design: A Systems Perspective*. Addison-Wesley, 1993.

[24] M. Lehman and N. Burla, "Skip techniques for high-speed carry propagation in binary arithmetic units," *IRE Transactions on Electronic Computers*, vol. EC-10, pp. 691–698, 1961.

[25] P. K. Chan et al., "Delay optmization of carry-skip adders and block carry-lookahead adders using multidimensional dynamic programming," *IEEE Transactions on Computers*, vol. 41, no. 8, pp. 920–930, August 1992.

[26] R. P. Brent and H. T. Kung, "A regular layout for parallel adders," *IEEE Transactions on Computers*, vol. C-31, no. 3, pp. 260–264, March 1982.

[27] O. J. Bedrij, "Carry-select adder," *IRE Transactions on Electronic Computers*, vol. EC-11, pp. 340–346, 1962.

[28] B. W. Kernighan and D. M. Ritchie, *The C Programming Language*. Englewood Cliffs, NJ: Prentice Hall, 1978.

[29] A. V. Oppenheim and R. W. Schafer, *Discrete-Time Signal Processing*. Englewood Cliffs, NJ: Prentice Hall, 1989.

[30] K. Hwang, *Computer Arithmetic: Principles, Architecture, and Design*. NY: Wiley, 1979.

[31] I. Koren, *Computer Arithmetic Algorithms*. Prentice Hall, NJ: Englewood Cliffs, 1993.

[32] M. D. Ercegovac and T. Lang, *Division and Square Root*. Norwell, MA: Kluwer Academic, 1994.

[33] J. E. Volder, "The CORDIC trigonometric computing technique," *IRE Transactions on Electronic Computers*, vol. EC-8, pp. 330–334, 1959.

[34] J. S. Walther, "A unified algorithm for elementary functions," *Spring Joint Computer Conference*, pp. 379–385, 1971.

[35] A. D. Booth, "A signed binary multiplication technique," *Q. J. mech. appl. math:4*, pp. 236–240, Oxford University Press, 1951.

[36] I. Chen and R. Willoner, "An O(n) parallel multiplier with bit-sequential input and output," *IEEE Transactions on Computers*, vol. C-28, pp. 721–727, October 1979.

[37] R. Gnanasekaran, "On a bit-serial input and bit-serial output multiplier," *IEEE Transactions on Computers*, vol. C-32, 1983.

[38] N. R. Strader and V. T. Rhyne, "A canonical bit-sequential multiplier," *IEEE Transactions on Computers*, vol. C-31, 1982.

[39] H. J. Sips, "Comments on "an o(n) parallel multiplier with bit-sequential input and output"," *IEEE Transactions on Computers*, vol. C-31, 1982.

[40] P. Hilfinger, "A high-level language and silicon compiler for digital signal processing," in *Proceedings of IEEE Customs Integrated Circuits Conference*, pp. 213–216, 1985.

[41] R. Lipsett, C. Schaefer, and C. Ussery, *VHDL: Hardware Description and Design*. Boston, MA: Kluwer Academic, 1989.

[42] R. Hartley et al., "A rapid prototyping environment for digit-serial processors," *IEEE Design and Test of Computers*, vol. 8, no. 2, pp. 11–25, June 1991.

[43] C. Mead and L. Connway, *Introduction to VLSI Systems*. Reading, MA: Addision-Wesley, 1980.

[44] R. Hartley and A. E. Casavant, "Optimizing pipelined networks of associative and commutative operators," *IEEE Transactions on Computer Aided Design*, vol. 13, no. 11, pp. 1418–1425, November 1994.

[45] D. M. Ritchie and K. Thompson, "The UNIX time-sharing system," *Communications of the ACM*, vol. 17, no. 7, pp. 365–375, July 1974.

[46] S. Kirkpatrick, C. D. Gelatt, and M. P. Vecchi, "Optimization by simulated annealing," *Science*, vol. 220, pp. 671–680, May 1983.

[47] S. N. et al. "A 30mhz chip set for graphics computations designed using a silicon compiler," in *Proc. of International Solid State Circuits Conference, ISSCC-87*, 1987.

[48] S. E. Noujaim, J. A. Mallick, and M. Wu, "Multichannel data acquisition system with on-chip digital processing," in *Proc. Custom Integrated Circuits Conference*, pp. 9.3.1 – 9.3.4, 1988.

[49] S. R. MacMinn and F. F. Yassa, "A single-chip vector rotator for ac drives," in *Proc. Conference on Applied Motion Control*, (Minneapolis, Minn), June 16-18 1987.

[50] P. L. Shaffer, "Implementation of a parallel extended kalman filter using a bit-serial silicon compiler," in *Proc. IEEE Fall Joint Computer Conference*, (Dallas), pp. 327–334, October 25-29 1987.

[51] F. Y. et al. "A multi-channel digital demodulator for lvdt and rvdt position sensors," in *Proc. IEEE Custom Integrated Circuits Conference*, (San Diego), pp. 20.5.1 – 20.5.4, May 15-18 1989.

BIBLIOGRAPHY

[52] F. F. Yassa and S. L. Garverick, "A multi-channel digital demodulator for lvdt/rvdt position sensors," *IEEE Journal of Solid State Circuits*, vol. 25, 1990.

[53] R. I. Hartley, P. Corbett, P. Jacob, and S. Karr, "An fir filter designed using a silicon compiler," in *Proc. IEEE Custom Integrated Circuits Conference*, (San Diego), pp. 20.2.1 – 20.2.4, May 15–18 1989.

[54] J. Tiemann and F. F. Yassa, "An iir filter architecture with programmable structure," in *Proc. VLSI Signal Processing, III*, (Monterey), pp. 30–38, November 2–4 1987.

[55] F. F. Yassa, J. R. Jasica, R. I. Hartley, and S. Noujaim, "A silicon compiler for digital signal processing: Methodology, implementation and applications," *IEEE Proceedings, Special Issue on Hardware and Software for DSP*, vol. 75, 1987.

[56] M. Potkonjak and J. M. Rabaey, "Fast implementation of recursive programs using transformations," in *Proceedings of IEEE Int. Conf. on Acoustics, Speech and Signal Processing*, (San Francisco, NJ), pp. V–569–572, March 1992.

[57] N. Park and A. Parker, "Sehwa: A software package for synthesis of pipelines from behavioral specifications," *IEEE Transactions on Computer Aided Design*, vol. 7, no. 3, pp. 356–370, 1988.

[58] J. D. Ullman, *Computational Aspects of VLSI*. Rockville, MD: Computer Science Press, 1984.

[59] K. K. Parhi and D. G. Messerschmitt, "Static rate-optimal scheduling of iterative data-flow programs via optimum unfolding," *IEEE Transactions on Computers*, vol. 40, no. 2, pp. 178–195, February 1991.

[60] K. K. Parhi, C.-Y. Wang, and A. P. Brown, "Synthesis of control circuits in folded pipelined DSP architectures," *IEEE Journal of Solid-State Circuits*, vol. 27, no. 1, pp. 29–43, January 1992.

[61] J. R. Jump and S. R. Ahuja, "Effective pipelining of digital systems," *IEEE Transactions on Computers*, vol. C-27, pp. 855–865, September 1978.

[62] C. Leiserson, F. Rose, and J. Saxe, "Optimizing synchronous circuitry by retiming," *Third Caltech Conference on VLSI*, pp. 87–116, March 1983.

[63] H. T. Kung, "Why systolic architectures?," *Computer*, pp. 37–46, 1982.

[64] S. Y. Kung, *VLSI Array processors*. Englewood Cliffs, NJ: Prentice Hall, 1988.

[65] P. R. Cappello and K. Steiglitz, "Unifying VLSI array design with linear transformations of space-time," *Advances in Computing Research*, vol. 2, pp. 23–65, 1984.

[66] H. V. Jagadish, S. Rao, and T. Kailath, "Array architectures for iterative algorithms," *Proceedings of IEEE*, vol. 75, no. 9, pp. 1304–1321, September 1987.

[67] G. Privat, "A novel class of serial-parallel redundant signed-digit multipliers," in *Proceedings of IEEE Symposium on Circuits and Systems*, (New Orleans, LA), pp. 2116–2119, May 1990.

[68] M. C. Chen, "The generation of a class of multipliers: Synthesizing highly parallel algorithms in vlsi," *IEEE Transactions on Computers*, vol. C-37, pp. 329–338, March 1988.

[69] C. R. Baugh and B. A. Wooley, "A two's-complement parallel array multiplication algorithm," *IEEE Transactions on Computers*, vol. C-22, pp. 1045–1047, December 1973.

[70] O. Rioul and M. Vetterli, "Wavelets and signal processing," *IEEE Signal Processing Magazine*, pp. 14–38, October 1991.

[71] M. Ohta, M. Yano, and T. Nishitani, "Entropy coding for wavelet transform of image and its applications for motion picture coding," in *Proceedings of SPIE Visual Communications and Image Processing: Visual Communications, Vol. 160*, (Boston, MA), pp. 456–466, November 1991.

[72] K. K. Parhi and T. Nishitani, "VLSI architectures for discrete wavelet transforms," *IEEE Transactions on VLSI Systems*, vol. 1, no. 2, pp. 191–202, June 1993.

[73] K. K. Parhi, "Systematic synthesis of DSP data format converters using life-time analysis and forward-backward register allocation," *IEEE Transactions on Circuits and Systems–II: Analog and Digital Signal Processing*, vol. 39, no. 7, pp. 423–440, July 1992.

[74] K. K. Parhi, "Calculation of minimum number of registers in arbitrary life time chart," *IEEE Transactions on Circuits and Systems–II: Analog and Digital Signal Processing*, vol. 41, no. 6, pp. 434–436, June 1994.

[75] J. Bae, V. K. Prasanna, and H. Park, "Synthesis of a class of data format converters with specified delays," in *Proceedings of IEEE Int. Conf. on Applications Specific Array Processors*, (San Francisco, CA), pp. 283–294, August 1994.

[76] P. Corbett and R. Hartley, "Designing systolic arrays using digit-serial arithmetic," *IEEE Transactions on Circuits and Systems-II: Analog and Digital Signal Processing*, vol. 39, no. 1, pp. 62–65, January 1992.

[77] A. Chandrakasan, S. Sheng, and R. Brodersen, "Low-power CMOS digital design," *IEEE Journal of Solid-State Circuits*, vol. 27, no. 4, pp. 473–484, April 1992.

[78] R. W. Reitwiesner, "Binary arithmetic," *Advances in Computers*, vol. 1, pp. 231–308, Academic, 1966.

[79] R. Jain, P. T. Yang, and T. Yoshino, "FIRGEN: A computer-aided design system for high-performance FIR filter integrated circuits," *IEEE Transactions on Signal Processing*, vol. 39, no. 7, pp. 1655–1668, July 1991.

[80] R. Hartley, "Optimization of canonic signed digit multipliers for filter design," in *Proc. of IEEE International Symposium on Circuits and Systems*, (Singapore), pp. 1992–1995, June 1991.

[81] A. V. Aho, R. Sethi, and J. D. Ullman, *Compilers: Principles, Techniques, and Tools*. Reading, MA: Addison-Wesley, 1986.

[82] A. Dempster and M. D. Macleod, "Constant integer multiplication using minimum adders," *IEE Proc. Circuits Devices Syst*, vol. 141, 1994.

[83] A. Chatterjee and R. K. Roy, "An architectural transformation program for optimization of digital systems by multi-level decomposition," in *Proc. 30th ACM/IEEE Design Automation Conference*, pp. 343–348, 1993.

[84] A. Chatterjee, R. K. Roy, and M. A. d'Abreu, "Greedy hardware optimization for linear digital circuits using number splitting and refactorization," *IEEE Transactions on VLSI Systems*, vol. 1, 1993.

[85] M. Potkonjak, M. B. Srivastava, and A. Chandrakasan, "Efficient substitution of multiple constant multiplications by shifts and additions using iterative pairwise matching," in *DAC-94, Proc. 31st ACM/IEEE Design Automation Conference*, pp. 189–194, 1994.

[86] T. G. Noll, "Carry-save architectures for high-speed digital signal processing," *Journal of VLSI Signal Processing*, vol. 3, no. 1/2, pp. 121–140, June 1991.

[87] C. S. Wallace, "A suggestion for a fast multiplier," *IEEE Transactions on Computers*, vol. EC-13, pp. 14–17, February 1964.

[88] T. Yoshino *et al.*, "A 100 MHz 64-tap FIR digital filter in 0.8 μm BiCMOS gate array," *IEEE Journal of Solid State Circuits*, vol. 25, no. 6, pp. 1494–1501, December 1990.

[89] *VAX Unix Macsyma Reference Manual, Version 11*. Symbolics Inc., 1985.

[90] A. Avizienis, "Signed digit number representation for fast parallel arithmetic," *IRE Transactions on Electronic Computers*, pp. 389–400, September 1961.

[91] M. D. Ercegovac and T. lang, "Fast multiplication without carry-propagate addition," *IEEE Transactions on Computers*, vol. 39, no. 11, pp. 1385–1390, November 1990.

[92] P. Montuschi and L. Ciminiera, "$n \times n$ carry-save multipliers without final addition," in *Proc. of 11th IEEE International Symposium on Computer Arithmetic*, (Windsor, Ontario), pp. 54–61, June 29 - July 2 1993.

[93] A. Guyot, Y. Herreros, and J. Muller, "Janus: An on-line multiplier/divider for manipulating large numbers," in *Proceedings of 9th IEEE Symposium on Computer Arithmetic*, (Santa Monica), September 1989.

[94] S. C. Knowles, J. G. McWhirther, R. F. Woods, and J. V. McCanny, "Bit-level systolic architectures for high-performance IIR filtering," *Journal of VLSI Signal Processing*, vol. 1, no. 1, pp. 9–24, 1989.

[95] A. Vandemeulebroecke, E. Vanzieleghem, T. Denayer, and P. G. A. Jespers, "A new carry-free division algorithm and its application to a single-chip 1024-b RSA processor," *IEEE Journal of solid state circuits*, vol. 25, no. 3, pp. 748–756, June 1990.

[96] M. J. Irwin and R. M. Owens, "Digit-pipelined arithmetic as illustrated by the paste-up system:a tutorial," *IEEE Computer*, pp. 61–73, April 1987.

[97] M. J. Irwin and R. M. Owens, "Design issues in digit serial signal processors," in *Proceedings of IEEE Int. Symp. on Circuits and Systems*, pp. 441–444, May 1989.

[98] H. R. Srinivas and K. K. Parhi, "High-speed VLSI arithmetic processor architectures using hybrid number representation," *Journal of VLSI Signal Processing*, vol. 4, pp. 177–198, 1992.

[99] H. R. Srinivas and K. K. Parhi, "A fast VLSI adder architecture," *IEEE Journal of Solid-State Circuits*, vol. 27, no. 5, pp. 761–767, May 1992.

[100] Y. S. Ming, L. C. Sung, C. C. Hsing, and L. J. Yien, "An efficient redundant-binary number to binary number converter," *IEEE Journal of Solid State Circuits*, vol. 27, no. 1, pp. 109–112, January 1992.

[101] M. D. Ercegovac and T. Lang, "On-the-fly conversion of redundant into conventional representations," *IEEE Transactions on Computers*, vol. C-36, pp. 895–897, July 1987.

[102] K. S. Trivedi and M. D. Ercegovac, "On-line algorithms for division and multiplication," *IEEE Transactions on Computers*, vol. C-26, no. 7, pp. 681–687, July 1977. reprinted in [133].

[103] S. H. Unger, "Tree realizations of iterative circuits," *IEEE Transactions on Computers*, vol. C-26, 1977.

[104] Y. Harata, Y. Nakamura, H. Nagase, M. Takigawa, and N. Takagi, "A high-speed multiplier using a redundant binary adder tree," *IEEE Journal of Solid State Circuits*, no. 1, pp. 28–34, February 1987.

[105] S. Kuninobu, T. Nishiyama, H. Edamatsu, T. Tanaguchi, and N. Takagi, "Design of high speed MOS multiplier and divider using redundant binary representation," in *Proc. of 8th Symposium on Computer Arithmetic*, (Como, Italy), pp. 80–86, 1987.

[106] N. Takagi, H. Yasuura, and S. Yajima, "High speed VLSI multiplication algorithm with a redundant binary addition tree," *IEEE Transactions on Computers*, vol. 34, pp. 789–796, September 1985.

[107] T. Rhyne and N. R. Strader, "A signed bit-sequential multiplier," *IEEE Transactions on Computers*, vol. C-35, 1986.

[108] M. D. Ercegovac and T. Lang, "Redundant and on-line cordic: Application to matrix triangularization and svd," *IEEE Transactions on Computers*, vol. 39, 1990.

[109] J. E. Robertson, "A new class of digital division methods," *IRE Transactions on Electronic Computers*, vol. EC-7, pp. 218–222, September 1958.

[110] T. D. Tocher, "Techniques of multiplication and division for automatic binary computers," *Quarter. J. Mech. App. Math.*, vol. 2, pt. 3, pp. 364–384, 1958.

[111] M. D. Ercegovac and T. Lang, "Fast radix-2 division with quotient-digit prediction," *Journal of VLSI Signal Processing*, pp. 169–180, 1989.

[112] D. E. Atkins, "Higher-radix division using estimates of the divisor and partial remainders," *IEEE Transactions on Computers*, vol. C-17, no. 10, pp. 925–934, October 1968.

[113] J. Fandrianto, "Algorithm for high-speed shared radix-4 division algorithm," in *Proc. of 8th IEEE Symposium on Computer Arithmetic*, (Como, Italy), pp. 73–79, May 1987.

[114] J. Fandrianto, "Algorithm for high speed shared radix-8 division and radix-8 square-root," in *Proc. of 9th IEEE Symposium on Computer Arithmetic*, (Santa Monica, CA), pp. 68–75, September 1989.

[115] P. Montuschi and L. Ciminiera, "Design of a radix-4 division unit with simple selection table," *IEEE Transactions on Computers*, vol. 41, no. 12, pp. 1606–1611, December 1992.

[116] A. Svoboda, "An algorithm for division," *Information Processing Machines*, vol. 9, pp. 183–190, March 1963.

[117] C. Tung, "A division algorithm for signed-digit arithmetic," *IEEE Transactions on Computers*, vol. 17, pp. 887–889, September 1968.

[118] M. D. Ercegovac and T. Lang, "Simple radix-4 division with operands scaling," *IEEE Transactions on Computers*, vol. C-39, no. 9, pp. 1204–1208, September 1990.

[119] N. Burgess, "A fast division algorithm for VLSI," in *Proc. of IEEE International Conference on Computer Design: VLSI in Computers and Processors*, (Boston, MA), pp. 560–563, October 1991.

[120] S. E. McQuillan, J. V. McCanny, and R. Hamill, "New algorithms and VLSI architectures for SRT division and square root," in *Proc. of 11th Symposium on Computer Arithmetic*, (Windsor, Ontario), pp. 80–86, June 29-July 2 1993.

[121] P. Montuschi and L. Ciminiera, "Over-redundant digit sets and the design of digit-by-digit division units," *IEEE Transactions on Computers*, vol. 43, no. 3, pp. 269–277, March 1994.

[122] H. R. Srinivas and K. K. Parhi, "A fast radix 4 division algorithm," in *Proc. of IEEE International Symposium on Circuits and Systems*, (London), pp. 311 – 314, May 30 - June 2 1994.

[123] G. Metze, "Minimal square rooting," *IEEE Transactions on Electronic Computers*, vol. EC-14, pp. 181–185, April 1965.

[124] J. C. Majithia, "Cellular array for extraction of squares and square roots of binary numbers," *IEEE Transactions on Computers*, vol. C-21, pp. 1023–1024, September 1972.

[125] L. Ciminiera and P. Montuschi, "Higher radix square rooting," *IEEE Transactions on Computers*, vol. 39, no. 10, pp. 1220–1231, October 1990.

[126] M. D. Ercegovac, "An on-line square rooting algorithm," in *Proc. of 4th IEEE Symposium on Computer Arithmetic*, (Santa Monica, CA), pp. 183–189, October 1978.

[127] V. G. Oklobdzija and M. D. Ercegovac, "An on-line square root algorithm," *IEEE Transactions on Computers*, vol. C-31, pp. 70–75, January 1982.

[128] T. Lang and P. Montuschi, "Higher radix square root with prescaling," *IEEE Transactions on Computers*, vol. 41, no. 8, pp. 996–1009, August 1992.

[129] M. D. Ercegovac and T. Lang, "Redundant and on-line CORDIC: Application to martix triangularization and svd," *IEEE Transactions on Computers*, vol. 39, no. 6, pp. 725–740, June 1990.

[130] N. Takagi, T. Asada, and S. Yajima, "Redundant CORDIC methods with a constant scale factor for sine and cosine computation," *IEEE Transactions on Computers*, vol. 40, no. 9, pp. 989–995, September 1991.

[131] J. A. Lee and T. Lang, "Constant-factor redundant cordic for angle calculation and rotation," *IEEE Transactions on Computers*, vol. 41, no. 8, pp. 1017–1025, August 1992.

[132] E. E. Swartzlander, Jr., ed., *Computer Arithmetic*, vol. 1. Los Alamitos, California: IEEE Computer Society Press, 1990.

[133] E. E. Swartzlander JR., ed., *Computer Arithmetic*, vol. 2. Los Alamitos, California: IEEE Computer Society Press, 1990.

[134] A. Chatterjee, R. K. Roy, J. A. Abraham, and J. H. Patel, "Efficient testing techniques for bit and digit-serial arrays," in *Proc. Fourth CSI/IEEE International Symposium on VLSI Design*, (New Delhi), pp. 142–147, Jan 4–8 1991.

Index

_for macro, 81
_for macro definition, 81
_prev macro, 81
_succ macro, 81
_while macro, 81
_while macro definition, 81

arithmetic operators, 72
associativity, 123, 124, 127

band matrix multiplication, 201
bit-serial adder, 67
bit-serial language, 63
bit-slicing, 24
block delay operator, 145
bound cell names, 76
bound cells, 77, 78
bound-cell names, 76
BSL, 63, 75, 76, 80, 81
BSSC, 4
built-in operators, 76

C language, 8, 72
CALL keyword, 71, 74, 76, 77, 80, 85
Calma, 66
canonic signed digit, 217, 218
carry-look-ahead adder, 5, 137, 241, 244
carry-save, 268
carry-save adder, 177, 241, 244
carry-save addition, 240, 244
carry-save arithmetic, 44, 225, 240, 241, 244, 245
carry-save multiplier, 177

carry-save subtraction, 242
carry-select adder, 5, 137, 241, 244
carry-skip adder, 5
Cathedral, 2, 4, 8, 63, 69
cell binding, 76
cell header, 64
CELL keyword, 64, 65, 67, 70, 78, 79, 81, 83, 84, 90–92
cells, 64
    composite, 64, 68
    GREATER_THAN, 65
    leaf, 64, 65
    stack, 64, 67, 79, 90
    symbolic, 64
choice operator, 73
circular graph, 190
clock signal, 2
common sub-expressions, 227, 228, 237, 238
    cost function, 236
complex arithmetic, 90
CONNECT keyword, 68, 79, 82, 91
CONST keyword, 71, 88
constant generation, 20
control structures, 80
convolution, 198
CORDIC, 37, 290
counting pairs, 239
CSD, 218–221, 226, 238–242
    asymptotic bit frequency, 249
    asymptotic frequency of pairs, 252
    conversion to, 218
    frequency of pairs, 251

multiplication, 220
number of non-zero bits, 247–249
number of values, 247
pairs, 249
uniqueness, 218
cyclic permutation, 37

data format conversion, 271
data-broadcast filter, 179
data-flow, 1, 8–10, 121, 124, 125, 129, 130, 144
decoder function, 85
define macro, 80
digit-serial adder, 18
digit-serial adder description, 67
digit-serial multiplication, 54
digit-serial multiplier, 74
digit-serial shifting, 35
digital signal processing, 1, 75
DigitSize keyword, 70, 76, 91–93
DIODES, 63
directory lists, 77
discrete wavelet transform, 183
division, 19, 34, 290
doubly-redundant multiplier, 275

EMULATION keyword, 67
END keyword, 64, 71, 76, 79, 82, 90–93
eval macro, 80

feedback loops, 2, 3, 117, 119, 120, 122
finite state machine, 23
FIR filter, 12, 131, 133, 144, 145, 179, 226, 227, 238, 244
FIRST, 4, 8, 63
FixedSize keyword, 70, 71, 76
folding, 44, 165
FOR keyword, 71, 73, 74, 76, 77, 79, 80, 85, 91–93
forward-backward allocation, 186, 191

frames, 83, 188
function call syntax, 75
function calls, 75
functions
adding new ones, 78
multiple output, 75

gate arrays, 244
generating functions, 246
generator, 21, 68, 74, 77, 79, 80, 82, 83, 85–87
command line syntax, 77
shift-left operator, 75

HEIGHT keyword, 65
heterogeneous styles, 2
high-level synthesis, 2
Hilo simulator, 67
Horner's rule, 51
hyper, 2

ifelse macro, 80
IN keyword, 64, 65, 67, 68, 70, 78, 79, 81, 83, 84
INSTANCE keyword, 68
IO-description, 64
iterative splitting, 231

Lager, 2
language
applicative, 69, 70, 72
prescriptive, 69
single assignment, 69
latch, 2, 84
insertion, 231
latch count, 231
latch function, 86
latch function call, 85–87
latency, 6, 7, 9, 10, 19, 46, 47, 50, 231
minimization, 133
layout, 24, 95
layout description, 65
LAYOUT keyword, 65

INDEX

LEFT keyword, 65
life-time analysis, 186
lifetime of a signal, 231, 234
logic operators, 73
low-latency multiplier, 56, 179

m4 macros, 80–82
macros
    built-in, 80
Mimic simulator, 67
MSD-first arithmetic, 253
MSD-first conversion, 273
multi-chip designs, 63
multiplication, 18, 43, 74

NEXT keyword, 66, 67
non-$TIME$-type inputs, 77
non-bound cells, 78
non-signal arguments, 75

online addition, 261
online arithmetic, 253
online conversion, 273
online multiplication, 275
operator binding, 73
operator precedence, 73
OUT keyword, 64, 65, 67, 70, 78,
    79, 81, 83, 84
overflow, 257, 264

parallel-serial converter, 16, 20, 22,
    82–84, 87, 143, 160
parameters, 75
Parsifal, 2, 4, 8, 18, 27, 67, 131
    generator, 80
    simulation description language,
        66
    adder, 75
    built-in operators, 76
    FOR statements, 73
    generator, 68, 74, 82, 87
    guiding principle, 73
    input language, 63, 71
    layout, 95

    typing, 72
    use of macros, 80
performance, 137
periodic function, 86, 87
periodic signals, 86
pipelined circuits, 231
pipelining, 167, 186, 191, 244
power consumption, 217
processor, 64
PROCESSOR keyword, 64, 76, 88,
    93
ps function call, 83, 84, 87

rapid prototyping, 2, 63
real-time, 1, 2
register allocation, 186, 188, 191
registers, 68
relational operators, 73
retiming, 167, 168
RIGHT keyword, 65
ripple-carry adder, 241, 243
ROM, 23, 85
rotation, 290

sample period, 14, 17, 19, 20, 22,
    65, 70, 83, 87
scaling, 51
scheduling, 9, 10, 12, 107
    earliest possible, 119–121, 130,
        233
    pipelined circuits, 231
serial-parallel converter, 67, 82–84,
    143
shimming delays, 127–132, 134, 135
signal arguments, 75
SIGNAL keyword, 86, 88, 92, 93
signals, 75
Silage, 63, 69
silicon compiler, 4
simulation description, 66
singly-redundant multiplier, 275
sp function call, 84
Specification keyword, 71, 76, 86,
    89, 91–93

square-root, 88, 290
STACK keyword, 68, 79, 81
standard libraries, 74, 82
static function, 86, 87
static signals, 86
STREAM, 82
Stream, 66
sub-expression sharing
    latch count, 234
SYMBOLIC keyword, 70, 76, 88, 91–93
synchronous circuit, 2, 3
systolic architectures, 165, 171, 175, 195

TIME signal, 6, 10, 14, 22, 32, 64, 65, 83, 108, 151
TIME signals, 159
TIME-type inputs, 77
timing information, 78
timing specification, 64, 65, 78
TOP keyword, 65
tree multiplication, 290
tree-height minimization, 73
tristate operator, 66
two-dimensional allocation, 186

unbound cells, 77
unbound-cells, 77
unfolding, 141, 147, 165, 193
UNIX, 77
USE keyword, 71, 73, 74, 76, 77, 79, 80, 85, 91–93

VAR keyword, 72
VHDL, 63

Wallace tree multiplier, 240, 241
WIDTH keyword, 65
width specification, 64
Wordsize keyword, 70, 71, 76, 88, 91–93

z-transform, 246